Technological Learning in the Transition to a Low-Carbon Energy System

Technological Learning in the Transition to a Low-Carbon Energy System

Conceptual Issues, Empirical Findings, and Use, in Energy Modeling

Edited by

Martin Junginger

Copernicus Institute of Sustainable Development, Utrecht University, Utrecht, The Netherlands

Atse Louwen

Copernicus Institute of Sustainable Development, Utrecht University, Utrecht, The Netherlands
Institute for Renewable Energy, Eurac Research, Bolzano, Italy

Academic Press is an imprint of Elsevier
125 London Wall, London EC2Y 5AS, United Kingdom
525 B Street, Suite 1650, San Diego, CA 92101, United States
50 Hampshire Street, 5th Floor, Cambridge, MA 02139, United States
The Boulevard, Langford Lane, Kidlington, Oxford OX5 1GB, United Kingdom

This book was published as part of the EU project REFLEX (Analysis of the European energy system under the aspects of flexibility and technological progress), which received funding from the European Union's Horizon 2020 research and innovation programme [GA-No. 691685]. For further information, see: http://reflex-project.eu/.

British Library Cataloguing-in-Publication Data
A catalogue record for this book is available from the British Library

Library of Congress Cataloging-in-Publication Data
A catalog record for this book is available from the Library of Congress

ISBN: 978-0-12-818762-3

For Information on all Academic Press publications
visit our website at https://www.elsevier.com/books-and-journals

Publisher: Brian Romer
Acquisition Editor: Graham Nisbet
Editorial Project Manager: Naomi Robertson
Production Project Manager: Surya Narayanan Jayachandran
Cover Designer: Mark Rogers

Typeset by MPS Limited, Chennai, India

Contents

List of contributors

Philipp Andres Perimeter Solar Inc., ON, Canada

Blanca Corona Copernicus Institute of Sustainable Development, Utrecht University, Utrecht, The Netherlands

Thijs de Groot Nouryon, Amsterdam, The Netherlands

David de Jager GROW Foundation (Growth Through Research, Development & Demonstration in Offshore Wind), Utrecht, The Netherlands

Oreane Y. Edelenbosch PBL Netherlands Environmental Assessment Agency, The Hague, The Netherlands; Department of Management and Economics, Politecnico di Milano, Milan, Italy

Matthew Fairlie Next Hydrogen Corporation, Mississauga, ON, Canada

Tobias Fleiter Fraunhofer Institute for Systems and Innovation Research ISI, Karlsruhe, Germany

Christoph Fraunholz Institute for Industrial Production (IIP), Karlsruhe Institute of Technology (KIT), Karlsruhe, Germany

Brian F. Gerke Lawrence Berkeley National Laboratory, Berkeley, CA, United States

Nilo Gomez Tuya Copernicus Institute of Sustainable Development, Utrecht University, Utrecht, The Netherlands

Jonatan J. Gómez Vilchez European Commission, Joint Research Centre, Ispra, Italy

Stephanie Heitel Fraunhofer Institute for Systems and Innovation Research ISI, Karlsruhe, Germany

Andrea Herbst Fraunhofer Institute for Systems and Innovation Research ISI, Karlsruhe, Germany

Eric Hittinger Rochester Institute of Technology, Rochester, NY, United States

Martin Jakob TEP Energy GmbH, Zurich, Switzerland

Gert Jan Kramer Copernicus Institute of Sustainable Development, Utrecht University, Utrecht, The Netherlands

Patrick Jochem Institute for Industrial Production (IIP), Karlsruhe Institute of Technology (KIT), Karlsruhe, Germany

Martin Junginger Copernicus Institute of Sustainable Development, Utrecht University, Utrecht, The Netherlands

Daniel M. Kammen Energy and Resources Group, UC Berkeley, Berkeley, CA, United States; Renewable and Appropriate Energy Laboratory, UC Berkeley, Berkeley CA, United States; Goldman School of Public Policy, UC Berkeley, Berkeley, CA, United States

Noah Kittner Group for Sustainability and Technology, ETH Zurich, Zürich, Switzerland; Department of Environmental Sciences and Engineering, Gillings School of Global Public Health, University of North Carolina at Chapel Hill, Chapel Hill, NC, United States; Department of City and Regional Planning, University of North Carolina at Chapel Hill, Chapel Hill, NC, United States; Environment, Ecology, and Energy Program, University of North Carolina at Chapel Hill, Chapel Hill, NC, United States

Subramani Krishnan Copernicus Institute of Sustainable Development, Utrecht University, Utrecht, The Netherlands

Juliana Subtil Lacerda Copernicus Institute of Sustainable Development, Utrecht University, Utrecht, The Netherlands

Atse Louwen Copernicus Institute of Sustainable Development, Utrecht University, Utrecht, The Netherlands; Institute for Renewable Energy, Eurac Research, Bolzano, Italy

David L. McCollum International Institute for Applied Systems Analysis (IIASA), Laxenburg, Austria; University of Tennessee, Knoxville, TN, United States

Dominik Möst Energy Economics, TU Dresden, Dresden, Germany

Hazel Pettifor Tyndall Centre for Climate Change Research, University of East Anglia (UEA), Norwich, United Kingdom

Ulrich Reiter TEP Energy GmbH, Zurich, Switzerland

Oliver Schmidt Grantham Institute for Climate Change and the Environment, Imperial College London, London, United Kingdom; Centre for Environmental Policy, Imperial College London, London, United Kingdom

Steffi Schreiber Energy Economics, TU Dresden, Dresden, Germany

Katrin Seddig Institute for Industrial Production (IIP), Karlsruhe Institute of Technology (KIT), Karlsruhe, Germany

Iain Staffell Centre for Environmental Policy, Imperial College London, London, United Kingdom

Dalius Tarvydas Joint Research Centre, European Commission, Petten, The Netherlands

Michael Taylor International Renewable Energy Agency, Bonn, Germany

Ioannis Tsiropoulos Joint Research Centre, European Commission, Petten, The Netherlands

Wilfried van Sark Copernicus Institute of Sustainable Development, Utrecht University, Utrecht, The Netherlands

Detlef P. van Vuuren Copernicus Institute of Sustainable Development, Utrecht University, Utrecht, The Netherlands; PBL Netherlands Environmental Assessment Agency, The Hague, The Netherlands

Ernst van Zuijlen WindWerk BV, Utrecht, The Netherlands

Eric Williams Rochester Institute of Technology, Rochester, NY, United States

Charlie Wilson International Institute for Applied Systems Analysis (IIASA), Laxenburg, Austria; Tyndall Centre for Climate Change Research, University of East Anglia (UEA), Norwich, United Kingdom

Ryan Wiser Lawrence Berkeley National Laboratory, Berkeley, CA, United States

Christoph Zöphel Energy Economics, TU Dresden, Dresden, Germany

PART I

Introduction and methods

CHAPTER 1

Introduction

Martin Junginger[1] and Atse Louwen[1,2]

[1]Copernicus Institute of Sustainable Development, Utrecht University, Utrecht, The Netherlands
[2]Institute for Renewable Energy, Eurac Research, Bolzano, Italy

Abstract

The ongoing energy transition is driven by a need to mitigate climate change and switch from fossil to low-carbon fuels and renewable energy. However, while technologies such as onshore and offshore wind energy, solar energy, and batteries have made significant progress over the past decades, and they can increasingly compete directly with fossil fuels, their deployment is still only a fraction of what is needed to fully decarbonize our economy—a process that is going to take at least several more decades and is going to require major investments. Also, the intermittent character of especially wind and solar energy will require major changes in, for example, storage of energy (both heat and electricity). How this transition will play out, and which technologies will ultimately become winners and losers are highly relevant questions which this book will help to answer by providing both the individual market deployment and cost reduction trends per technology and the results of modeling a portfolio of energy technologies in various sector models and overall energy models.

Chapter outline

1.1 Introduction

1.1.1 Background and rationale

It is clear that the further development of various energy technologies is crucial to reduce the emission of greenhouse gases (GHGs), achieve other environmental targets, limit growing global energy demand, and ultimately enable the transition to a low-carbon society—preferably at low costs. These aims can only be achieved when a large number of technologies to supply renewable energy and to save energy become commercially available and thus are at the core of most energy and climate policies worldwide. Important scenario analyses of the world's future energy system and climate change mitigation

Technological Learning in the Transition to a Low-Carbon Energy System.
DOI: https://doi.org/10.1016/B978-0-12-818762-3.00001-7

scenarios illustrate that technological progress is key to minimizing costs of such development pathways. Given the need for drastic decarbonization, and related substantial investment needs, the political and public debate about the societal costs of this transition is increasing, making it even more important to point out possible cost reductions of novel energy technologies and ultimately the benefits of a low-carbon energy system. Furthermore, the speed of development is essential in order to meet required reductions and supply contributions on time. Many scenarios also highlight the positive economic and security impacts of strong support for research, development, demonstration and deployment of such technologies. Lastly, developing and deploying such energy technologies is seen as a major opportunity for development, (sustainable) industrial activity, and (high-quality) employment. Many (national) policies support both research and development (R&D) and market deployment of promising new energy technologies.

The latter, in particular, will require substantial investment. However, designing such policies effectively (e.g., timing and amount of incentives) has proved to be a challenge. The energy sector and manufacturing industry need strategic planning of their R&D portfolio and have to identify key market niches for new technologies (with or without policy support). Taken together, this situation makes an improved understanding of technological learning pivotal. Currently, most strategies and policies are only based to a limited extent on a rational and detailed understanding of learning mechanisms and technology development pathways. The conditions that provide efficient development routes are subject to much research, for example, in the innovation sciences. However, in addition to what may provide the optimal conditions and settings to achieve technological progress and rapid market deployment, it is clear that a detailed understanding of specific technologies, their performance, and factors influencing their performance are essential in order to design and implement effective policies and strategies. Historically, technological learning has resulted in the improvement of many technologies available to mankind, subsequent efficiency improvements and reduction of production costs, and has been an engine of economic development as a whole. Many of the conventional technologies in use today have already been continually improved over several decades, sometimes even over a century (e.g., most bulk chemical processes, cars, ships, and airplanes). Specifically for the electricity sector, coal-fired power plants have been built (and improved) for nearly a century now, while nuclear plants and gas-fired power plants have been built and developed since the 1960s and 1970s on a large commercial scale. Note that these well-established technologies are also still continuously improved, though this mainly leads to incremental improvements and concomitant cost reductions. Due to this long-term development, the established fossil fuel technologies have relatively low production costs. However, they also have a number of negative externalities, especially the emission of GHGs.

In contrast, many renewable/clean fossil fuel—energy technologies and energy-saving technologies used to have higher production costs, but lower fuel demands and GHG emissions. A few examples are electricity from biomass, offshore wind, and photovoltaics

(PVs), and energy-efficient lighting and space-heating technologies. For many of these new technologies the potential for further technological development and resulting production cost reductions is deemed substantial, and relatively high-speed cost reduction occurs compared to the conventional technologies. In the past 10 years, the gap between conventional and new technologies has been (largely) closed, and in some cases breakeven points have been reached. Electricity from onshore and offshore wind parks and large amounts of PV systems already today push out fossil generation units in Germany, as these technologies have no fuel costs. Crucial questions are, however, what will happen when intermittent electricity technologies will gain an even larger market share, making backup capacity and various forms of electricity storage (and associated additional investments) a necessity. Also, many renewable heating and transport technologies cannot yet compete with their fossil counterparts, so for these technologies, further technological progress is essential.

Thus the past and future development in time of production costs of (renewable) energy technologies (and the linked cost of CO_2 equivalent emission reduction) are of great interest, as the information allows policy makers to develop strategies for cost-effective implementation of these new technologies.

One approach to analyzing the reduction in production costs employs the so-called experience curve. It has been empirically observed for many different technologies that production costs tend to decline by a fixed percentage with every doubling of the cumulative production. As a rule of thumb, this cost reduction lies between 10% and 30%. To date, the experience curve concept has been applied to (renewable) energy technologies with a varying degree of detail.

The importance of progress in technological development of energy technologies is evident. Many (national) policies support R&D and provide the usually costly incentives for market deployment of targeted energy technologies. However, timing of incentives, the specific design of policy measures, and the amount of support that may be effective for success are very hard to determine. The resulting situation makes an improved understanding of technological learning extremely important. The relevance is clear from the urgency to achieve significant changes in the energy system (both in efficiency and in supply) at a rapid pace, to minimize costs and at the same time achieve competitive performance as soon as possible.

In recent years, much more insight has been gained into how learning regarding energy technologies has been acquired and also how their vital, further improvement can continue in the future. Many of these insights are derived from studies that have employed the experience curve approach. In 2009, Junginger, van Sark, and Faaij compiled a first comprehensive overview of experience curves for various energy technologies (both fossil and renewable). Since then, however, the energy transition has further progressed, and technologies have for the first time been commercially deployed on a significant scale (e.g., LED lamps and electric vehicles), which further matured (e.g., offshore wind and heat pumps) or regained new interest (such as green hydrogen production). In this book, less emphasis than previously has been put on the incumbent fossil energy technologies, and the focus has been put on these new technologies, which are likely to play an important role in the coming decades.

Also, the future energy system is challenged by the intermittent nature of renewables and requires therefore several flexibility options. Still, the interaction between different options, the optimal portfolio, and the impact on environment and society are unknown. It was the core objective of the H2020 REFLEX project to analyze and evaluate the development toward a low-carbon energy system with focus on flexibility options in the EU. The analysis was based on a modeling environment that considered the full extent to which current and future energy technologies and policies interfere and how they affect the environment and society while considering technological learning of low-carbon and flexibility technologies. By assessing the competitiveness of technologies and their interrelation, the cost-effectiveness of the whole system for future years as well as the systemic context demands for a well-founded energy system analysis including an appraisal of technological learning. Within REFLEX, this challenge was addressed by the integration of experience in an integrated energy models system. The main findings and lessons of integrating experience curves in the various sector models are of importance for modelers, but also policy makers and industry.

The renewed need for comprehensive overview of the technological learning progress of various technologies needed for the energy transition and the increasing interlinkage of the electricity, heat and transport sectors and the need to jointly model are the rationales for this book.

1.1.2 Objectives and structure

This book aims to provide an overview of the technological development and cost reductions achieved by the major energy technologies that are expected to be deployed as part of the ongoing energy transition. At the same time, it shows how future cost reductions and subsequent deployment of these technologies may shape the future mix of the electricity, heat and transport sectors. The central concept in this book is the experience curve, which quantifies the (past) cost reductions that occur together with the cumulative deployment of a technology.

The book first explains the concept in detail, including possible pitfalls and a new pedigree approach of mapping the quality of experience curves, discusses how this tool is currently implemented in models (and associated methodological challenges and solutions), and explores new applications of the concept, such as the ex ante assessment of environmental impacts of energy technologies (e.g., reduction of energy use of associated GHG emissions with increasing deployment). In a second part, nine chapters will focus on specific technologies that are relevant for the energy transition (e.g., PV and concentrated solar power, onshore and offshore wind, batteries and electric vehicles, heat pumps, power-to-hydrogen technologies, and space-heating technologies). For each technology the current market trends, past cost reductions and underlying drivers, available experience curves, and future prospects are treated in detail. In the third part of the book the results of various electricity supply, energy demand, and transport sector models play a key role. They also show how the future deployment of these technologies (and their associated costs) will determine whether ambitious decarbonization climate targets can be reached (and at what costs). The final chapter focuses on general lessons and recommendations for policy makers, industry, and academics, focusing on among others what technologies may require further policy support on the short term to have a major impact later on, which investments will be needed, and what scientific knowledge gaps remain for future research.

CHAPTER 2

The experience curve: concept, history, methods, and issues

Atse Louwen[1,2] and Juliana Subtil Lacerda[1]

[1]Copernicus Institute of Sustainable Development, Utrecht University, Utrecht, The Netherlands, [2]Institute for Renewable Energy, Eurac Research, Bolzano, Italy

Chapter outline

2.1 Introduction

Especially since the Industrial Revolution, impressive technological progress has been made. This progress has resulted in the development of many novel technologies, but there

Technological Learning in the Transition to a Low-Carbon Energy System.
DOI: https://doi.org/10.1016/B978-0-12-818762-3.00002-9

are also abundant examples of technologies that have remained in essence the same (at least performing the same function) but which have seen gradual but strong improvements over time. Examples that come to mind are the airplane and passenger car and, more recently, technologies in the computer industry, such as processors and memory chips. Learning is considered a key driver in these examples of endogenous technological change (Junginger et al., 2010). In this chapter, we introduce the concepts of the learning curve and experience curve, mathematical relationships that describe the technological progress for a technology—measured in unit cost reductions—as a result of increases in cumulative production of this technology. We discuss their origins, definitions, key applications, and finally present some main methodological issues and drawbacks.

2.2 Learning and experience curves

Within the context of technological learning, it is important to distinguish two concepts: (1) learning and (2) experience curves. Both refer in some ways to the same phenomenon that, as producers, gain more *experience* with manufacturing of a product, the costs of production will decrease. However, the exact parameters that the curves describe differ between the two concepts.

The phenomenon of the learning curve was first observed and documented in the 19th century by a German psychologist Hermann Ebbinghaus. He described that learning is an exponential process, meaning that the fastest learning occurs in the beginning and that exponentially more effort is required for subsequent increases in learning (Ebbinghaus, 1885). Ebbinghaus was the first researcher to mathematically document the learning process in an experiment he conducted upon himself. He measured the number of repetitions it took to memorize lists of words and found that those declined in an exponential manner (Ebbinghaus, 1885), as shown in Fig. 2.1.

The first well-documented quantified example of the learning curve in the context of technology costs was published by Wright (1936). When examining the manufacturing of airplanes, Wright stated that the time required (measured in unit labor costs) for each airplane built decreased with a constant percentage every time the cumulative number of airplanes produced doubled. The relation was described by Wright with the equation:

$$F = N^X \tag{2.1}$$

where F is the observed variation in labor costs, N the quantity of airplanes produced. For an 80% reduction in labor costs the exponent X has the value of 0.322. The unit labor costs are then the reciprocal of F. Wright attributed the unit labor cost reductions to a well-known theory that states that as assembly line workers gain more experience, they become more efficient in their work.

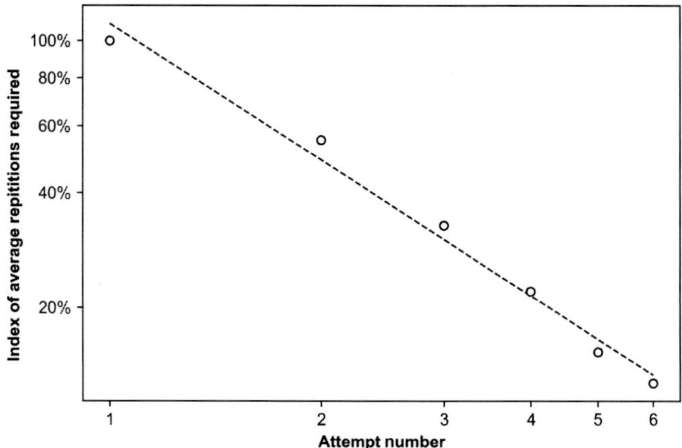

Figure 2.1
Learning curve made from Hermann Ebbinghaus' experiments on the number of repetitions required to memorize certain lists of words. *Data from: Ebbinghaus (1885).*

Wright also examined the relation between quantities of airplanes produced and unit material costs, and observed different mechanisms that led to cost decreases. First, by increasing the production quantity, relative amounts of waste decrease. Second, higher quantities allow for more economical purchasing of materials from external suppliers. By combining the curves for labor and materials, and also including overhead costs, Wright stated that the total airplane costs follow a curve that has a steeper slope in the beginning due to the higher contribution of labor costs and which gradually has a less steep slope, as the proportion of material costs to overall costs increases. He saw the value in this concept and the mathematical representation of the cost decrease resulting from the cumulated production experience to be able to assess the cost developments for very large numbers of production.

2.2.1 The single factor experience curve

After Wright, it took quite some time for this theory to become more mainstream. As discussed in Junginger et al. (2010), it was not until the RAND Corporation revisited the subject to study the possibility of cost reductions in production of war materials that gained more prominence. In the 1960s, the concept was broadened and introduced into the field of economics by Arrow (1962), and it was developed further by the Boston Consulting Group (1970) into the concept of the experience curve. Boston Consulting Group (BCG) expanded Wright's learning curve concept to describe the total cost of products and used it to describe unit cost of a product across a whole industry, rather than within a single company, and called this concept the experience curve to distinguish it from the previous learning curve. BCG included in this theory the combined effects of learning-by-doing, learning-by-researching (more commonly research and development, R&D), scale, and investment. Taking all this into account, the experience curve got the following form:

$$C_Q = C_1 \cdot Q^b \tag{2.2}$$

In this equation, C_Q is the cost of the product at cumulative production Q, C_1 is the cost of the first unit ($Q = 1$) produced, and b is the experience parameter. In essence, this is the same formula as posed by Wright, with the distinction that C_Q gives the unit cost (not cost reduction) as Q^b is multiplied with C_1, and so

$$\frac{C(n)}{C_1} = \frac{1}{F} \tag{2.3}$$

where F is as in Eq. (2.1). As Wright already showed, this power law shows a straight line when plotted on a double-logarithmic scale. With this in mind the equation can also be expressed as a linear equation by expressing it in a logarithmic form:

$$\log\ C_Q = \log\ C_1 + b \cdot \log\ Q \tag{2.4}$$

The experience curve parameter b thus represents the slope of the linear representation of the experience curve in a double-logarithmic graph. Since the slope of this line indicates the rate at which a technology's cost decreases, two terms have been connected to the experience parameter b: the progress ratio (PR) and the learning rate (LR):

$$PR = 2^b \tag{2.5}$$

$$LR = 1 - 2^b \tag{2.6}$$

For an LR of 20% (PR of 80%), the cost of a product decreases by 20% for every doubling of cumulative production n. Hence, these parameters are a more intuitive expression of the experience parameter b. An example of an experience curve, for solar photovoltaic (PV) modules, is given in Fig. 2.1. Shown in Fig. 2.2 is the raw, empirical data collected, the derived experience curve, and an example of plotting this data on normal, linear scales and on double-logarithmic or log–log scales. For this dataset, an LR of 23.9% was derived, indicating a decline in price of 23.9% for every doubling of cumulative production of PV modules.

2.2.2 Two and multifactor experience curves

As discussed in the previous section, the unit cost reductions observed for a technology, are the result of a combination of learning drivers. In addition, developments in input material prices can also have a large effect on the cost development of a technology. The experience curve as discussed above, which we from now refer to as the one-factor experience curve

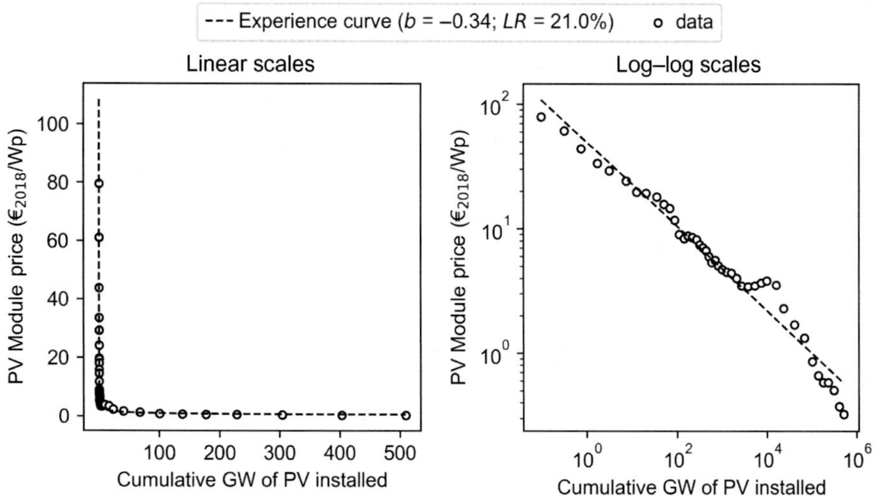

Figure 2.2
Example of experience curves on two different graph scales: normal, linear scales (left), and double-logarithmic or log–log scales (right). *Source: Data from Louwen et al. (2018).*

(OFEC), treats all different drivers and developments essentially as a black box, and thus only describes the observed empirical trend of decreasing unit costs but gives little to no insight in the underlying mechanisms driving these cost reductions. By trying to separate the different drivers of cost reductions, invaluable insight can be obtained on how to influence cost reductions (e.g., from a policy maker's point of view) and how to explain cost developments that seem to deviate from the long-term trend that is given by the OFEC.

To be able to include these considerations in the experience curve concept, several extensions have been made to the OFEC. The first example, we give here, is the extension of the OFEC to include separately the effects of learning-by-doing and learning-by-researching (R&D), in a two-factor experience curve (TFEC). By taking into account R&D expenditures, either directly or using some proxy metric, these drivers can theoretically be separated and measured:

$$C_Q = C_1 Q^b KS^c \tag{2.7}$$

$$\log C_Q = \log C_1 + b \log Q + c \log KS \tag{2.8}$$

$$LR_{LBD} = 1 - 2^b \tag{2.9}$$

$$LR_{LBR} = 1 - 2^c \tag{2.10}$$

Here, we obtain two separate *LR*s, LR_{LBD}, being the *LR* for learning-by-doing, or cumulative production, and LR_{LBR} being the *LR* for learning-by-researching. The latter describes the unit cost reductions of the studied technology, as a function of doubling the R&D effort, in this formula measured by the parameter *KS*, the "knowledge stock." The parameter *KS* takes into account the fact that knowledge gained from R&D expenditures generally does not directly result in cost reductions (there is a certain time lag), and that the knowledge stock depreciates over time when no further research is conducted:

$$KS_t = (1-\delta)KS_{t-1} + RD_{t-x} \tag{2.11}$$

In this equation, *KS* in year t is based on the *KS* of the previous year, with addition of the R&D expenditures *RD* in year $t-x$, where x is the time lag for implementation of R&D-related learning in technology improvements. The parameter δ describes the depreciation of the knowledge stock, based on the underlying assumption that without further research, the value of knowledge gained from R&D efforts gradually declines, along with its ability to drive down unit costs.

An example of application of the TFEC in the currently developing technology is the study by Kittner et al. (2017), who studied the development of energy storage technology. By taking into account both production volume (as opposed to cumulative production) and "innovation activity," they analyzed the effect of economies of scale and learning-by-researching on declining lithium-ion battery prices and found that this TFEC is better able to describe the observed price trends than an OFEC based on either production volume or cumulative production (Kittner et al., 2017). Further detail on this study is given in Chapter 8.

Further examples of extensions of the OFEC were presented by Yu et al. (2011) for PV technology. Yu et al. analyzed the effect of input prices for silver and silicon, as well as increasing manufacturing scale for PV plants on the cost developments of PV technology. The TFEC was expanded to a multifactor experience curve (MFEC), with different parameters for input prices $(P_1, P_2, \ldots, P_i)$ and multiple learning parameters $(q_1, q_2, \ldots, q_i)$ in the following generalized equation:

$$C_{\text{cum}} = aQ_x^{(1-r)/r}\left(\prod_{i=1}^{m}\left(q_i^{\sigma_i}\right)\right)^{1/r}\left(\prod_{i=1}^{n}\left(P_i^{\delta_i}\right)\right)^{1/r} \tag{2.12}$$

where Q_x is the instantaneous production, δ_i is the elasticity of input prices P_i, and a is defined as $a = r\left(\prod_{i=1}^{n}\delta_i^{\delta_i}\right)^{-(1/r)}$, with r being the returns-of-scale parameter. The first product $\prod_{i=1}^{m}\left(q_i^{\sigma_i}\right)$ in Eq. (1.12) represents the effects of technological changes, for instance as a result of learning-by-doing or learning-by-researching, while the second product $\prod_{i=1}^{n}\left(P_i^{\delta_i}\right)$ represents the effect of prices of inputs, while $Q_x^{(1-r)/r}$ shows the scale effect. The number of input learning variables and input prices is denoted by parameters m and n, respectively.

In addition to the TFEC approach discussed earlier, Kittner et al. also studied a MFEC for lithium technology, incorporating raw material prices in addition to economies-of-scale and innovation activity. Although a high correlation was observed for this MFEC, no significant improvement compared to the TFEC was observed.

2.3 Empirical data collection for experience curves

The value of the experience curve concepts stems for a large part from the use of empirical data, giving us an evidence-based description of technology cost reduction as a function of cumulated production experience. In the following sections, we describe guidelines for data collection and harmonization, experience curve parameter derivation, and assessment of data quality.

2.3.1 Cost versus price data

To be able to derive the experience curve parameters as shown in Eqs. (1.2)–(1.6), at least two datasets are required: the development of unit technology costs over time and the development of cumulative production of this technology over time. Until now in our discussions, both the learning and experience curves refer to the development of production *cost*. However, for most technologies, cost data is not readily available. Hence, most studies of experience curves rather make use of the unit *price* of technologies.

Market prices of technologies depend not only on the production cost and some fixed margin but are also affected by marketing strategy, supply and demand of the product, competitiveness of the market for this product, available subsidies, etc. The Boston Consulting Group (1970) proposed four stages of cost development as a result of pricing policy relative to market and product maturity:

- Development: prices below costs to compete with existing alternatives
- Umbrella: early commercialization leads to price increase above costs
- Shakeout: strong price reductions due to increasing competition
- Stability: costs and prices move parallelly in mature markets

During the development phase a manufacturer introduces a new technology at a price point below that which does not cover production costs (e.g., with a negative margin). With this strategy the manufacturer can compete with incumbent technologies that are already in a further stage of development and can create a market for the new product (forward pricing). When cumulative production increases, unit *production* costs decline rapidly but prices are (kept) stable, until a profit margin becomes viable. This “umbrella” phase shows that the product is commercially viable, and hence should attract competitors to enter the market. The dominant market players can commonly determine the market price for a certain time, until a shakeout occurs where prices decline rapidly in a short period of time. After the shakeout phase the market would reach stability, where both

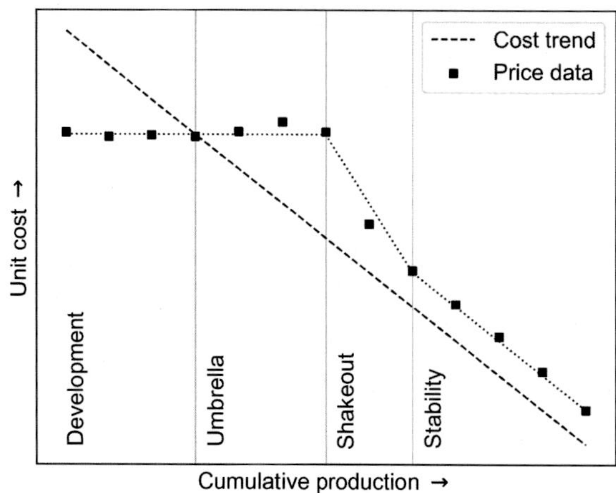

Figure 2.3
Illustration of cost versus price dynamics during the development and introduction of a new technology. *Source: Adapted from Boston Consulting Group (1970); Junginger et al. (2010).*

costs and prices decline at the same stable rate, with constant relative profit margins (Fig. 2.3).

In the context of experience curves, price data is a good proxy for cost only in this final stage of stability, since the slope of cost and price curves should be identical. Experience curves made with data points from early market deployment stages can be highly variable as more data comes in and do not likely reflect the *LR* in a stable, mature market situation. Even in the phase of stability, certain market developments might severely affect the cost versus price dynamics. In particular, when price data is used, a period of at least 10 years' worth of historical data or two orders of magnitude of cumulative output should be available for price trends to be reliably reflective of cost trends (Junginger et al., 2010; Gross et al., 2013).

Although the decision to use price rather than cost data is usually made out of necessity, since cost data is hardly available, there are still reasons to justify the use of price data. When making investment decisions, informed consumers or businesses consider the unit price, total cost of ownership or levelized cost of electricity (LCOE) of a technology, rather than taking into account production costs for the manufacturers. Since experience curves are commonly used to make projections about future investment decisions between a number of competing technologies, it makes sense to consider prices rather than costs. This does mean that when analyzing experience curves, particular care needs to be taken to take into account market dynamics.

2.3.2 Functional unit

When gathering unit cost or price data for experience curve analyses, it is important to determine for which unit this data is being collected. This unit should relate to the function of the product, as it is likely that a manufacturer will try to optimize production so that the economic performance of the products' function will improve. Hence, data should be collected for the unit that best describes the technology's function, for example, the *functional unit*. The technology characteristics determine the functional unit.

For instance, when purchasing solar PV systems, consumers (or commercial installers) often make a comparison between different offered PV systems based on the cost per unit of rated capacity (kW) and use this metric to make an investment decision. Although in the end, the LCOE (EUR/kWh) is most likely the most suitable metric to make an investment decision; the cost in EUR/kW is the most convenient and hence the most likely method in making this decision.

For other technologies or products, for instance hydrogen production by means of electrolysis, costs are often reported per unit of electrolysis stack capacity (EUR/kW), while an end consumer of hydrogen would be more interested in the cost per unit of output product (EUR/kg hydrogen). In the end the application of the experience curve determines the functional unit that is required, but technology characteristics will affect whether this unit is suitable for deriving an experience curve. An example of this issue is found in Chapters 6 and 7, where experience curves for on- and offshore wind are discussed. While earlier studies show reasonable experience curves of unit capacity cost (EUR/kW) versus cumulative production, certain developments in wind technology have led to an increase in the unit capacity cost. One of the reasons behind this is that manufacturers and installers of wind farms seem to optimize this technology to improve LCOE, rather than unit capacity cost (Williams et al., 2017).

Related to the topic of the functional unit is the issue whether it is reasonable to assume that a technology learns as a whole, or as a function of learning in several separate components. Datasets for solar PV-system components (see Chapter 5) indicate that different components have different *LR*s; hence, the overall cost decline of the PV systems is more accurately described by taking into account these separate *LR*s, rather than a single *LR* for the complete systems.

2.3.3 Data harmonization

When collecting data, it is likely that datasets gathered will be consisting of different source datasets, each in its own currency and/or currency year. When deriving an experience curve, it is important to make sure the complete dataset is harmonized to a single currency and currency year. For this harmonization the following steps are necessary: correction for inflation of the currency and conversion to a single currency using exchange rates. We suggest the following approach:

- Correct for inflation using a GDP deflator for the local currency
- Convert the currency to EUR using exchange rate of the given currency year

Other means of inflation correction are those using consumer price indices (CPI) or producer price indices, and additionally, one could argue that the conversion furthermore needs to take into account purchasing power parity. Depending on the technology or product under study, it might be more reasonable to use CPI for inflation correction, especially if the technology is primarily a consumer product. However, for most of the technologies discussed in this book, using the GDP deflator approach is deemed most suitable.

2.3.4 Common issues with data collection

When collecting data for derivation of experience curves, a number of issues often arise. In Table 2.1, an overview is given of common issues, based on the experiences in the REFLEX project (Louwen et al., 2018). In addition, several ways to solve these issues are listed and are discussed next. In Chapters 5–13, wherever applicable, these data collection and quality issues are discussed.

Arguably, the most common issue with data collection is the unavailability of production cost data. Hence, the majority of experience-curve studies are based on price data. In the previous section, we have discussed the implications of this issue. Out of necessity, the solution is to use price data, but another important recommendation is to thoroughly analyze market dynamics and market diffusion stages for the technology under study. Connected to the issue of cost versus price data is one of its implications: that supply and demand dynamics can affect prices significantly. An option would be to use MFECs that account for, for example, supply/demand balance (Gan and Li, 2015), which is also recommended if there is evidence that input material prices affect the price or cost to a substantial degree.

Data that can be used to derive experience curves is often gathered on a country basis. Hence, price data could be very specific for the local market conditions. In addition, when deriving the *LR* based on, for example, local cumulative production data, the value of the *LR* will be very different from that based on global cumulative production data. A trivial solution is to collect more data from different regions in order to form an aggregate dataset that could more closely reflect average global prices. For specific technologies, it might be necessary to adjust the data for purchasing power.

Table 2.1: Overview of possible data gathering issues.

Issue	Resolution
Data is not for cost but for price	Use price data as indicator for costs
Data not available for desired cost unit	Convert data to desired unit if possible Use available data as a proxy
Data is valid for limited geographical scope	Convert currency and adjust for PPP if necessary Combine with other datasets from various geographical scopes
Cumulative production figures not available	Calculate from annual production figures Calculate from annual sales figures
Data is in incorrect currency or currency year	Convert currency and correct for inflation and PPP if necessary
Early cumulative production figures are not clear or available	Restrict the dataset to time horizon for which reasonable cumulative production figures are available
Supply/Demand affecting costs significantly	Correct using multifactor experience curves (if required data is available) Otherwise, decide whether to discard this data, or keep data as is
Lack of empirical (commercial scale) data	Use proxy technologies, use expert estimates

PPP, Purchasing power parity.

Just like technology production costs are often difficult to collect, cumulative *production* data is normally also not readily available. In many cases cumulative *sales* or *installation* data is much more readily available and can serve as an adequate proxy for cumulative production data, as long as there is no clear evidence of large discrepancies between production and sales or installation. A related issue is the availability or accuracy of early cumulative production data for the beginning of the dataset. In this case it might be advisable to restrict the dataset timeframe to a range for which there is sufficient confidence in the accuracy of the data.

One of the most difficult issues when deriving experience curves, especially apparent for very new technologies, is a complete lack of empirical data. A very prominent example of such a technology is carbon capture and storage (CCS). Although there are a (small) number of pilot plants installed worldwide, and there is cost data for some of these pilot plants, the dataset is too limited and there is too much of heterogeneity between the pilot plants to derive an accurate *LR*. Given the importance of CCS in many energy scenarios, it is thus nearly impossible to use experience curves based on empirical data from CCS installations. In this case a possible solution is to make use of proxy technologies that are closely related to the technology under study. Rubin et al. (2007) made use of this approach for CCS, where they used experience curve data for a number of related proxy technologies to derive a nominal *LR* for power plants equipped with CCS systems.

2.4 Estimation of experience curve parameters

2.4.1 Regression method: linear or nonlinear

To be able to estimate the values of the experience curve parameters, the empirical data is to be used to perform a regression fit. Given the expected relation of $y = a{\cdot}x^b$, one of the options would be to perform a nonlinear regression of the raw, harmonized dataset. However, most commonly for this type of relation, a linear regression is performed of a logarithmic transformation of both x and y data. In general, transformation of the input data is recommended when the resulting fit of a nonlinear regression results in nonnormally distributed fit residuals, or when these fit residuals show heteroscedasticity. The choice between the two methods is also related to the assumption of the type of error in the dependent variable. Assuming an additive error in the original power law function, the equation to be fitted becomes

$$C_Q = C_1 \cdot Q^b + \epsilon \tag{2.13}$$

where ϵ is the error term. Given the power law relation and the developments of unit costs from very high initial values to much lower current values, it seems fair to assume that the error term in Eq. (1.13) would be multiplicative, rather than additive, for example:

$$C_Q = C_1 \cdot Q^b \cdot \exp(\epsilon) \tag{2.14}$$

which becomes, after log-transformation, an equation with an additive error:

$$\log C_Q = \log C_1 + b \log Q + \epsilon \tag{2.15}$$

The difference between the two regression methods is exemplified in Fig. 2.4. In this figure it becomes apparent that there is a large difference between the experience curves derived with the two regression procedures. Visually, the linear regression of log-transformed data gives much better results. In Fig. 2.4 (middle and right panels), the fit residuals are plotted against the "measured" cost data. It is apparent that the nonlinear regression shows large heteroscedasticity, when compared to the fit residuals of the log-transform linear regression. This suggests that using the latter method would be recommended.

2.4.2 Determining fit accuracy and experience curve parameter errors

When performing the regression of the empirical experience curve data, it is recommended to present the accuracy of this regression. Commonly, the authors present the accuracy in terms of the coefficient of determination R^2, which shows that in the context of experience curves, the proportion of observed variance in the costs can be attributed to variation in the cumulative production. Although the R^2 value can be considered a good measure to show goodness of fit, it is important to realize that it does not say anything about causality in the

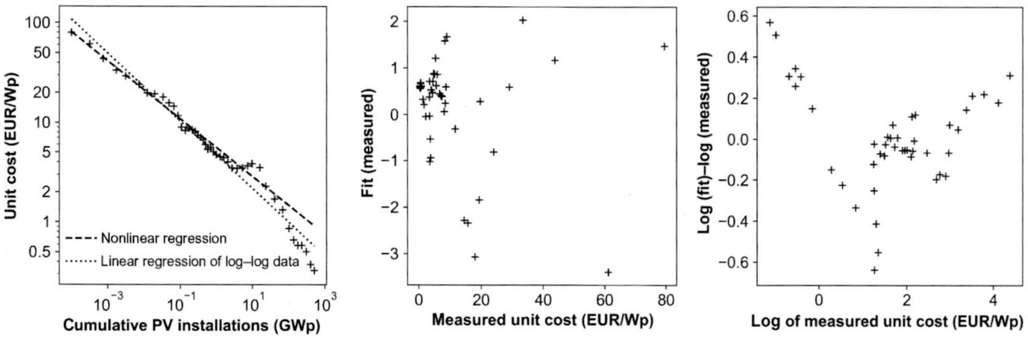

Figure 2.4

Comparison of nonlinear regression of raw data and linear regression of log-transformed data. The example dataset used here is for PV modules. *Data from: Fraunhofer ISE (2019); Louwen et al. (2018).*

observed correlation, if the used dataset is sufficiently large for statistical validity. Furthermore, it does not indicate whether the chosen regression model is the appropriate one, and as Kittner et al. (2017) discuss, it gives no information about omitted-variable bias.

A partial solution to some of these issues, is to furthermore present the *P*-values of the linear regression. Statistical software commonly gives *P*-values for the regression coefficients (intercept and slope). These values for the regression coefficients indicate the probability that the null-hypothesis is valid, with the null-hypothesis stating that intercept and slope of the regression are zero. Aside from the *P*-value for statistical significance, it is also informative to present the errors or the confidence intervals of the fitted regression coefficients, especially for the slope of the regression.

In addition to the *P*-value for the regression coefficients, it is also useful to present the overall significance of the linear regression. This can be calculated by performing an *F*-test, and the associated *P*-value. The *F*-test compares explained and unexplained variance and tests against the null-hypothesis whether the dataset can be described by a model with only an intercept, and no additional coefficients. A *P*-value can be attributed to the *F*-test result, allowing us to reject this null-hypothesis at different levels of significance. It can be said that while the R^2 shows how strong the correlation is between the x and y data, the *F*-test allows us to judge whether this correlation is statistically significant.

2.5 Applications of experience curves

As discussed in the previous sections, the experience curve is a broadly applied tool, and its validity has been empirically demonstrated for many different technologies (Junginger et al., 2010). The experience curve enables to forecast future changes in the technology costs by assuming that these costs decline as experience with a technology is gained through production and use. A large body of empirical evidence has been already built demonstrating a strong negative correlation between experience and cost for different technologies. The experience curve approach has been used increasingly in economic modeling to endogenize cost developments by accounting for the interrelationship between a technology's cost and its deployment (Löschel, 2002; Gillingham et al., 2008; Della Seta et al., 2012; Wei et al., 2017a,b).

There are many examples of literature of the past decades of experience curves applied to maturing and emerging energy technologies, such as solar PV modules (Bhandari, 2018), wind turbines (Williams et al., 2017), hydropower (Chang et al., 2016a,b), energy storage (Kittner et al., 2017; Schmidt et al., 2017), fuel cells (Wei et al., 2017a,b), and CCS (Riahi et al., 2004).

The experience curve is thriving as a key analytical approach due to its fundamental role in forecasting trajectories of energy system transformation, which is required to tackle challenges such as climate change, energy security, and poverty. By enabling to foresee most likely cost trajectories of different energy technologies, the experience curve facilitates strategy design and policy making as it provides insight into which technological mix is most efficient from economic, environmental, and social points of view. The International Energy Agency (IEA) (200AD) published one of the first comprehensive overviews of experience curves for renewable energy technologies. This assessment only covered a limited set of renewable energy technologies. In 2006 the NEEDS project (Neij et al., 2006) resulted in experience curves for a variety of energy technologies, and also included comparisons with bottom-up engineering estimates. The results of the NEEDS project were updated and extended in 2010 in Junginger et al. (2010), including the addition of demand-side technologies and lessons for policy makers. In this book, we update the results of Junginger et al. with up-to-date datasets, and now include more novel technologies such as energy storage (Chapter 8), electric vehicles (Chapter 9), and heat pumps (Chapter 11). This work is the result of the REFLEX project, where a large set of energy models are coupled and technology costs are modeled using experience curves.

Experience curves can be applied by companies, analyzing the speed at which the manufacturing costs of the products they sell may decline, by energy modelers as a tool to forecast future investment costs, and by policy makers as a tool to design policy measures.

For companies, business digitalization together with electrification and large amounts of data available has made energy a common variable for a range of business decisions almost regardless of the sector. As energy consumption becomes a key cost variable and GHG emissions a fundamental factor for a business reputation, companies are increasingly moving toward forecasting the most economically and environmentally efficient energy sources, giving rise to another application of experience curves in business.

For policy makers, we can identify at least three key applications. First, they can apply experience curves directly as a tool that allows for monitoring and quantification of investments required for technologies to attain a competitive level. Second, experience curves can be used more indirectly, as a means for energy models to estimate future prices of technologies and to forecast the deployment of several competing technologies. Third, experience curves inform climate and energy policy by providing data on the best combination of energy technologies to reduce GHG emissions, on how to increase energy security and on how to reduce energy poverty.

2.5.1 Direct applications of experience curves for policy makers

From a policy maker's point of view, the experience curve can inform policy design at different dimensions such as policy instrument choice (Kivimaa and Kern, 2015), level and type of financial support (Cárdenas Rodríguez et al., 2015), mix of command-and-control and market-based instruments (Jaffe et al., 2005), tackling uncertainty (Guivarch et al., 2017), adaptation to support or regulation needs of a specific technological sector (Malerba, 2005), and efficient pace of phasing out subsidies (Rezai and van der Ploeg, 2016). A key element on policy guidance based on the experience curve relates to the sources of learning. Cost reductions, derived from the experience curve, result from learning gained mainly from three different sources, namely, learning-by-doing (Arrow, 1962), learning-by-using (Chang et al., 2016a,b; Morstyn et al., 2018), and learning-by-interacting.

Learning-by-doing, related to the concept originally developed by Arrow (1962), refers to the cost reductions due to higher efficiency achieved through increasing experience with the production process. Here, deeper knowledge through work specialization and repletion can lead to process improvements such as work pace, waste reduction, and higher labor safety.

Learning-by-using comes from the demand side where efficiency gains come from the users' side, who, by using a technology, can learn how to operate it more efficiently. With the increasing decentralization of energy supply (Kainiemi et al., 2019) and growing role of prosumers (Olkkonen et al., 2017; Masson et al., 2018), learning-by-using is bound to have a stronger impact on the experience curve. Moreover, the spreading number of user groups tends to generate networking effects (Beermann and Tews, 2017; Li et al., 2017) that may strengthen this kind of learning.

Finally, learning-by-interacting is related to the open innovation idea (Bogers et al., 2016) where, along a production chain, actors (from product developers to final users passing by regulators, suppliers, and grid managers) exchange information regarding possible improvements or problems related to the use of a technology (Kreitlein et al., 2015). This information exchange, either formal or informal, opens up further learning opportunities as a broader view of technology performance becomes available to all actors involved. The data collected for experience curve studies show that experience curves are rational and systematic tools that can be used to describe historical developments in cost and performance of technologies. Furthermore, they can be applied to forecast future cost developments. By increasing cumulative production, market prices of a technology are driven down. This means that experience curves are a valuable tool to design policies that aim to increase deployment especially of low-carbon technologies.

The experience curve has been applied to policy making mainly in two areas. First, by looking at the experience curve of a technology (solar PV panels) or even of a technological system (e.g., smart grids), policy makers can forecast future price developments that tend to increase policy performance due to a more efficient design of policy measures as well as the optimization of public funds allocation to guide technology deployment and energy system transformation. For example, price changes, as forecasted by the experience curve, were a key factor to design the change of subsidies to solar energy in Germany and plan the pace of phasing out such incentives whereby controlling for side effects on the industry such as on-job creation and on-international competitiveness (Cherp et al., 2017). In addition, experience curves can serve as an indicator of technological maturity, which helps guide R&D investment both for private and public agents.

Second, experience curves can be applied to estimate the required investments for novel technologies to attain a certain competitive price level. To illustrate this, Fig. 2.5 takes a 20-year step back in history and indicates how technological learning as a result of increasing cumulative production reduced the cost of PV up to 1998. For PV, two levels of competitiveness are shown: socket parity (PV is competitive with the retail price of electricity, for consumers) and grid parity (PV is competitive with electricity spot prices).

The shaded areas of Fig. 2.5 show the total "learning investments" that would be required to reach certain levels of competitiveness. These required investments can be estimated by taking the integral of the experience curve between the current price level and the competitive price level. Taking 1998 as a starting point, and assuming grid parity at a price level of €2.1/Wp (which was roughly the level at which grid parity was achieved in the Netherlands), we can calculate that the total learning investments required are around €13 billion, compared to a total investment in PV modules of €6.52 up to 1998. To achieve grid parity (in this example at €0.37/Wp) would have required a total learning investment of just €925 billion. Current module prices are shown in Fig. 2.5.

The experience curve thus shows the total investments that are required for a technology to reach a competitive price level, but it gives us no information about the timing of reaching this break-even point. This timing depends on technology deployment rates, which policy

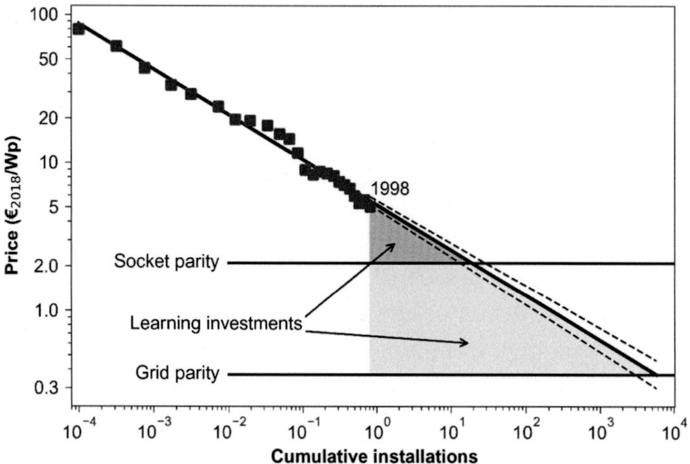

Figure 2.5
Visual example of the calculation of required investment in a technology to reach a certain break-even level of cost competitiveness with an incumbent technology. The dashed lines indicate the 95% confidence interval of the experience curve regression line.

makers can foster through demand pull and technology push incentives. Demand pull focuses on transforming market conditions to incentivize investments in innovation (Dosi and Grazzi, 2009); it is commonly directed toward diffusion improvements and justified by economies of learning obtained through gains of scale and incremental innovations (Nemet, 2009a,b). On the other hand, technology push emphasizes advances in knowledge as the main driver of technological change (Dosi and Grazzi, 2009), directed mostly to early stages of the innovation process; it is typically formulated as public funding of R&D activities (Peters et al., 2012) and performs better for the development of radical innovations (Costantini et al., 2015).

It should be noted that the learning investments only include the direct expenditures toward the technology and thus do not include other societal costs that are not necessarily included in the learning investments as estimated from the experience curve. They are primarily a result of market mechanisms, for example, consumers buying technologies thereby sustaining their markets. In the case of, for instance, government-funded research, development, and demonstration (RD&D), there might be an overlap with learning expenditures, as at least a part of the government funding is used for demonstration projects, and thus used for manufacturing the technology. For technologies very early in their market diffusion, government-funded RD&D might constitute significantly to the ongoing learning investments, but for more mature technologies (such as solar PV and wind), the majority of investments will come from the market. Even in these cases, government funding might be necessary to incentivize deployment.

Thus in order to reach competitiveness, the technology needs to be "driven down the experience curve," requiring that scaling up takes place at the technology domain level. To this end, the classic approach adopted so far by policy makers to foster the deployment of energy technologies in general—but environment-friendly ones in particular—is to adopt a policy mix that combines demand and supply side measures so that all types of learning are benefited. There is consensus that optimal policies to support environmental energy innovation should combine environmental regulation and technology policy (Popp et al., 2010; Acemoglu et al., 2012; Mowery, 2012). Environmental regulation focuses on pricing emissions, seeking to foster diffusion of environment-friendly technologies; whereas technology policy focuses on new technologies through support for knowledge creation. The combined implementation of such policies, addressing both market failures simultaneously, offers several advantages. For example, it enables the use of more precise, detailed, policy instruments, lowering the cost of emission reductions (Fischer et al., 2013). It also favors synergies between policy goals, since public research funding is more effective to foster environment-friendly innovation when accompanied by emission reduction policies (Newell, 2010). In addition, this averts escape routes as the green paradox (emission pricing could avoid increases in energy consumption due to cost-reducing innovations), and environmental rebound, emission pricing tends to stimulate the selection of technologies with least environmental impacts. Also, since emission control policy commonly involves assuring future demand for renewable energy generation, it provides a positive investment stimulus on innovation in this field, reducing uncertainty (Peters et al., 2012) and fossil fuel lock-in (Lehmann and Söderholm, 2017).

2.5.2 Indirect application of experience curves for policy: energy and integrated modeling

In Fig. 2.5 we showed how a single renewable energy technology rides down the experience curve toward a price level that is competitive with an incumbent technology. When thinking of the layout of future energy systems, there are of course many different technologies to consider that compete with each other and fossil fuel alternatives. To be able to analyze possible future energy systems, many research groups have developed energy models in the recent decades. Possibly the most prominent application of these models for policy makers is for the Intergovernmental Panel on Climate Change assessment reports, for which integrated assessment models are used to analyze, among others, how climate targets can be achieved with a mix of renewable and fossil technologies, and what the associated costs are (IPCC, 2014). These models, and many other energy models as well, very commonly use experience curves in some form to be able to model future technology price developments and technology deployment. Especially in endogenous applications, there is a feedback loop in these models between deployment and price of technologies, allowing these models to simulate and optimize future energy system layouts.

Thus by using these modeling results, policy makers gain insight in pathways toward a low-carbon energy system, and the associated costs required for market diffusion of clean but currently noncompetitive technologies. In Chapter 3, we discuss model implementation of experience curves in more detail.

2.6 Main issues and drawbacks of experience curves

As discussed in Section 2.3, a key issue of experience curves is data availability and the relation between cost and price of technologies and its implications for deriving experience curves. Aside from these issues, there are a number of factors that can hamper the applicability or accuracy of price extrapolations from experience curves. In this section, we highlight a number of key issues and drawbacks of experience curves.

2.6.1 Nonconstant learning rates and learning rate uncertainty

Since experience curves are most commonly applied as a tool to make extrapolations on future costs, the value of the *LR* is the key factor that determines how technology costs develop as cumulative production increases. The cost projections are thus extremely sensitive to the chosen *LR* value and its uncertainty. As Nemet (2009a,b) shows, there can be significant variation in the derived *LR* when taking into account datasets of different time periods. In his study of a dataset for PV module prices from 1976 to 2006, Nemet shows that when analyzing all periods in this dataset of 10 years or greater, the value of the derived *LR* can range from 14% to 25% (from 5th to 95th percentile). As a result, when analyzing the break-even year to attain a certain competitive cost level (see also Section 2.5.1), the break-even year ranges from 2017 to 2057 and requires learning investments from US$33 to US$903 billion (both from 5th to 95th percentile).

Aside from this nonconstant behavior of the *LR*, variance in the dataset away from the estimated experience curve trend results in uncertainty of the derived *LR* value. The result is generally an experience curve with relatively low R^2 values and high *LR* errors. When applying the *LR* for cost extrapolations, this *LR* error should be taken into account, for instance by a stochastic model formulation that explicitly calculates the impact of *LR* uncertainty (Junginger et al., 2010).

2.6.2 Technology systems and components

Depending on the technology, the dataset that is collected might represent a technology as an aggregate of components with a single price level. For instance, lithium battery storage system is composed of several components, including the lithium-ion cells, an inverter, and battery management system, and total system costs that can include or exclude installation costs. PV systems comprise PV modules and so-called balance-of-system components, which is shown in Chapter 5, to have different *LR*s. To accurately model and project the total costs of such technology systems, the use a component-based experience curve would be more suitable (Yeh and Rubin, 2012).

Only if a technology system is based on components that all have equal *LR*s based on one single cumulative production value, outcomes of projections will have the same results as those based on one single experience curve. In all other cases, projections might be under- or overestimating the future technology costs.

A similar issue lies in definition of the technology and its system boundaries. Again considering lithium battery storage, we know that lithium batteries exist in many different forms, with different materials, battery layouts, etc. In an ideal situation an experience curve would be based on a clearly defined single-battery technology, but unfortunately, this increases the data requirements substantially.

2.6.3 No explanation for cost reductions and causality

A fundamental issue of experience curves relates to the statistical correlation between cumulative production and observed cost declines. Many authors have argued that the correlation between these two variables offer little explanation of the reasons of the cost declines, as well as the causality between those variables (Yeh and Rubin, 2012). Although the theory on mechanisms of technological progress, such as learning-by-doing, learning-by-searching, and upscaling, is well documented and said to give a realistic representation of how technology costs decline, they are often difficult to separate in experience curve studies. Still, given the often strong correlation between cumulative production volumes and technology cost decline, the representation of this process in a single parameter remains attractive (Junginger et al., 2010). The value of experience-curve studies, however, relies not only on the datasets collected and the derived *LR*s, but equally as much on a thorough examination of the underlying mechanisms of the observed cost decline, as this gives valuable information on (limits to) future prospects of cost development.

2.6.4 Radical innovations

Since the experience curve is based on empirical historical data and is in its simplest form a trend characterized by one parameter that describes its slope, it cannot account for future (or historical) innovations that lead to a step change in technology costs. Substitutions of key materials or other innovations can lead to a drastic change in the technology cost, deviating from the long-term cost trend, and possibly resulting in a change in the observed *LR*. With sufficient data, it is possibly to identify such structural breaks in the experience curve trend, but for future projections, it is difficult to account for such trend breaks when using experience curves. This means that in addition to a thorough examination of historical technology developments (as discussed in the previous section), it is also recommended to make an inventory of possible future innovations of the technology under study.

2.6.5 Technology quality

Another downside to describing technologies with a single parameter, such as the *LR*, is that the quality of the technology is not necessarily taken into account. As discussed in Chapter 6, and by Williams et al. (2017), the unit capacity costs of wind turbines have varied substantially over the past decades. Up to 2005, onshore wind farm costs declined, but between 2005 and 2011, they increased substantially. This rise has been attributed to a variety of factors, but several changes to wind turbine and wind farm quality have contributed to this cost increase. For PV systems, experience curves are mostly also based on unit capacity costs and as such do not take into account variables such as system lifetime or capacity factors. In both cases the issues can be partly solved by analyzing experience curves for LCOE, but as is discussed in Chapter 6, this does not necessarily solve the issue completely. In any case it is important that studies explicitly analyze and discuss the roles of heterogenous quality attributes (Junginger et al., 2010).

References

Acemoglu, D., et al., 2012. The environment and directed technical change. Am. Econ. Rev. 102 (1), 131–166. Available from: https://doi.org/10.1257/aer.102.1.131.

Arrow, K.J., 1962. The economic implications of learning by doing. Rev. Econ. Stud. 29 (3), 155–173.

Beermann, J., Tews, K., 2017. Decentralised laboratories in the German energy transition. Why local renewable energy initiatives must reinvent themselves. J. Cleaner Prod. 169, 125–134. Available from: https://doi.org/10.1016/j.jclepro.2016.08.130.

Bhandari, R., 2018. Riding Through the Experience Curve for Solar Photovoltaics Systems in Germany. Available from: https://doi:10.1109/IESC.2018.8439945.

Bogers, M., et al., 2016. The open innovation research landscape: established perspectives and emerging themes across different levels of analysis. Ind. Innov. 24 (1), 8–40. Available from: https://doi.org/10.1080/13662716.2016.1240068.

Boston Consulting Group, 1970. Perspectives on Experience. Boston Consulting Group.

Cárdenas Rodríguez, M., et al., 2015. Renewable energy policies and private sector investment: evidence from financial microdata. Environ. Res. Econ. 62 (1), 163–188. Available from: https://doi.org/10.1007/s10640-014-9820-x.

Chang, Y.S., Jo, S.J., Jeon, S., 2016a. Using experience curve to project net hydroelectricity generation—in comparison to EIA's projection. SSRN Electr. J. Available from: https://doi.org/10.2139/ssrn.2732485.

Chang, Y.S., et al., 2016b. Is the 2040 projection on net electricity generation by Energy Information Administration too conservative? In comparison to alternate projection using experience curve. SSRN Electr. J. Available from: https://doi.org/10.2139/ssrn.2732488.

Cherp, A., et al., 2017. Comparing electricity transitions: a historical analysis of nuclear, wind and solar power in Germany and Japan. Energy Policy 101, 612–628. Available from: https://doi.org/10.1016/j.enpol.2016.10.044.

Costantini, V., et al., 2015. Demand-pull and technology-push public support for eco-innovation: the case of the biofuels sector. Res. Policy 44 (3), 577–595. Available from: https://doi.org/10.1016/j.respol.2014.12.011.

Della Seta, M., Gryglewicz, S., Kort, P.M., 2012. Optimal investment in learning-curve technologies. J. Econ. Dyn. Control 36 (10), 1462–1476. Available from: https://doi.org/10.1016/j.jedc.2012.03.014.

Dosi, G., Grazzi, M., 2009. On the nature of technologies: knowledge, procedures, artifacts and production inputs. Cambridge J. Econ. 34 (1), 173–184. Available from: https://doi.org/10.1093/cje/bep041.

Ebbinghaus, H., 1885. Über das Gedächtnis: Untersuchungen zur experimentellen Psychologie. Duncker & Humblot, Leipzig, Germany.

Fischer, C., Newell, R.G., Preonas, L., 2013. Environmental and technology policy options in the electricity sector: interactions and outcomes. SSRN Electr. J. Available from: https://doi.org/10.2139/ssrn.2432161.

Fraunhofer ISE, 2019. Photovoltaics Report. Freiburg, Germany: Fraunhofer ISE. Available at: https://www.ise.fraunhofer.de/en/publications/studies/photovoltaics-report.html.

Gan, P.Y., Li, Z., 2015. Quantitative study on long term global solar photovoltaic market. Renew. Sustain. Energy Rev. 46, 88–99. Available from: https://doi.org/10.1016/j.rser.2015.02.041.

Gillingham, K., Newell, R.G., Pizer, W.A., 2008. Modeling endogenous technological change for climate policy analysis. Energy Econ. 30 (6), 2734–2753. Available from: https://doi.org/10.1016/j.eneco.2008.03.001.

Gross, R., et al., 2013. Presenting the Future—An Assessment of Future Costs Estimation Methodologies in the Electricity Generation Sector. UK Energy Research Center, London.

Guivarch, C., Lempert, R., Trutnevyte, E., 2017. Scenario techniques for energy and environmental research: an overview of recent developments to broaden the capacity to deal with complexity and uncertainty. Environ. Modell. Softw. 97, 201–210. Available from: https://doi.org/10.1016/j.envsoft.2017.07.017.

International Energy Agency (IEA), 2000. Experience Curves for Energy Technology Policy. International Energy Agency, Paris.

IPCC, 2014. Zwickel, T., Minx, J.C., Edenhofer, O., Pichs-Madruga, R., Sokona, Y., Farahani, E., et al., (Eds.), Mitigation of Climate Change. Contribution of Working Group III to the Fifth Assessment Report of the Intergovernmental Panel on Climate Change. Cambridge University Press, Cambridge, United Kingdom and New York.

Jaffe, A.B., Newell, R.G., Stavins, R.N., 2005. A tale of two market failures: technology and environmental policy. Ecol. Econ. 54 (2–3), 164–174. Available from: https://doi.org/10.1016/j.ecolecon.2004.12.027.

Junginger, M., van Sark, W., Faaij, A., 2010. Technological Learning in the Energy Sector: Lessons for Policy, Industry and Science. Edward Elgar Publishing, Cheltenham.

Kainiemi, L., Eloneva, S., Levänen, J., 2019. Transition towards a decentralised energy system: analysing prospects for innovation facilitation and regime destabilisation in Finland. Technol. Anal. Strateg. Manage. 1–13. Available from: https://doi.org/10.1080/09537325.2019.1582765.

Kittner, N., Lill, F., Kammen, D.M., 2017. Energy storage deployment and innovation for the clean energy transition. Nat. Energy 2 (9), 17125. Available from: https://doi.org/10.1038/nenergy.2017.125.

Kivimaa, P., Kern, F., 2015. Creative destruction or mere niche creation? Innovation policy mixes for sustainability transitions. SSRN Electr. J. Available from: https://doi.org/10.2139/ssrn.2743173.

Kreitlein, S., et al., 2015. Green factories Bavaria: a network of distributed learning factories for energy efficient production. Procedia CIRP 32, 58–63. Available from: https://doi.org/10.1016/j.procir.2015.02.219.

Lehmann, P., Söderholm, P., 2017. Can technology-specific deployment policies be cost-effective? The case of renewable energy support schemes. Environ. Res. Econ. 71 (2), 475–505. Available from: https://doi.org/10.1007/s10640-017-0169-9.

Li, Y., et al., 2017. A combined forecasting approach with model self-adjustment for renewable generations and energy loads in smart community. Energy 129, 216–227. Available from: https://doi.org/10.1016/j.energy.2017.04.032.

Löschel, A., 2002. Technological change in economic models of environmental policy: a survey. Ecol. Econ. 43 (2–3), 105–126. Available from: https://doi.org/10.1016/s0921-8009(02)00209-4.

Louwen, A., et al., 2018. Comprehensive Report on Experience Curves. Utrecht, The Netherlands. Deliverable of the project REFLEX. Available on http://reflex-project.eu/.

Malerba, F., 2005. Sectoral systems of innovation: a framework for linking innovation to the knowledge base, structure and dynamics of sectors. Econ. Innov. New Technol. 14 (1–2), 63–82. Available from: https://doi.org/10.1080/1043859042000228688.

Masson, G., et al., 2018. A snapshot of global PV markets—the latest survey results on PV markets and policies from the IEA PVPS programme in 2017. In: 2018 IEEE Seventh World Conference on Photovoltaic Energy Conversion (WCPEC) (A Joint Conference of 45th IEEE PVSC, 28th PVSEC & 34th EU PVSEC). IEEE. Available from: https://doi.org/10.1109/pvsc.2018.8547794.

Morstyn, T., et al., 2018. Using peer-to-peer energy-trading platforms to incentivize prosumers to form federated power plants. Nat. Energy 3 (2), 94–101. Available from: https://doi.org/10.1038/s41560-017-0075-y.

Mowery, D.C., 2012. Defense-related R&D as a model for "grand challenges" technology policies. Res. Policy 41 (10), 1703–1715. Available from: https://doi.org/10.1016/j.respol.2012.03.027.

Neij, L., et al., 2006. Cost Development: An Analysis Based on Experience Curves, Deliverable 3.3—RS1A of the NEEDS Project, co-funded by the European Commission Within the Sixth Framework Programme.

Nemet, G.F., 2009a. Interim monitoring of cost dynamics for publicly supported energy technologies. Energy Policy 37 (3), 825–835. Available from: https://doi.org/10.1016/J.ENPOL.2008.10.031.

Nemet, G.F., 2009b. Demand-pull, technology-push, and government-led incentives for non-incremental technical change. Res. Policy 38 (5), 700–709. Available from: https://doi.org/10.1016/j.respol.2009.01.004.

Newell, R.G., 2010. The role of markets and policies in delivering innovation for climate change mitigation. Oxford Rev. Econ. Policy 26 (2), 253–269. Available from: https://doi.org/10.1093/oxrep/grq009.

Olkkonen, L., Korjonen-Kuusipuro, K., Grönberg, I., 2017. Redefining a stakeholder relation: finnish energy "prosumers" as co-producers. Environ. Innov. Societal Trans. 24, 57–66. Available from: https://doi.org/10.1016/j.eist.2016.10.004.

Peters, M., et al., 2012. The impact of technology-push and demand-pull policies on technical change – does the locus of policies matter? Res. Policy 41 (8), 1296–1308. Available from: https://doi.org/10.1016/j.respol.2012.02.004.

Popp, D., Newell, R.G., Jaffe, A.B., 2010. Energy, the environment, and technological change. In: Handbook of the Economics of Innovation, vol. 2. Elsevier, pp. 873–937. Available from: http://doi.org/10.1016/s0169-7218(10)02005-8.

Rezai, A., van der Ploeg, F., 2016. Second-best renewable subsidies to de-carbonize the economy: commitment and the green paradox. Environ. Res. Econ. 66 (3), 409–434. Available from: https://doi.org/10.1007/s10640-016-0086-3.

Riahi, K., et al., 2004. Technological learning for carbon capture and sequestration technologies. Energy Econ. 26 (4), 539–564. Available from: https://doi.org/10.1016/J.ENECO.2004.04.024.

Rubin, E.S., et al., 2007. Use of experience curves to estimate the future cost of power plants with CO_2 capture. Int. J. Greenhouse Gas Control 1 (2), 188–197. Available from: https://doi.org/10.1016/s1750-5836(07)00016-3.

Schmidt, O., et al., 2017. The future cost of electrical energy storage based on experience rates. Nat. Energy 2 (8). Available from: https://doi.org/10.1038/nenergy.2017.110.

Wei, M., Smith, S.J., Sohn, M.D., 2017a. Experience curve development and cost reduction disaggregation for fuel cell markets in Japan and the US. Appl. Energy 191, 346–357. Available from: https://doi.org/10.1016/j.apenergy.2017.01.056.

Wei, M., Smith, S.J., Sohn, M.D., 2017b. Non-constant learning rates in retrospective experience curve analyses and their correlation to deployment programs. Energy Policy 107, 356–369. Available from: https://doi.org/10.1016/J.ENPOL.2017.04.035.

Williams, E., et al., 2017. Wind power costs expected to decrease due to technological progress. Energy Policy 106, 427–435. Available from: https://doi.org/10.1016/J.ENPOL.2017.03.032.

Wright, T.P., 1936. Factors affecting the cost of airplanes. J. Aeronaut. Sci. 3, 122–128.

Yeh, S., Rubin, E.S., 2012. A review of uncertainties in technology experience curves. Energy Econ. 34 (3), 762–771. Available from: https://doi.org/10.1016/J.ENECO.2011.11.006.

Yu, C.F., Van Sark, W.G.J.H.M., Alsema, E.A., 2011. Unraveling the photovoltaic technology learning curve by incorporation of input price changes and scale effects. Renew. Sustain. Energy Rev. 15 (1), 324–337. Available from: https://doi.org/10.1016/j.rser.2010.09.001.

CHAPTER 3

Implementation of experience curves in energy-system models

Atse Louwen[1,2], Steffi Schreiber[3] and Martin Junginger[1]

[1]*Copernicus Institute of Sustainable Development, Utrecht University, Utrecht, The Netherlands,* [2]*Institute for Renewable Energy, Eurac Research, Bolzano, Italy,* [3]*Energy Economics, TU Dresden, Dresden, Germany*

Abstract

To analyze the transition toward a low-carbon energy system and to develop appropriate policy measures toward this goal, large efforts are currently taking place to model future layouts of our energy system. In this context, it is important to consider the technological progress in energy-system technologies to take into account how this progress affects technology cost and deployment. In this chapter, we discuss the implementation of experience curves in energy modeling. Experience curves allow for endogenous modeling of cost reductions resulting from technological progress and are, therefore, widely applied in energy modeling. Several issues with implementation of experience curves in energy modeling are discussed, as well as possible solutions.

Chapter outline

Technological Learning in the Transition to a Low-Carbon Energy System.
DOI: https://doi.org/10.1016/B978-0-12-818762-3.00003-0

3.1 Introduction

In the past few decades the development of climate change mitigation and adaptation strategies are key issues of national and international discussions in policy, economy, and science. Therefore large efforts are currently taking place to model the energy transition and future pathways toward a low-carbon energy system across several sectors. In this context, assessing the impact of technological improvements is important to determine which technologies will increase their expansion and which technologies will be phased out in coming years (Louwen et al., 2018). Thus future cost developments of incumbent and new or premature technologies are influencing results of energy-system models significantly. To consider technology cost reductions with increased experience in energy models, mathematical formulations as learning curves or experience curves are implemented in the modeling code. The approach of incorporating the correlation between technology deployments and costs provides a framework for evaluating whole-system effects caused by and initiating further technology cost reductions (Heuberger et al., 2017). Hence, with consideration of technological learning in energy-system models through experience curves, scientists and policy makers can identify least-cost pathways and alternative pathways to encourage a low-carbon energy system and achieve CO_2 reduction levels at low costs. Furthermore, the consideration of cost-learning effects shifts periods for optimal investments to earlier planning years, which influences the competitiveness of technologies (Junginger et al., 2010). However, the implementation of experience curves in energy-system models still has some disadvantages and therefore chances and barriers encountered for modelers need to be discussed.

3.2 Energy modeling approaches: bottom-up versus top down

In general, energy-system models can be distinguished between top-down models (macroeconomic models) and bottom-up models (detailed techno-economic or process-oriented models). For both types of energy models, main aims are to examine deployment of energy technologies, the effects of energy policies, and the interplays between the economy, environment, and energy system.

Top-down models are applied to depict the whole economy on a national or regional level. Therefore effects of energy as well as climate change policies are generally assessed in monetary units. Further, macroeconomic models equilibrate market developments by maximizing consumer welfare, applying feedback loops between economic growth, employment, and welfare as well as by using production factors (Herbst et al., 2012). As shown in Fig. 3.1, a variety of modeling approaches exist under the umbrella of top-down models. Commonly in top-down modeling, general equilibrium modeling is applied (Junginger et al., 2010).

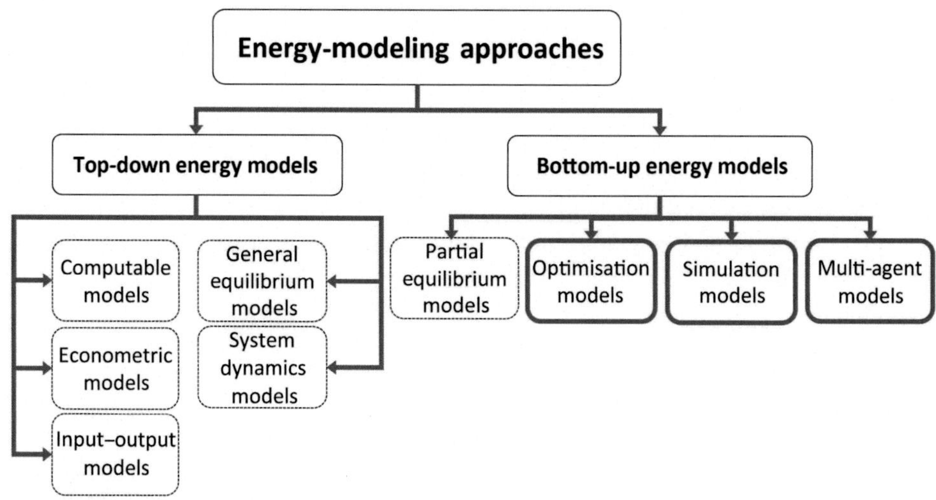

Figure 3.1
Overview of energy system modeling approaches. *Source: Based on Herbst et al. (2012). Credit: Steffi Schreiber.*

In contrast, bottom-up models are applied to depict energy sectors and the economy in an aggregated perspective by simulating economic developments, energy demand and supply as well as employment. Bottom-up models are much more detailed in terms of technological parameters, as compared to top-down models, and are often focused on separate sectors of the energy system. In the REFLEX project, bottom-up models separately model the transport sector, industry, and residential-energy demand, the electricity sector, and the heat sector. Bottom-up models can be generally distinguished into optimization models, simulation models, and multiagent or agent-based models.

Optimization models generally aim to minimize the cost of supplying some exogenous energy demand, while taking into account the available portfolio of energy technologies, including their technical and economic performance. Investment decisions are made based on, for example, total cost of ownership (TCO) or levelized cost of energy, but often, the whole timeframe of the modeling scope is optimized at once, meaning the models have perfect foresight.

Simulation models have a substantially different approach. Rather than finding a cost-optimal solution for a whole sector, over the whole modeling timeframe, simulation models attempt to more realistically capture behavior of actors in energy systems. Starting from a set of preexisting conditions, different actors in the modeled energy system make investment decisions based on TCO principles at each point in time. Often, these types of models employ algorithms that prevent technologies gaining a 100% market share, to ensure heterogeneity in market shares and simulate nonrational behavior (Herbst et al., 2012).

Finally, agent-based models, which can be considered a type of simulation model, are comprised of a set of autonomous "agents," who individually make decisions about deployment of technologies and their activities in the energy system. In contrast with other simulation models, there are several market players interacting in agent-based models, each making decisions and technology choices from a portfolio of available technologies, rather than an overall bottom-up simulation of an energy system as a whole. With this in mind, agent-based modeling is said to give a more natural description of (energy) systems, has the ability to include emergent phenomena, and should be more flexible compared to other modeling techniques (Bonabeau, 2002).

3.2.1 Integrated assessment models

A different category of models is the integrated assessment models (IAMs). These models, as their name indicates, integrate a variety of natural and economic processes. The aim of IAMs is to assess the effects of human activities on the Earth's system (van Sluisveld et al., 2018), with outcomes being effects on economy, greenhouse gas emissions, the energy system, and land use. IAMs are often applied to assess the effect of policy measures on climate change mitigation (Weyant, 2017; van Sluisveld et al., 2018). The models aim to inform policy makers about the requirements and consequences of limiting global-temperature increase (van Sluisveld et al., 2018).

IAMs, thus, extend their scope far beyond energy sectors, are often global models, and include representations of the economy, the land and climate system, and the energy system. Key model drivers include population growth, economic developments, policies, resources, and technological change. The prominent IMAGE model uses experience curves to model technological change in especially energy supply technologies (Stehfest et al., 2014), and many models incorporate technological learning endogenously (Stanton et al., 2009). Well-known models aside from IMAGE are (van Sluisveld et al., 2018): AIM/Enduse (Hibino et al., 2003), GCAM (Calvin et al., 2019), MESSAGE (Huppmann et al., 2019), REMIND (Luderer et al., 2015), and WITCH (Bosetti et al., 2006).

3.3 Implementation of experience curves in energy models

Modeling transitioning pathways for future energy systems requires precise cost estimations for several technologies across different sectors. The costs for technologies are changing over time regarding their technological improvements that can result in higher efficiency, reliability, or lower investment, operation, and maintenance costs (Junginger et al., 2010). By miscellaneous methods the decrease in technology costs due to learning mechanism as learning-by-doing, learning-by-researching, product upscaling (larger products), or production upscaling (economies of scales) can be estimated. One of these few methods are experience curves, which consist of empirical data that derive in mathematical functions to relate cumulative production experiences to cost decreases of technologies (Louwen et al., 2018). This chapter points out why it is necessary to implement experience curves into energy-system models. Further, the technical implementation of experience curves in energy models is described, followed by practical implications in different types of models. As experience curves have some limitations, the issues encountered for both endogenous and exogenous implementation are discussed at the end of this section.

3.3.1 Technical discussion on model implementation of experience curves

Costs are the key drivers for technology diffusion; however, estimating future costs is difficult and afflicted by uncertainties. Therefore experience curves are one of the few methods to allow evidence-based cost projections. In order to devise experience curves, two empirical datasets need to be gathered: data about the development of technology-production costs and about the development of cumulative production of the technology over a certain period (Louwen et al., 2018). The unit of the datasets depends on different types of applications. For instance, the production-cost data for energy supply technologies are given per unit of electrical capacity (e.g., EUR/MW_{el}) and the cumulative production in terms of total capacity (e.g., MW_{el}). According to these gathered data, the devised experience curves are the reduction of total product costs as a function of cumulative production (Boston Consulting Group, 1970).

$$C(Q) = C_1 \cdot Q^b \tag{3.1}$$

where $C(Q)$ is the cost C of a technology at cumulative production Q. Here, C_1 is the cost of the first unit produced and b is the experience curve parameter. The experience curve can be formulated in a linear equation by expressing it in a logarithmic form:

$$\log C(Q) = \log C_1 + b \cdot \log Q \tag{3.2}$$

The experience curve parameter b is the incline of the linear function represented in a double-logarithmic graph. The parameter b indicates at which rate the technology's costs decrease. Two parameters are connected to the experience curve parameter b: the learning rate (LR) and the progress rate (PR).

$$\mathrm{LR} = 1 - 2^b \tag{3.3}$$

$$\mathrm{PR} = 2^b \tag{3.4}$$

These two parameters are more meaningful than the experience curve parameter b since the LR (= 1 − PR) describes the decrease in costs of a product for every doubling of cumulative production Q (Louwen et al., 2018).

The equations above describe *single-factor experience curves*. However, technology cost reductions are influenced by different learning mechanisms (e.g., learning-by-researching and economies of scales) as well as by multiple, independent input material prices. Consequently, single-factor experience curves have to be extended to *multifactor experience curves* to describe the cost developments of a technology in more detail and devise the learning curves accurately (Yu et al., 2011; Heuberger et al., 2017; Kittner et al., 2017). According to the consideration of multiple, independent input variables, the extension to multifactor experience curves requires more empirical data input compared to single-factor experience curves. Furthermore, technologies can consist of several components with different costs and LRs, which also lead to higher data requirements. Energy-system models generally do not supply the required input variables for multifactor experience curves and thus normally use only single-factor experience curves (Louwen et al., 2018).

The implementation of experience curves in energy-system models by including the equations mentioned above directly into the modeling code allows the endogenous modeling of technological progress. For the endogenous implementation the model needs to implement the development of cumulative production. However, the endogenous implementation of experience curves is not feasible for all types of energy-system models. Amongst others, possible reasons are that the mathematics or optimization approach of a model does not allow for endogenous implementation, another reason can be that the geographical scope is limited or that the model does not calculate the cumulative production data in the required unit (Louwen et al., 2018). For models that do not use experience curves for endogenous learning, the exogenous implementation of experience curves can be an alternative by taking future cost reductions into account. The technology costs are changing over time by following an autonomous and exogenous cost-decline path (Junginger et al., 2008).

The technical implementation of endogenous and exogenous experience curves in energy-system models is presented in Fig. 3.2 in a simplified overview, where the gray boxes with the dotted frame represent external data sources, the blue boxes illustrate the model functions, and the transparent boxes stand for model-produced data. The direct *endogenous implementation* (left) is indicated by calculating the required data of the cumulative production in the energy-system model. The data is transferred into the experience curve function. Consequently, the technology costs are calculated, feedback looped, and applied in the energy model. With the dotted arrows, alternative routes of endogenous implementation of experience curves are illustrated. The endogenous calculated technology demand or technology penetration is converted into the required data and unit of cumulative production and further transferred into the experience-curve function. This can be defined as a direct approximation that can still be considered as endogenous calculation. Following the *exogenous implementation* route (right), the model provides only the year for which the

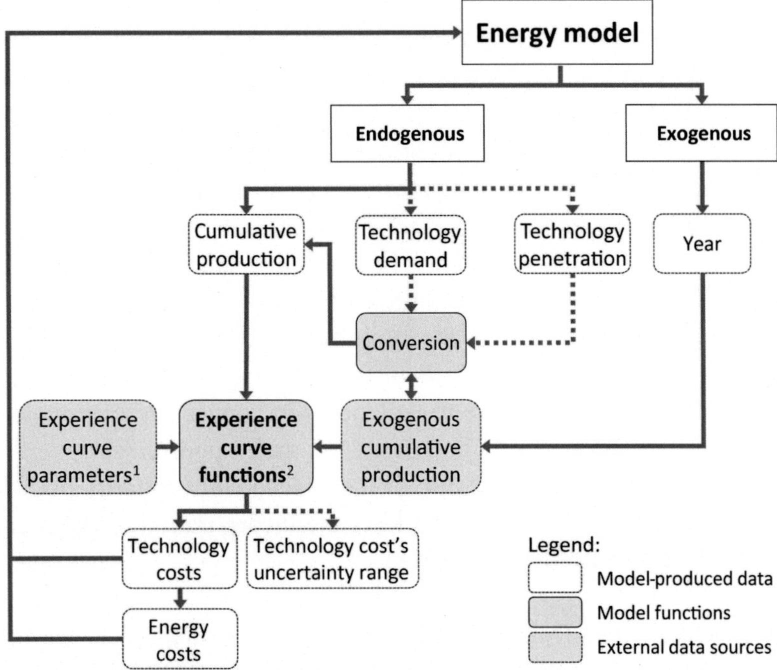

[1] Empirical data on Q (cumulative production) and C (cost)
[2] $C(Q) = C_1 \cdot Q^b$

Figure 3.2
Overview of possible endogenous and exogenous experience-curve implementation routes in energy-system models. *Source: Based on Louwen et al. (2018). Credit: Steffi Schreiber.*

technology costs should be calculated, while external data are describing the cumulative production over time, which is converted into the right unit and implemented into the experience curve function.

3.4 Practical implications in different types of models

3.4.1 Endogenous technological learning

The importance of considering endogenous technological learning in energy-system models is addressed by Weyant and Olavson (1999). The authors highlight that the response of technological learning to economic incentives is of crucial importance for designing appropriate energy and environmental policy measures. A comprehensive review of large-scale models employing endogenous learning curves is given in the Fourth Assessment Report of the IPCC (2007) in Junginger et al. (2010) by Lensink et al. (2010). While energy-system models often display long-term perspectives, the effect of technological learning and innovation can have a large effect on the cost-competitiveness of different technologies and therefore a significant effect on the general model results. Endogenous technological learning is widely used in macroeconomic top-down models, for example, in RICE (Castelnuovo et al., 2005), MIND (Edenhofer et al., 2006), or E3MG (Barker et al., 2006) as well as in bottom-up energy models, for example, MESSAGE (Messner, 1997), MARKEL–TIMES (Loulou et al., 2004), or ESO–XEL (Heuberger et al., 2017). In general, two types of models can be defined that employ experience curves endogenously—the general equilibrium models (top-down model) and the partial equilibrium models (bottom-up model). In the case of top-down models, usually the general equilibrium models are used because of their simple representation of the energy sector including all other sectors of the economy. Hence, the general equilibrium models are able to estimate the relationship between research and development investments in energy technologies and their opportunity costs. The partial equilibrium models are appropriate because of their reasonable quantity of technological detail and as its mathematical formulation allows the incorporation of the nonlinear experience-curve function (Rubin et al., 2015).

The endogenous technological-learning model produces internally consistent technology-cost trajectories, which enables the evaluation of policy measures on realizable future cost reductions. However, with increasing complexity of a model, the interpretation of the model results becomes more difficult (Junginger et al., 2010). Thus the endogenous incorporation of technological learning in energy-system models has some threats to mention. While the relationship between investment costs and installed capacity is nonlinear, binary variables have to be implemented in energy models, which lead to a significant increase of computational burden. Furthermore, as technological learning occurs in almost all technologies, learning asymmetries have to be avoided by applying endogenous learning consistently among all relevant technologies (DeCarolis et al., 2017). Seebregts et al. (2000) and Anandarajah et al. (2013) are using clustering of technological learning for similar technology modules, whereas the learning is applied across a set of technologies with similar components. In addition, technological improvements can be driven by a modeled country or region, but indeed, technological learning is a global phenomenon. Therefore modelers and policy makers should be careful while structuring the model and interpreting its results that are influenced by endogenous experience curves (DeCarolis et al., 2017). Another caveat related to the implementation of endogenous learning is that LRs are not trivial to estimate and that they are not remaining constant over time (McDonald and Schrattenholzer, 2001). Further, LRs for a certain technology vary between studies as different datasets are used, for example, different gross domestic product (GDP) deflator rates (Rubin et al., 2015). The variation of LRs over time must also be faced by considering exogenous cost assumptions, as a small change in the assumptions leads to substantial different optimal investment decisions (DeCarolis et al., 2017). Energy-system models with perfect foresight assumptions can place enormous investments in nonmature technologies with high LRs without failure. Hence, the investment patterns can differ significantly from the probable reality.

Several studies indicate that models with endogenous learning curves demonstrate benefits from the early adoption of a new technology as it encourages greater cost reductions over the long term (Mattsson and Wene, 1997; van der Zwaan et al., 2002; Nordhaus, 2009). Hence, in models with endogenous experience curves the cost of delays in introducing a new or enhanced technology can be extremely high compared to models that are not considering technological learning (Bosetti et al., 2011). Applying the MESSAGE–MACRO model, Riahi et al. (2004) investigated the effect on the global carbon dioxide abatement levels with and without technological learning for carbon capture and storage (CCS) technologies. The findings show that in the scenario with endogenous technological learning, the overall CCS costs are lower, resulting in higher CO_2 abatement levels compared to other mitigation methods and in lower opportunity costs of global CO_2 abatement in contrast to the scenario without endogenous technological learning. Thus the results evince that CO_2 mitigation policies are less costly in models with endogenous technological learning than in models without learning. Divergent from the expected reality, CCS technologies are playing a crucial role in the future energy system when experience curves are implemented in the analyses (Heuberger et al., 2017).

To summarize, the implementation of endogenous learning curves in energy-system models is complex but of crucial importance and enables modelers to determine the effectiveness of technology improvements and the supporting policy measures. However, the interpretation of modeling results should be done carefully, as the models with endogenous experience curves are not predicting policy impacts but achievable outputs. Thus for some specific cases, it can be more transparent to identify changes in technology costs exogenously over time and verify it by sensitivity analyses (DeCarolis et al., 2017).

3.4.2 Exogenous technological learning

A large part of energy-system models are considering constant or exogenously driven technology-cost reductions as a time-dependent input parameter (Gillingham et al., 2008; Green and Staffell, 2016). Three ways exist to employ technology performance and cost trajectories exogenously. The first method changes future technology costs and/or the technology efficiency by an annual rate from a reference year, that is, $x\%$ per year decrease in capital costs and/or $y\%$ increase in technology efficiency. The second method would be to directly estimate the absolute technology costs or performance parameter over time, that is, in EUR per capacity and net plant efficiency (Rubin et al., 2015). As a third option, experience curves can be derived by empirical data and can emerge the future cost developments over a specific time period, followed by the exogenous implementation of the data in absolute values (e.g., EUR/kW$_{el}$) into the energy-system model. The exogenous implementation of experience curves by the first two methods with its assumed quantitative values implies the judgment of modelers that may derive from data analyses and/or expertise. Implementing the cost reduction and performance improvements as time-dependent variables lead to the fact that investments in new technologies are avoided until the costs are decreasing significantly and the technology becomes competitive. Thus policy implications are delayed to later time periods. In contrast, with endogenous technological learning, the early adoption of a technology helps to drop down costs and makes the deployment more attractive (Rubin et al., 2015). Hence, the choice of method to implement endogenous or exogenous learning can have substantially different policy implications.

The fundamental difference between exogenous and endogenous technological learning is that the exogenous technological change is only time dependent, while the endogenous technological improvements can be influenced in several ways from past, present, and/or future expected policies and prices (Gillingham et al., 2008).

3.5 Issues, caveats, and drawbacks of experience-curve implementation in energy models

The implementation of technological learning in energy-system models is complex, related to many uncertainties and, therefore, connected with some burden. In the following sections, we will discuss several issues that can be encountered when implementing experience curves in energy models.

3.5.1 Geographical scope of model

Only few models exist that are displaying worldwide developments. As the majority of energy-system models are limited in their geographical scale, the technological learning outside the system boundaries are not considered endogenously. But technological learning is a global process. In the REFLEX project, only the developments in the European Union are considered. However, technologies as CCS or battery storages are used and will be used in future years worldwide. Hence, to derive consistent and reliable experience curves, worldwide learning has to be taken into account. This could be realized by assuming that technological learning outside the system boundaries (e.g., outside the EU) advances with the same velocity as inside the EU. An alternative would be to base technological learning on global energy scenarios as the World Energy Outlook. In addition, the technological learning could partly be exogenized by modeling developments outside the model scope with a different global model. The learning curve function will, therefore, be enhanced by parameters as n_{glob}—the global cumulative developments in the global model and by $n_{\text{glob}_{\text{in}}}$—the cumulative developments in the global model for countries in the local model.

Hence, the experience curve function to project the model costs considering exogenous global learning in a local model is as follows (Louwen, 2018):

$$C(Q) = C_1 \cdot (Q_{\text{in}} + Q_{\text{glob}} - Q_{\text{glob}_{\text{in}}})^b \tag{3.5}$$

where Q_{in} is the cumulative developments *in local model*, Q_{glob} is the global cumulative developments *in global model*, and $Q_{\text{glob}_{\text{in}}}$ is the cumulative developments *in global model* for countries *in local model.*

However, this approach requires that the external global model has a compatible geographical subdivision. Further, the level of technology detail can be insufficient in a global model. Therefore the *S*-curve approach (Fleiter and Plötz, 2013; Schmidt et al., 2017) can be applied if the global model does not consider the required technology.

3.5.2 Technological learning in an energy modeling system

Considering technological learning in an energy modeling system, where several models are soft-linked with each other (e.g., as in the REFLEX project), can lead to model inconsistencies as some technologies are considered by different models (e.g., heat pumps) and thus each model represents the technology deployment individually. Hence, technology penetration is often modeled based on different settings and criteria in each model, sometimes even exogenously. As experience curves and cost reductions are a mathematical function of cumulative deployment, strategies are needed to keep the deployment levels and costs synchronized between the models (Louwen, 2018).

With the incorporation of endogenous experience curves in top-down or bottom-up models a cost floor may be implemented to prevent the technology costs from falling below a specific value. This would help to estimate a feasible solution and avoid that costs decrease in an extreme path where technologies become unrealistically cheap within a certain time period (Rubin et al., 2015).

3.5.3 Technical issues

Another issue encountered regarding learning curves in energy-system models is that especially optimization models are not compatible with the nonlinearity and nonconvexity of experience-curve functions. Nonconvexity can lead to local maxima or multiple global equilibria (Messner, 1997). Therefore a global optimum cannot be guaranteed as it is required for linear optimization problems. The development of a piece-wise linear approximation of the exponential experience-curve function can be a solution for optimization models as presented by Barretto (2001) and Heuberger et al. (2017).

Another issue, especially with optimization models, is the tendency of these models to prefer technologies with high LRs, as these models have "perfect foresight" due to their technical definition. On the other hand, technologies with high initial costs can hardly be overcome in simulation models, which can suffer from their myopic nature. Only when taking into account, for example, government subsidies or other incentives do they enter market deployment, and otherwise stagnate in end-consumer prices and thus do not enter the market.

Aside from these technical issues, implementation of experience curves in modeling also requires increased processing time, especially in endogenous implementations that create certain feedback loops in the modeling computations.

3.5.4 Technology deployment constrained by modeling scenario and policy targets

Furthermore, technology diffusion can be constrained by policy targets and assumptions set in the scenarios, which will lead to a prompt stoppage of technological learning in the energy models. If this is the case, an ex post check of the experience curves and the results should be performed to understand which assumptions have what influence on the experience curves (Louwen, 2018). Another threat can occur if models are not producing the data required for the needed input of experience curves (e.g., no calculation of cumulative capacity for a specific technology). If this is the case, a proxy method shall be used to convert the data into the desired input data, for example, data gathering on relation between TWh and TW, or the data shall be used from an external model with similar scenarios (Louwen, 2018).

3.5.5 Other issues

The most common caveat of deriving and implementing learning curves in energy-system models is the lack of empirical data, especially for innovations and new technologies as CCS. If empirical data is lacking proxy technologies, expert elicitations or a simplified estimation of LRs can be used. Further, cross-sectoral or spillover effects are difficult to take into account. Spillover effects can, for instance, describe how the technological learning of lithium-ion batteries influences the reduction in prices of electric vehicles, which for a large part depend on the battery system. Conversely, large deployment of battery electric vehicles could lead to sharp price declines in stationary battery-storage applications. To consider these effects the use of component-based experience curves are a possible solution. However, this approach is more complex, requiring models to produce a larger set of input parameters, and enlarging the threat of missing empirical data (Rubin et al., 2015).

Price developments of technologies can severely be affected by market dynamics, including among others, different levels of market diffusion (see Chapter 2), supply and demand balance, and input material prices. Multifactor learning curves can partly solve these issues of market dynamics. Multifactor experience curves can take factors such as policies, competition, R&D spending, and prices of input materials into account. Nevertheless, usually one model does not produce all of the required multiinput parameters and considering multifactor experience curves will thus increase modeling complexity rapidly (Louwen, 2018). Multifactor learning curves are not widely used; therefore further research is required in context of endogenous model implementation of multifactor experience curves.

Obviously, there exists a trade-off between increasing the accuracy of modeling results and limiting additional model complexity. Therefore each modeler should find a balance between modeling accuracy and complexity, and analyze as well as interpret modeling results very carefully.

3.6 Concluding remarks

Implementation of technological learning processes in energy modeling is an essential part in analyses of future energy systems. It is critical to take into account continuous development and improvement of new and incumbent energy-system technologies when designing policy measures and analyzing energy-transition pathways laid out to achieve climate targets. By using experience curves in energy modeling, technology-cost trajectories can be modeled endogenously, creating a direct feedback between technology deployment and associated learning processes and cost reductions.

That being said, model implementation of experience curves is not without its issues and drawbacks. As we have discussed, a variety of model characteristics can hamper endogenous implementation. Many energy models are restricted in geographical scope, while technological progress is most often considered a global process. In these cases, it is likely that an (at least partly) exogenous cost trajectory based on experience curves is necessary, but a feedback between development within the model under study and technological learning is in this case only possible to a limited extent. Other issues encountered relate to technical or practical considerations in energy modeling, such as the mathematical layout of the model, computation time, or the ability of models, to produce the required input parameters for endogenous (multifactor) experience curves.

References

Anandarajah, G., McDowall, W., Ekins, P., 2013. Decarbonising road transport with hydrogen and electricity: long term global technology learning scenarios. Int. J. Hydrogen Energy 38 (8), 3419–3432. Available from: https://doi.org/10.1016/j.ijhydene.2012.12.110.

Barker, T., et al., 2006. Decarbonizing the global economy with induced technological change: scenarios to 2100 using E3MG. Energy J. 27, 241–258. Available from: <http://www.jstor.org/stable/23297066>.

Barretto, L., 2001. Technological Learning in Energy Optimization Models and Deployment of Emerging Technologies. Swiss Federal Institute of Technology, Zurich.

Bonabeau, E., 2002. Agent-based modeling: methods and techniques for simulating human systems. Proc. Natl. Acad. Sci. U.S.A. 99 (Suppl. 3), 7280–7287. Available from: https://doi.org/10.1073/pnas.082080899.

Bosetti, V., et al., 2006. WITCH—a world induced technical change hybrid model. SSRN Electron. J. Available from: https://doi.org/10.2139/ssrn.948382.

Bosetti, V., et al., 2011. What should we expect from innovation? A model-based assessment of the environmental and mitigation cost implications of climate-related R&D. Energy Econ. 33 (6), 1313–1320. Available from: https://doi.org/10.1016/j.eneco.2011.02.010.

Boston Consulting Group, 1970. Perspectives on Experience. Boston Consulting Group.

Calvin, K., et al., 2019. GCAM v5.1: representing the linkages between energy, water, land, climate, and economic systems. Geosci. Model Dev. 12 (2), 677–698. Available from: https://doi.org/10.5194/gmd-12-677-2019.

Castelnuovo, E., et al., 2005. Learning-by-doing vs. learning by researching in a model of climate change policy analysis. Ecol. Econ. 54 (2–3), 261–276. Available from: https://doi.org/10.1016/J.ECOLECON.2004.12.036.

DeCarolis, J., et al., 2017. Formalizing best practice for energy system optimization modelling. Appl. Energy 194, 184–198. Available from: https://doi.org/10.1016/J.APENERGY.2017.03.001.

Edenhofer, O., Lessmann, K., Bauer, N., 2006. Mitigation strategies and costs of climate protection: the effects of ETC in the hybrid model MIND. Energy J. 27, 207–222. Available from: <http://www.jstor.org/stable/23297064>.

Fleiter, T., Plötz, P., 2013. Diffusion of energy-efficient technologies. In: Encyclopedia of Energy, Natural Resource, and Environmental Economics. Elsevier, pp. 63–73. Available from: http://10.1016/b978-0-12-375067-9.00059-0.

Gillingham, K., Newell, R.G., Pizer, W.A., 2008. Modeling endogenous technological change for climate policy analysis. Energy Econ. 30 (6), 2734–2753. Available from: https://doi.org/10.1016/j.eneco.2008.03.001.

Green, R., Staffell, I., 2016. Electricity in Europe: exiting fossil fuels? Oxford Rev. Econ. Policy 32 (2), 282–303. Available from: https://doi.org/10.1093/oxrep/grw003.

Herbst, A., et al., 2012. Introduction to energy systems modelling. Swiss J. Econ. Stat. 148 (2), 111–135. Available from: https://doi.org/10.1007/bf03399363.

Heuberger, C.F., et al., 2017. Power capacity expansion planning considering endogenous technology cost learning. Appl. Energy 204, 831–845. Available from: https://doi.org/10.1016/j.apenergy.2017.07.075.

Hibino, G., et al., 2003. A guide to AIM/Enduse model. In: Climate Policy Assessment. Springer Japan, Tokyo, pp. 247–398. Available from: https://doi.org/10.1007/978-4-431-53985-8_15.

Huppmann, D., et al., 2019. The MESSAGEix integrated assessment model and the ix modeling platform (ixmp): an open framework for integrated and cross-cutting analysis of energy, climate, the environment, and sustainable development. Environ. Modell. Softw. 112, 143–156. Available from: https://doi.org/10.1016/J.ENVSOFT.2018.11.012.

Junginger, M., van Sark, W., Faaij, A., 2010. Technological Learning in the Energy Sector: Lessons for Policy, Industry and Science. Edward Elgar Publishing, Cheltenham.

Kittner, N., Lill, F., Kammen, D.M., 2017. Energy storage deployment and innovation for the clean energy transition. Nat. Energy 2 (9), 17125. Available from: https://doi.org/10.1038/nenergy.2017.125.

Lensink, S., Kahouli-Brahmi, S., van Sark, W., 2010. The use of experience curves in energy models. In: Junginger, M., van Sark, W., Faaij, A. (Eds.), Technological Learning in the Energy Sector: Lessons for Policy, Industry and Science 2. Edward Elgar Publishing, Cheltenham and Northampton, MA, pp. 48–62.

Loulou, R., Goldstein, G., Noble, K., 2004. Documentation for the MARKAL Family of Models. Available from: <https://iea-etsap.org/MrklDoc-I_StdMARKAL.pdf>.

Louwen, A., 2018. Technological learning in energy modelling – implementation of experience curves. In: EMP-E 2018 Conference, Session on Technological Learning. Brussels, Belgium.

Louwen, A., et al., 2018. Comprehensive Report on Experience Curves. Utrecht, The Netherlands.

Luderer, G., et al., 2015. Description of the REMIND model (version 1.6). SSRN Electron. J. Available from: https://doi.org/10.2139/ssrn.2697070.

Mattsson, N., Wene, C.-O., 1997. Assessing new energy technologies using an energy system model with endogenized experience curves. Int. J. Energy Res. 21 (4), 385–393. Available from: https://doi.org/10.1002/(SICI)1099-114X(19970325)21:4 < 385::AID-ER275 > 3.0.CO;2-1.

McDonald, A., Schrattenholzer, L., 2001. Learning rates for energy technologies. Energy Policy 29 (4), 255–261. Available from: https://doi.org/10.1016/s0301-4215(00)00122-1.

Messner, S., 1997. Endogenized technological learning in an energy systems model. J. Evol. Econ. 7 (3), 291–313. Available from: https://doi.org/10.1007/s001910050045.

Nordhaus, W., 2009. The Perils of the Learning Model for Modeling Endogenous Technological Change. National Bureau of Economic Research. Available from: https://doi.org/10.3386/w14638.

Riahi, K., et al., 2004. Technological learning for carbon capture and sequestration technologies. Energy Econ. 26 (4), 539–564. Available from: https://doi.org/10.1016/J.ENECO.2004.04.024.

Rubin, E.S., et al., 2015. A review of learning rates for electricity supply technologies. Energy Policy 86, 198–218. Available from: https://doi.org/10.1016/j.enpol.2015.06.011.

Schmidt, O., et al., 2017. The future cost of electrical energy storage based on experience rates. Nat. Energy 2 (8). Available from: https://doi.org/10.1038/nenergy.2017.110.

Seebregts, A., et al., 2000. Endogenous learning and technology clustering: analysis with MARKAL model of the Western European energy system. Int. J. Global Energy Issues 14 (1/2/3/4), 289. Available from: https://doi.org/10.1504/ijgei.2000.004430.

van Sluisveld, M.A.E., et al., 2018. Comparing future patterns of energy system change in 2°C scenarios to expert projections. Global Environ. Change 50, 201–211. Available from: https://doi.org/10.1016/J.GLOENVCHA.2018.03.009.

IPCC, 2007. In: Solomon, S., et al., (Eds.), Climate Change 2007: The Physical Science Basis. Contribution of Working Group I to the Fourth Assessment Report of the Intergovernmental Panel on Climate Change. Cambridge University Press, Cambridge and New York, 996 pp.

Stanton, E.A., Ackerman, F., Kartha, S., 2009. Inside the integrated assessment models: four issues in climate economics. Clim. Dev. 1 (2), 166–184. Available from: https://doi.org/10.3763/cdev.2009.0015.

Stehfest, E., et al., 2014. Integrated Assessment of Global Environmental Change With IMAGE3.0. Model Description and Policy Applications. PBL Netherlands Environmental Assessment Agency, Hague, The Netherlands.

van der Zwaan, B.C.C., et al., 2002. Endogenous technological change in climate change modelling. Energy Econ. 24 (1), 1–19. Available from: https://doi.org/10.1016/s0140-9883(01)00073-1.

Weyant, J., 2017. Some contributions of integrated assessment models of global climate change. Rev. Environ. Econ. Policy 11 (1), 115–137. Available from: https://doi.org/10.1093/reep/rew018.

Weyant, J.P., Olavson, T., 1999. Issues in modeling induced technological change in energy, environmental, and climate policy. Environ. Model. Assess. 4, 67–85.

Yu, C.F., Van Sark, W.G.J.H.M., Alsema, E.A., 2011. Unraveling the photovoltaic technology learning curve by incorporation of input price changes and scale effects. Renew. Sustain. Energy Rev. 15 (1), 324–337. Available from: https://doi.org/10.1016/j.rser.2010.09.001.

CHAPTER 4

Application of experience curves and learning to other fields

Atse Louwen[1,2], Oreane Y. Edelenbosch[3,4], Detlef P. van Vuuren[1,3], David L. McCollum[5,6], Hazel Pettifor[7], Charlie Wilson[5,7] and Martin Junginger[1]

[1]Copernicus Institute of Sustainable Development, Utrecht University, Utrecht, The Netherlands, [2]Institute for Renewable Energy, Eurac Research, Bolzano, Italy, [3]PBL Netherlands Environmental Assessment Agency, The Hague, The Netherlands, [4]Department of Management and Economics, Politecnico di Milano, Milan, Italy, [5]International Institute for Applied Systems Analysis (IIASA), Laxenburg, Austria, [6]University of Tennessee, Knoxville, TN, United States, [7]Tyndall Centre for Climate Change Research, University of East Anglia (UEA), Norwich, United Kingdom

Abstract

The concept of experience curves is normally applied to analyze cost or price developments of technologies, but studies have shown that the concept can also be extended to other applications and fields. In this chapter, we show the application of experience curves to analyze energy use in industrial processes, energy demand of household appliances and discuss its application to describe and project developments in environmental impact of technologies. A second part of the chapter shows a case study of applying the concept of technological learning to the field of social behavior in what is called social learning (SL). This part of the chapter shows how the diffusion of electric vehicles is affected by different SL mechanisms.

Chapter Outline

Technological Learning in the Transition to a Low-Carbon Energy System.
DOI: https://doi.org/10.1016/B978-0-12-818762-3.00004-2

4.1 Introduction

Until now, we have discussed the main concepts: technological learning (TL) and a mathematical representation of its effects: the experience curve. In this context of TL, we consider *technological* improvements to production processes that result in a decline of the *cost* of technologies. In this chapter, we will discuss several ways to expand on this concept, by discussing learning processes related to social influence social learning (SL) rather than technological progress, and by including improvement due to technological progress in other metrics than cost (energy-efficiency and environmental learning), as well as application of the experience curve concept in environmental impact analysis.

4.2 Energy experience curves

4.2.1 Experience curves for energy efficiency in industrial processes

Given the value of experience curves to describe the cost reductions in technologies, and to project future costs of the studied technologies, several authors have attempted to apply this concept to analyze industrial processes.

Ramírez and Worrell (2006) analyzed the production of ammonia and urea in the United States, in order to determine experience curves for the cost of these products. During their analyses, the authors found that although there is clear evidence for technological improvements in the production processes of these fertilizers, the historical price developments of these products is affected by two major parameters: supply and demand relations, and the price of natural gas, which is used as a key input in the production of both ammonia and urea. The natural gas that is required was found to account for 70%–90% and 70%–75% of the total costs for ammonia and urea, respectively. Ramirez and Worrell argued that since these contributions are so high, technological progress would be likely to focus on reducing the amount of natural gas used and thus modified the experience curve concept to analyze not overall costs, but rather the specific energy consumption (SEC) of the production process. Furthermore, since this process is constrained by a physical minimum, they added this minimum to the experience curve:

$$\mathrm{SEC} = \mathrm{SEC}_{\min} + \mathrm{SEC}_0 \cdot \mathrm{Cum}^m$$

where SEC is the specific energy consumption of the production process, $\mathrm{SEC}_{\min}$ is the minimum specific energy consumption, Cum is the cumulative production, and m is the experience curve parameter. The parameter $\mathrm{SEC}_{\min}$ represents the theoretical physical minimum for this conversion process and serves as the asymptote of the experience curve function. Ramirez and Worrell found learning rates of SEC of 23% for ammonia, and 9% for urea production.

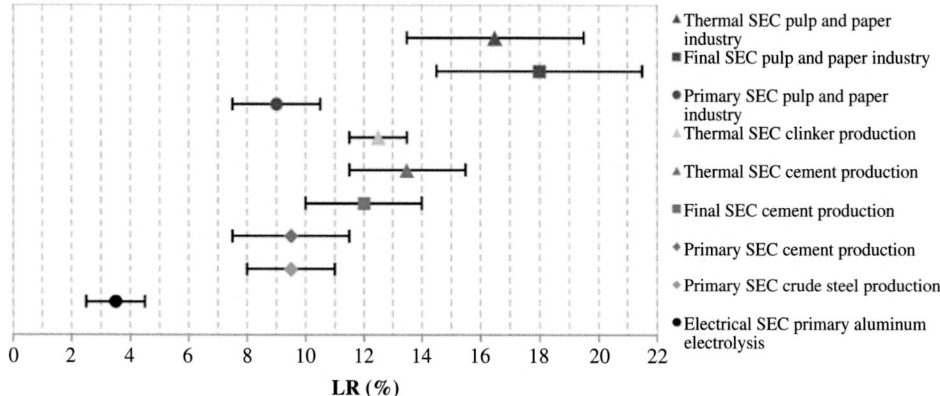

Figure 4.1
Overview of estimated learning rates for various industrial processes, as devised by Brucker et al. (2014). *Source: From Brucker et al. (2014).*

In similar work, Brucker et al. (2014) analyzed the specific energy consumption of several other industrial processes. The authors collected SEC data and built time series based on sector averages, or data on best-available-technology processes for paper and pulp industry, cement, clinker and crude steel production, and primary aluminum electrolysis. As shown in Fig. 4.1, a range of learning rates was found for these different products and industries, from 3.5% to 12.5% for primary energy SEC, while the coefficient of determination was 0.79 or higher, showing good agreement of the devised experience curve with the empirical data collected. The authors note that the learning rates were generally higher as the system boundary increased, seemingly confirming the notion that overall learning rates in industries are a combination of TL across a product's supply chain. The learning rates that were derived from the data for the different products were applied to analyze the development of SEC values up to 2035, based on bottom-up modeling results of the industrial sector (Fleiter et al., 2013), and note that although the results nicely agree with business-as-usual scenarios for developments of energy consumption, some results seemed too optimistic due to past leap improvements in SEC that would not be expected to occur again in the future.

4.2.2 Experience curves for energy efficiency in energy demand technologies

In a study of TL of energy demand technologies, Weiss et al. (2010) analyzed the learning rates of a variety of technologies, including dishwashers, washing machines, laundry dryers, refrigerators, and freezers. Table 4.1 gives an overview of the technologies analyzed, and the associated learning rates for different noncost parameters. Weiss et al. collected data from "The Consumentenbond," a Dutch nonprofit organization for consumer protection, over a time frame ranging from 1965 to 2008.

Table 4.1: Overview of results presented by Weiss et al. (2010) on the energy efficiency of large household appliances.

Technology	Learning rate	R^2	Learning rate parameter
Washing machines	35% ± 3%	0.92	Specific energy consumption in kWh_e/kg laundry
	28% ± 8%	0.57	Specific water consumption in L/kg laundry
Laundry dryers	20% ± 6%	0.84	Specific energy consumption in kWh_e/kg laundry
Dishwashers	18% ± 3%	0.89	Energy efficiency index
	31% ± 5%	0.84	Water consumption in liters
Refrigerators	17% ± 2%	0.87	Energy efficiency index
Freezers	13% ± 3%	0.79	Energy efficiency index

As shown in Table 4.1, especially the energy consumption of the household appliances reviewed show strong correlation with cumulative production of these appliances, and learning rates for energy consumption range from 13% to 35%. For water consumption of washing machines and dishwashers, strong learning rates are also observed, although especially for the former, the coefficient of determination is quite low, and the estimated error of the learning rate is quite large.

Weiss et al. list a variety of reasons that have led to the improvements of energy and water consumption in the household appliances. Improvements in refrigerators and freezers were the result of improved insulation and various optimizations to the cooling equipment (such as compressors, condensers, and heat exchangers). For dishwashers and laundry machines the reduction in energy consumption was stated to be for a large part of the result of decreased water consumption, as well as better heat recovery and progress in in other areas, such as improvements in detergent quality.

An interesting insight from the study by Weiss et al. was found when the authors analyzed the implementation of several energy efficiency policies in The Netherlands. By comparing the learning rate for energy efficiency before and after the implementation of these measures, the authors found that after the implementation, the learning rate increased significantly for dishwashers, refrigerators, and freezers. Especially for the freezers and refrigerators, the increase was large, going from 17% to 49% for refrigerators, and from 11% to 47% for freezers. These findings seem to indicate that learning rates can be influenced by policy makers, at least on the short term.

4.3 Experience curves for environmental impacts and life cycle assessment

As the previous section already made clear, the experience curve concept can be applied to several other metrics than just costs of technologies. The application of experience curves in energy modeling already has a long history, being applied to aid in the exploration of future energy systems that allow for a transition toward a low-carbon energy supply, in order to meet climate targets. Taking into account and quantifying the effects of technological progress is however also an important aim in studies of environmental impact of (energy) technologies, such as ex ante or prospective life cycle assessment (LCA).

Most of the examples of using experience curves for assessment of environmental impacts are related to photovoltaic (PV) technology. Louwen et al. (2016) collected LCA results of PV technology over a range of several decades. These LCA results were used to derive experience curves for the cumulative energy demand (CED) and greenhouse gas (GHG) emissions of mono- and polycrystalline–based PV systems. The study showed that these technologies exhibit learning rates of 12%–13% for CED and 17%–24% for GHG emissions. The derived experience curves were applied to calculate the total cumulative energy use and GHG emissions from the PV industry, in order to compare these environmental impacts with the environmental benefits gained as a result of electricity generation from the global installed PV system capacity. The authors found that in their most likely scenario, breakeven was achieved in terms of net environmental impact, and that after this breakeven point, the PV industry as a whole has had a net positive environmental impact, in terms of both GHG emissions avoided and energy output. Due to a decline in energy and GHG payback times of PV systems, strong growth (around 50% per year) of installed PV capacity is possible without creating an energy sink or increasing global GHG emissions on the short term.

Other authors have also applied experience curves in the domain of LCA. Sandén and Karlström (2007) applied experience curves to develop a method for consequential LCA to assess the environmental impacts of investing in alternative technologies. They argue that due to the concept of TL, marginal investments in alternative technologies help "drive technologies down" the experience curves; and hence, these contribute not only marginally to the current system but also make contributions to more radical system changes.

A more direct application of experience curves for prospective LCA would be to use them to generate foreground or life cycle inventory (LCI) data. As discussed by Arvidsson et al. (2018), experience curves could be used to generate this LCI data and could include technological parameters (such as those discussed in Section 4.2), or material inputs required to produce such technologies.

Bergesen and Suh (2016) used experience curves to generate these future material inputs for a prospective LCA on cadmium telluride (CdTe) PVs. By analyzing historical developments in the supply change of CdTe PV, they established learning rates for several of the foreground technology parameters, such as the amount of CdTe required, the efficiency of the PV modules, and the amount of electricity required in manufacturing. By combining these different learning rates, they were able to build a model that estimates the environmental impact of this technology as a function of increases in cumulative production.

Caduff et al. (2012) also applied experience curve concepts in order to assess the environmental effects of the increasing scale of wind turbines. The authors report that due to a combination of upscaling and TL the environmental footprint of electricity produced with wind turbines decreases as they are upscaled and improved.

4.4 Social learning

N.B. this section is a summary of Edelenbosch et al. (2018).

Besides technology costs, other behavioral factors, related to for example preferences, habits and values play a key role in technology choice (Mundaca et al., 2010; Gifford et al., 2011). Modeling behavioral influences on technology choice is however complex. There are a large number of factors that could be represented and are not easy to quantify (Stern et al., 2016). Behavioral factors also tend to be highly heterogeneous across different consumer groups (Laitner et al., 2000). In a technology transition, this is important because market heterogeneity itself can affect the choices of others through SL. We use "SL" in this context to indicate the change in individuals' understanding and preferences toward new technologies as a result of interactions within social networks (Rogers, 2003; Young, 2009; Reed et al., 2010). As an example, early adopters (EAs) moving to a new technology can impact others' preferences and decision-making processes by changing their perspectives on the status, reliability, and safety of a new vehicle (Axsen and Kurani, 2012; McShane et al., 2012). SL by users about the benefits and risks of new technologies is a key process in technology diffusion. In his seminal work on "diffusion of innovations," Rogers (2003) defines diffusion as the process by which an innovation is communicated over time among the members of a social system. These members are heterogeneous in their preferences, particularly toward risk and uncertainty. Earlier adopters are risk-tolerant or risk-seeking, preferring new and relatively untested technologies that offer novel attributes. Later adopters are risk-averse, preferring to wait until perceived technology risks are lowered by observing the experiences of EAs. Heterogeneous adopters are therefore interdependent, connected through social communication processes. Although the specific mechanisms of SL are diverse—ranging from word of mouth to visible "neighborhood effects" and compliance with social norms—the basic insight that heterogeneous consumers exchange information through social networks (Rogers, 2003) has been repeatedly confirmed both in general terms (e.g., Peres et al., 2010; McShane et al., 2012) and in studies specific to vehicle choice (e.g., Grinblatt et al., 2007; Axsen and Kurani, 2012).

A novel modeling approach developed by Pettifor et al. (2017) represents consumer heterogeneity in a global context and the dynamic nature of SL processes. Pettifor et al. (2017) compiled and synthesized empirical data on risk aversion to new vehicle technologies among different consumer groups. Following diffusion of innovations theory (Rogers, 2003), they then translated differing adoption propensities in to a single aggregated "risk premium (RP)," which declined as a result of social influence effects between the heterogeneous adopter groups.

That work was advanced by Edelenbosch et al. (2018) exploring how a dynamic representation of *both* SL *and* TL influences the long-term transition to battery electric vehicles (BEVs). The "SL" process has a clear analogy with TL as a process by which costs or barriers are reduced. Both types of learning effect impact how technologies diffuse, and both are processes that unfold over time. Both for TL as well as for SL it is not time per se that decreases perceived risks or costs but rather the experience of others (SL) and the experience of manufacturing and using technologies (TL).

4.4.1 Social learning in the transport sector

The transport sector represents one of the fastest growing sources of GHG emissions (Sims et al., 2014). Empirical studies show that in the transport sector behavioral factors such as esthetics, performance, attitude, lifestyle, and social norms have a strong effect on the choice of technology (Tran et al., 2012; Stephens, 2013; McCollum et al., 2017). The discussed recent modeling efforts have explored whether the behavioral realism of integrated assessment models can be improved, focusing on consumer choices for light duty vehicles (LDVs). LDVs are of particular interest as they account for approximately half of current energy consumption in the transport sector (Sims et al., 2014), and the sector is a very heterogeneous, with many users.

4.4.2 Modeling risk premiums and social and technological learning

Rogers (2003) distinguishes consumer segments along a normal distribution of adoption propensities. EAs have high initial adoption propensities and so high risk tolerance; early majority (EM), late majority (LM), and laggards (LG) are increasingly risk averse and have low initial adoption propensities. Based on this conceptualization, Pettifor et al. (2017) calculate initial RPs as a measure of adoption propensity for each of the four different adopter groups. Their RP estimates are based on discrete choice experiments that provide willingness-to-pay (WTP) estimates for new technologies, such as BEVs, for which limited market data is available. Pettifor et al. (2017) use a normal distribution of WTP point estimates from discrete choice studies to calculate a mean RP ($\overline{x}$RP) with associated standard deviation ($\overline{\sigma}$RP) for different adopter groups. Negative initial RPs indicate attraction to new technologies (risk-seeking), and high positive initial RPs indicate aversion to new technologies (risk-aversion). Following Rogers (2003), the EAs[1] occupy a 16% market share; both the EM and LM account for 34% of the market; and the LG the final 16%.

[1] Our Early Adopter (EA) group contains the both the early adopters and innovators described by Rogers, E.M., 2003. Elements of diffusion. Diffus. Innovations 5, 1–38.

Pettifor et al. (2017) also use a metaanalysis of 21 empirical studies to measure the effect of social influence on vehicle purchase propensities. They find that for every one standard deviation increase in market share, RPs decrease by 0.241 standard deviations, which increases vehicle adoption propensities [95% CI (0.157, 0.322), $Z = 5.505$, $|P| < .000$]. In other words, RPs decline as market share grows, using market share as a proxy for social influence. In the vehicle choice model of IMAGE the RPs (in $/passenger km) for each consumer group have been added to the travel cost. More details on the empirical analysis and the implementation in IMAGE are provided in "Supplementary Materials" of the study by Edelenbosch et al. (2018).

Besides SL dynamics, this study focuses on TL of the battery costs and distinguish between exogenous and endogenous learning scenarios. Battery costs in electric vehicles (EVs) have declined rapidly over recent years (Nykvist and Nilsson, 2015); therefore the battery costs start from a cost estimate of 300 US$/kWh in 2014 (Nykvist and Nilsson, 2015). In the exogenous cost scenario, we assume that battery costs could reach 125 $/kWh by 2025 (Faguy, 2015) and decline further to 100 US$/kWh over the course of the century. In the endogenous cost scenario, we use a learning rate of 7.5% (uncertainty range from 6% to 9%) in line with estimates from the literature (Nykvist and Nilsson, 2015). We also assume a floor price of 50 $/kWh, affecting the purchase cost of plug-in electric vehicles (PHEVs), BEVs, and fuel cell vehicles (FCVs). More widely used components of cars such as the car frame or engine are not assumed to be influenced by learning after many years of experience and so follow the same path as in the exogenous scenario.

4.4.2.1 Model setup

Consumer heterogeneity, TL, SL, and policy measures, can all influence vehicle choice. Fig. 4.2 demonstrates schematically how these processes are related in the model setup within the IMAGE transport vehicle choice module, a global integrated assessment model used for this analysis (Girod et al., 2012). Increased market share affects SL and TL for different adopter groups: EA, EM, LM, and LG.

4.4.2.2 Scenario framework

We use a set of four scenarios to explore the effects of SL and TL, and how they dynamically interact. In the reference scenario (labeled "Ref"), technology costs decline exogenously over time, and RPs are frozen for the four adopter groups. In the TL scenario (labeled "TL"), RPs are also frozen, but technology cost reductions occur endogenously based on a learning curve. In the reference + SL scenario (labeled "Ref + SL"), SL is included but with exogenous technology cost assumptions. Finally, in the technological and SL scenario (labeled "TL + SL"), both TL and SL occur endogenously.

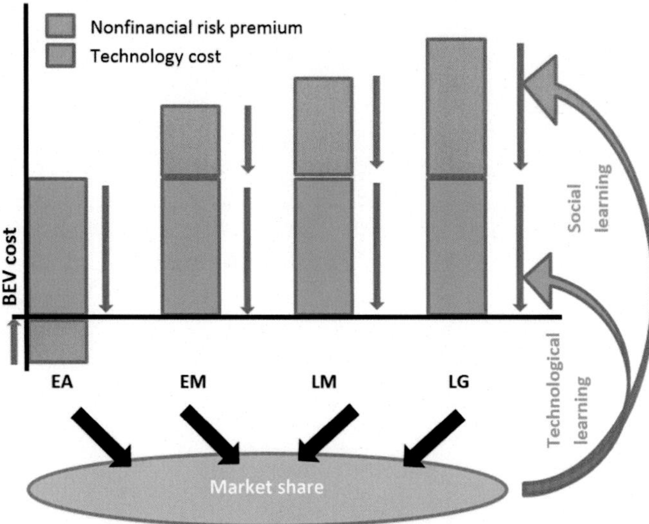

Figure 4.2
Schematic overview of the dynamic relationship between technological learning, social learning, and market deployment of new technologies. Four adopter groups are distinguished: EA, EM, LM, and LG. At a given time point, all four groups face the same technology cost but different monetized risk premiums. Net perceived costs therefore differ per group, with the lowest perceived cost vehicle selected by the cost-minimizing decision algorithm, resulting in changes to market share, which in turn stimulates further technological and social learning. *EAs*, Early adopters; *EM*, early majority; *LG*, laggards; *LM*, late majority. *Source: From Edelenbosch et al. (2018).*

4.4.3 Model results

4.4.3.1 Technological learning scenarios

Fig. 4.3 depicts market shares of the global vehicle fleet under endogenous and exogenous TL assumptions in the absence of SL. In the TL scenario the EA group shifts to PHEVs in the first half of the century given their preference for new technologies (represented by a negative RP that remains constant as there is no SL). Although EAs are also attracted to BEVs, this new technology remains too expensive through the first half of the century (Fig. 4.3 right panel). The deployment of PHEVs leads to reduction of both PHEV and BEV costs through TL in battery costs (Fig. 4.3 left panel). In the Ref (reference) scenario, BEV costs are projected to reduce rapidly in this period as well, based on exogenous assumptions. Once a certain BEV cost threshold has been passed, depending heavily on the learning rate (indicated by the TL range), EAs shift from PHEVs to BEVs. This shift leads to faster BEV cost reductions (Fig. 4.3 left panel). Under high learning rate assumptions, the EM group also adopts BEVs by the end of the century, by which point a small group of EAs move on to FCVs that have become more cost competitive.

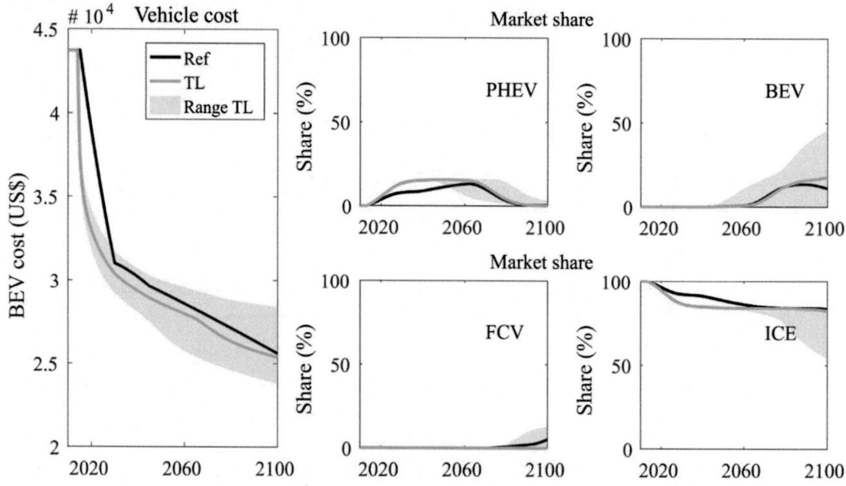

Figure 4.3

BEV cost over time in the Ref and TL scenarios (left panel), with resulting BEV, PHEV, FCV, and ICE market shares of the global vehicle fleet (middle and right panels). Shaded colors indicate the scenario range depending on assumed TL rates. *BEV*, Battery electric vehicle; *FCV*, fuel cell vehicle; *ICE*, internal combustion engine; *PHEV*, plug-in electric vehicle; *TL*, technological learning. *Source: From Edelenbosch et al. (2018).*

The EA group and TL play an important role in this initial phase of a technology transition. With slower learning rates, BEVs remain relatively expensive, and EV adoption might not take place at all. Even though the technology is competitive in terms of costs, if RPs remain at current levels purchasing a BEV is not an attractive option for the EM, LM, and LG.

4.4.3.2 Social learning and technological learning scenarios

In the SL scenarios the market deployment of BEVs drives down the RPs of the EM, LM, and LG, whereas for EAs the reduced novelty of BEVs makes them less attractive as RPs become less negative. Fig. 4.4 shows how the BEV RPs change over time for all four adopter groups in the Ref + SL and TL + SL scenarios.

The effect of SL can be seen in the diffusion of BEVs from EAs to the EM (Fig. 4.4 top right panel, compared to the reference scenario). The risk decline leads to higher BEV deployment that again leads to more risk decline (SL). As BEVs become mainstream, EAs become more attracted to distinctive alternatives, such as FCVs (seen previously in Fig. 4.3). Similarly, PHEVs become less attractive to EAs, which leads to an increase in the BEV share in the first half of the century compared to those scenarios where social influence is not represented. The Ref + SL scenario range shows that social influence effect size has little impact on the initial phase of the transition but significantly affects the speed of diffusion from EAs to other groups.

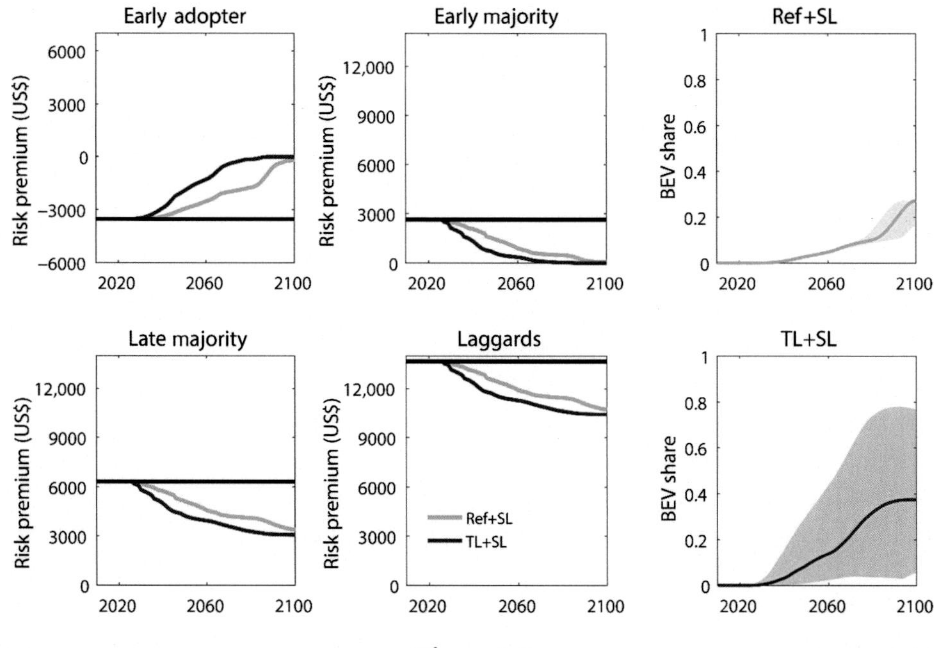

Figure 4.4

Risk premiums toward BEVs in scenarios with SL for the early adopter, early majority, late majority, and laggards (*left* and *middle panels*), and resulting market shares of the global vehicle fleet for BEVs (*right panels*). Shaded colors indicate the scenario range depending on technology learning rates and social influence effect size (*right panel*). *BEV*, Battery electric vehicle; *SL*, social learning. *Source: Figure adapted from Edelenbosch et al. (2018).*

The lower right panel of Fig. 4.4 shows how the combined effect of technological and SL leads to a faster technology transition and higher market penetration under assumptions of average learning rates and social influence effects. There are different phases during the technology transition in this scenario. First PHEV use by EAs leads to battery learning reducing BEV costs. The EAs then shift to BEVs, which results in increased TL and risk decline for the other adopter groups. The EM starts to adopt BEVs enlarging both types of learning effect. TL has occurred faster in the beginning and now starts to level off. RPs continue to decrease for the LM and LG. But additional policy is still needed to overcome the RP barrier for these groups. Clearly, these results are highly dependent on the social influence effect size and the learning rate, indicated by the colored area.

The importance of SL and TL during the different phases of the technology transition—with TL affecting the initial phase, and SL affecting further diffusion—can be traced back to their equational forms. The social influence effect equals the reduction in RP after an increase in market share, whereas the TL rate equals the cost reduction per doubling of cumulative battery production in EV application. Given the exponential form of the learning rate equation with its floor price to limit ever falling costs, the fastest learning happens in the initial deployment phase. In contrast, social influence has a linear relationship with deployment.[2]

While both processes impact vehicle choice in expected ways, their interaction is interesting and revealing. This new modeling approach demonstrates the different phases of a technology transition and its relevant dynamics. It shows how niche or EA groups can drive technology innovation by stimulating market demand.

BEVs can reach a larger market share if TL and SL processes work to mutually reinforce each other. Through SL and TL, new technologies can become more attractive to consumers. Generally speaking, TL affects the timing of adoption by EAs, whereas SL affects diffusion to other adopter groups. The two learning processes can stimulate each other in a positive feedback loop. Policy incentives stimulating EV deployment, such as a carbon tax, information campaigns, or dedicated transport sector policies, can spark positive learning feedbacks. However, the size of this effect depends strongly on the assumed TL rate and social influence effect size, which are key future uncertainties.

4.5 Conclusion

As we have shown earlier, the concept of the experience curve has many applications beyond its original aim to describe cost or price reductions of technologies in the global economy. Experience curves can be used to describe the effects of technological progress on the development of energy use of (industrial) production processes, energy demand of appliances, and environmental impact of technologies. Therefore experience curves have the potential to make valuable additions to for instance energy system modeling activities, broadening their application from "only" technology costs to a set of other metrics. Since only a limited number of studies have investigated these types of applications, further research should analyze whether they are as generally applicable as the original cost-based experience curves are regarded to be.

The second part of the chapter highlights the value of considering learning in other than technical domains. By taking into account SL processes, the modeling of behavior of consumers toward new technologies can be improved, allowing for a more realistic representation of this behavior in energy and integrated assessment models. Furthermore, it was discussed that while TL improves the competitiveness of new technologies, SL can increase the consumer acceptance of these technologies; hence, the market diffusion of such technologies benefits from both learning mechanisms, and both learning mechanisms can interact in a feedback loop.

[2] This linear relation has varying slope coefficients in specific periods of adoption due to the varying size of a market share corresponding to a standard deviation.

References

Arvidsson, R., et al., 2018. Environmental assessment of emerging technologies: recommendations for prospective LCA. J. Ind. Ecol. 22 (6), 1286–1294. Available from: https://doi.org/10.1111/jiec.12690.

Axsen, J., Kurani, K.S., 2012. Interpersonal influence within car buyers' social networks: applying five perspectives to plug-in hybrid vehicle drivers. Environ. Plann. A: Econ. Space 44 (5), 1047–1065. Available from: https://doi.org/10.1068/a43221x.

Bergesen, J.D., Suh, S., 2016. A framework for technological learning in the supply chain: a case study on CdTe photovoltaics. Appl. Energy 169, 721–728. Available from: https://doi.org/10.1016/J.APENERGY.2016.02.013.

Brucker, N., Fleiter, T., Plötz, P., 2014. What about the long term? Using experience curves to describe the energy-efficiency improvement for selected energy-intensive products in Germany. In: ECEEE Industrial Summer Study Proceedings. ECEEE, Arnhem, The Netherlands, pp. 341–352. Available from: <https://www.eceee.org/library/conference_proceedings/eceee_Industrial_Summer_Study/2014/3-matching-policies-and-drivers-policies-and-directives-to-drive-industrial-efficiency/what-about-the-long-term-using-experience-curves-to-describe-the-energy-efficien> (accessed 28.05.19.).

Caduff, M., et al., 2012. Wind power electricity: the bigger the turbine, the greener the electricity? Environ. Sci. Technol. 46 (9), 4725–4733. Available from: https://doi.org/10.1021/es204108n.

Edelenbosch, O.Y., McCollum, D.L., et al., 2018. Interactions between social learning and technological learning in electric vehicle futures. Environ. Res. Lett. 13 (12), 124004. Available from: https://doi.org/10.1088/1748-9326/aae948.

Edelenbosch, O.Y., Hof, A.F., et al., 2018. Transport electrification: the effect of recent battery cost reduction on future emission scenarios. Clim. Change 151 (2), 95–108. Available from: https://doi.org/10.1007/s10584-018-2250-y.

Faguy, P., 2015. Overview of the DOE Advanced Battery R&D Program. U.S. Department of Energy. Available from: <https://www.energy.gov/sites/prod/files/2015/06/f23/es000_faguy_2015_o.pdf>.

Fleiter, T., Schlomann, B., Eichhammer, W., 2013. Energieverbrauch und CO_2-Emissionen industrieller Prozesstechnologien—Einsparpotenziale, Hemmnisse und Instrumente. Fraunhofer Verlag, Karlsruhe, Germany.

Gifford, R., Kormos, C., Mcintyre, A., 2011. Behavioral dimensions of climate change: drivers, responses, barriers, and interventions. Wiley Interdiscip. Rev. Clim. Change 2, 801–827.

Girod, B., van Vuuren, D.P., Deetman, S., 2012. Global travel within the 2°C climate target. Energy Policy 45, 152–166. Available from: https://doi.org/10.1016/J.ENPOL.2012.02.008.

Grinblatt, M., Keloharju, M., Ikaheimo, S., 2007. Social influence and consumption: evidence from the automobile purchases of neighbors. SSRN Electron. J. Available from: https://doi.org/10.2139/ssrn.995855.

Laitner, J.A., De Canio, S.J., Peters, I., 2000. Incorporating behavioural, social, and organizational phenomena in the assessment of climate change mitigation options. In: Society, Behaviour, and Climate Change Mitigation. Springer, The Netherlands, pp. 1–64. Available from: https://doi.org/10.1007/0-306-48160-x_1.

Louwen, A., et al., 2016. Re-assessment of net energy production and greenhouse gas emissions avoidance after 40 years of photovoltaics development. Nat. Commun. 7, 13728. Available from: https://doi.org/10.1038/ncomms13728.

McCollum, D.L., et al., 2017. Improving the behavioral realism of global integrated assessment models: an application to consumers' vehicle choices. Transp. Res., D: Transp. Environ. 55, 322–342. Available from: https://doi.org/10.1016/J.TRD.2016.04.003.

McShane, B.B., Bradlow, E.T., Berger, J., 2012. Visual influence and social groups. J. Mark. Res. 49 (6), 854–871. Available from: https://doi.org/10.1509/jmr.11.0223.

Mundaca, L., et al., 2010. Evaluating energy efficiency policies with energy-economy models. Annu. Rev. Environ. Res. 35, 305–344.

Nykvist, B., Nilsson, M., 2015. Rapidly falling costs of battery packs for electric vehicles. Nat. Clim. Change 5 (4), 329–332. Available from: https://doi.org/10.1038/nclimate2564.

Peres, R., Muller, E., Mahajan, V., 2010. Innovation diffusion and new product growth models: a critical review and research directions. Int. J. Res. Market. 27 (2), 91–106. Available from: https://doi.org/10.1016/j.ijresmar.2009.12.012.

Pettifor, H., et al., 2017. Modelling social influence and cultural variation in global low-carbon vehicle transitions. Global Environ. Change 47, 76–87. Available from: https://doi.org/10.1016/J.GLOENVCHA.2017.09.008.

Ramírez, C.A., Worrell, E., 2006. Feeding fossil fuels to the soil: an analysis of energy embedded and technological learning in the fertilizer industry. Resour. Conserv. Recycl. 46 (1), 75–93. Available from: https://doi.org/10.1016/J.RESCONREC.2005.06.004.

Reed, M.S., et al., 2010. What is social learning? Ecol. Soc. 15 (4). Available from: https://doi.org/10.5751/es-03564-1504r01.

Rogers, E.M., 2003. Elements of diffusion. In: Diffusion of Innovations, fifth ed. Simon & Schuster, pp. 1–38.

Sandén, B.A., Karlström, M., 2007. Positive and negative feedback in consequential life-cycle assessment. J. Cleaner Prod. 15 (15), 1469–1481. Available from: https://doi.org/10.1016/J.JCLEPRO.2006.03.005.

Sims, R., et al., 2014. Transport. In: Edenhofer, O., et al., (Eds.), Climate Change 2014: Mitigation of Climate Change. Cambridge University Press, Cambridge. Available from: https://doi.org/10.1017/CBO9781107415416.005.

Stephens, T., 2013. Non-Cost Barriers to Consumer Adoption of New Light-Duty Vehicle Technologies. Transportation Energy Futures Series. U.S. Department of Energy, Argonne, IL.

Stern, P.C., Sovacool, B.K., Dietz, T., 2016. Towards a science of climate and energy choices. Nat. Clim. Change 6 (6), 547–555. Available from: https://doi.org/10.1038/nclimate3027.

Tran, M., et al., 2012. Realizing the electric-vehicle revolution. Nat. Clim. Change . Available from: https://doi.org/10.1038/nclimate1429.

Weiss, M., et al., 2010. Analyzing price and efficiency dynamics of large appliances with the experience curve approach. Energy Policy. Available from: https://doi.org/10.1016/j.enpol.2009.10.022.

Young, H.P., 2009. Innovation diffusion in heterogeneous populations: contagion, social influence, and social learning. Am. Econ. Rev. 99 (5), 1899–1924. Available from: https://doi.org/10.1257/aer.99.5.1899.

PART II

Case studies

CHAPTER 5

Photovoltaic solar energy

Atse Louwen[1,2] and Wilfried van Sark[1]

[1]Copernicus Institute of Sustainable Development, Utrecht University, Utrecht, The Netherlands, [2]Institute for Renewable Energy, Eurac Research, Bolzano, Italy

Abstract

Solar photovoltaics are currently the fastest growing energy generation technology in terms of capacity additions. Over 500 gigawatt (GW) of capacity was installed at the end of 2018, and by the end of 2019, there will most likely be a total global installed capacity of well over 600 GW. In this chapter, we analyze the price trends of solar photovoltaic systems, modules, and balance-of-system components and use these to derive a set of experience curves. We give an overview of several mechanisms that underlie the observed reductions in solar photovoltaic module and system prices. Finally, using the derived experience curves, future solar photovoltaic systems, and modules, price trajectories in a selection of energy technology deployment scenarios are shown.

Chapter outline

5.1 Introduction

Photovoltaic (PV) solar energy is a renewable energy technology that has been under development for more than 60 years. The technology makes use of the PV effect, in which a current and voltage is generated in a semiconductor material upon exposure to light. Currently, about 95% of PV systems produced are multi ($\sim$60%) or monocrystalline

Technological Learning in the Transition to a Low-Carbon Energy System.
DOI: https://doi.org/10.1016/B978-0-12-818762-3.00005-4

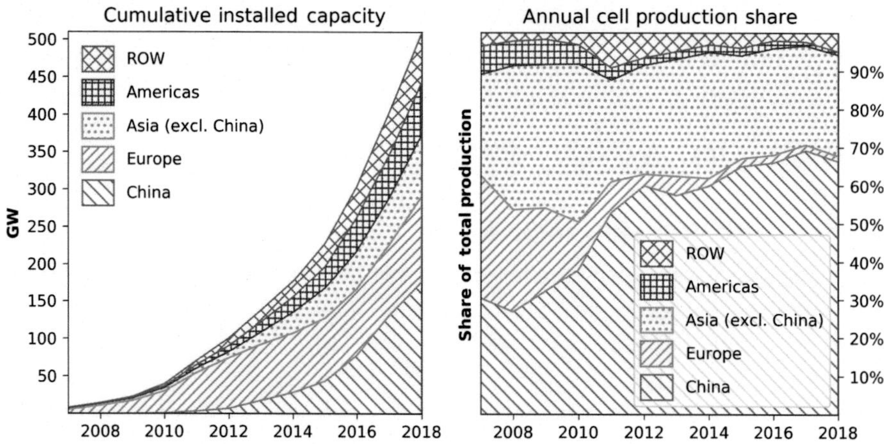

Figure 5.1
Development of installed capacity and cell production. *ROW,* Rest of world. *Data from: IEA PVPS (2018); Jäger-Waldau (2019).*

(~35%) silicon-based, while thin film technologies, such as cadmium telluride (CdTe) and copper-indium-[gallium-]diselenide (CI(G)S), only account for about 5% of PV production (Fraunhofer ISE, 2019).

In recent years the share of mono versus multicrystalline silicon has been increasing and is expected to increase further. Although the first implementation of the currently most common technology was already developed in 1954 by Bell Laboratories, over 95% of currently installed capacity was installed in the last 9 years, and almost 50% in the last 3 years. By the end of 2018, over 500 GWp[1] of PV system capacity was installed worldwide, while for 2019 it is estimated that 128 GWp will be installed (SolarPower Europe, 2019).

The development of the global PV market is shown in Fig. 5.1. In Fig. 5.1 the development of installed capacity is shown. During 2018 around 102 GWp is estimated to have been installed, representing a growth rate of cumulative capacity of 25% (SolarPower Europe, 2019), compared to an average annual growth rate in the last 5 years of almost 30%. In terms of annual net additions of power generation capacity, solar PV is currently the fastest growing of any electricity generation technologies (SolarPower Europe, 2019). The total global installed PV capacity produces an estimate of around 2.6% of our total electricity supply, which increases with roughly 30% each year. With the forecasted crossing of the

[1] The capacity of PV modules and systems is commonly expressed in Watt-peak (Wp), referring to the power of a PV module or system at standard testing conditions (STC), for example, solar irradiance of 1000 W/m^2, a solar spectrum in accordance with an air mass of AM1.5, and a module temperature of 25°C. These STC conditions are used to give PV modules a power rating.

1 TWp mark during 2021 (SolarPower Europe, 2019), around 5% of our global electricity demand will be produced from solar PV.

Currently, the country with both greatest installed PV capacity and the largest annual growth is China, with a cumulative installed capacity (end of 2018) of 176 GWp. As shown in Fig. 5.1, China is also the main producer of PV cells and modules, with a market share of around 70%. Other large markets are Europe, the United States, and Asia-Pacific (excluding China). Although Europe was the largest solar PV market up to about 2016, growth of installed capacity has decreased substantially there, predominantly due to declining feed-in tariffs and market saturation in Germany, Europe's main PV market (IEA PVPS, 2014; Rekinger et al., 2015). For the next 5 years, main growth markets are expected to be again China, but also India, the United States, and Australia, in terms of absolute numbers of installed capacity, while the highest compound annual growth rates for the period 2019–23 are expected in especially Saudi Arabia and other Arab countries such as Egypt and the United Arab Emirates (SolarPower Europe, 2019).

Since PV is a modular technology, its application ranges from small-scale, off-grid applications with installed capacities in the order of kilowatts of peak power, to large, utility-scale applications with capacities of over 1.5 GWp for the largest currently operating solar park.[2] A typical PV system, regardless of the scale, consists of a number of PV modules (an array) linked together and connected to an inverter. The inverter converts the DC power output of the PV system into AC power that can be fed into the electricity grid but also performs power optimization of the PV module array by maximum power point tracking.

In some cases the so-called module-level power electronics can perform the maximum power point tracking as well, optionally also including conversion to AC power. Commonly the cost of PV systems is shown as complete system costs per unit of rated power and broken down into module costs and so-called balance of system (BOS) costs, the latter referring in this case to all components aside from the PV modules (inverter, mounting structure, and components). Sometimes, this also includes installation and procurement costs and other so-called soft costs.

Concurrent with the fast deployment of PV systems, prices for PV system components have decreased substantially. Prices for PV modules have decreased drastically in the past decades, dropping roughly 90% in price from 2007 to 2017 (Fraunhofer ISE, 2019). Some datasets start tracking PV module prices from the 1970s, when the average selling price of a PV module was over 50€/Wp (2018 euros), hence almost a factor 200 as expensive compared to today's PV module prices of around 0.26€/Wp for mainstream modules and as low as 0.20€/Wp for low-cost modules (PV Magazine International, 2019).

---

[2] <https://www.power-technology.com/features/the-worlds-biggest-solar-power-plants/>.

For complete systems, prices vary substantially when comparing different markets. In Germany a large PV rooftop system (10–100 kWp) can be installed for around 1.07€/Wp, while a smaller scale residential rooftop system can be installed for around 1.40€/Wp (Fraunhofer ISE, 2019). In 2017 residential systems in the United States costed on average 2.20€/Wp, while large commercial systems (> 500 kW) were found to be installed for on average of 0.81€/Wp (Barbose and Darghouth, 2018).

The drop in PV system prices has also resulted in record-low bids in several large-scale PV tenders of recent years. In 2016 the lowest bid awarded was 22.3€/MWh for a part of a 1.18 GW plant in Abu Dhabi, while in 2018 a bid of 19.3€/MWh was accepted for a 300 MW plant in Saudi Arabia (SolarPower Europe, 2019). In locations with high solar irradiance, stable political climate, and well-developed financing environments, large-scale solar PV installations are thus competitive with fossil-based electricity generation. Even in Germany, where the solar irradiance is just over half that of countries such as Saudi Arabia, tenders for solar PV systems with capacities around 200 MWp are less than 50€/MWh already since August 2017 (Fraunhofer ISE, 2019).

5.2 Methodological issues and data availability

In the context of energy technologies, PV is one of the prime examples of a technology for which experience curves have been shown to work very well, and as a result, many publications are available that discuss experience curves for PV modules and systems, even including some studies that apply multifactor experience curves (MFECs) to take into account the effect of different parameters affecting the production cost of PV modules, such as R&D expenditures, economies-of-scale, but also prices of input parameters such as silicon and silver (Yu et al., 2011). In Table 5.1, we present an overview of the reviewed literature studies on PV experience curves that have been published in the last 10 years.

As can be discerned from this table, a majority of studies report experience curves for *price* of PV modules (most common), systems, and/or BOS components. As was reported previously (van Sark et al., 2010), experience curves for production cost of PV are hardly reported due to the difficulty of obtaining production cost data from manufacturers or otherwise. Since this is the case in general, not only for PV, we focus on the development of price and experience curves for prices of PV systems and their components. Most commonly, studies show learning curves for the price per unit of capacity (€/Wp), as a function of cumulative installations in units of capacity (Wp), although some studies analyze experience curves for, for example, LCOE as a function of cumulative electricity generation (kWh). For the latter, it must be noted that the levelised cost of electricity (LCOE) of a PV system is highly dependent on solar irradiance conditions and as such can show very large (roughly a factor 2) geographical variation. For PV the costs of rated capacity are highly reflective of the cost of generated electricity. Applying a learning curve for LCOE would thus only be more

Table 5.1: Overview of reviewed studies on experience curves of PV modules, systems, and BOS components.

Study	Learning rate (%)	R^2	Period	Region	n	Comments
Elshurafa et al. (2018)	10.2		1983–2015	20 Countries	Variable	BOS only, value reported is average across 20 countries studied
Bhandari (2018)	30		2007–15	Germany		Short time period
Trappey et al. (2016)	9.7	0.881	1999–2014	Taiwan	16	OFEC
	12.3	0.975				Hierarchical linear model including silicon, steel, oil price
Sampedro and Gonzalez (2016)	8.43	0.846	2001–12	Spain		Complete PV systems
Haysom et al. (2015)	18	0.72	2007–13	Global	16	Concentrator PV systems
Hong et al. (2015)	3.1	0.934	2004–11	Korea	8	Learning-by-doing rate for LCOE in OFEC
	2.33	0.958				Learning-by-doing rate for LCOE in TFEC
	5.13	0.958				Learning-by-searching rate for LCOE in TFEC
Gan and Li (2015)	14.2	0.8652	1988–2006	Global	29	OFEC
	8.1	0.9688				TFEC including silicon prices
	8.9	0.9671				MFEC including silicon prices and supply/demand
de La Tour et al. (2013)	20.1		1990–2011	Global	22	Learning rate in TFEC including silicon prices
Haysom et al. (2013)	22	0.54	2007–12	Global	5	Annual average prices for Concentrator PV systems
	18	0.83	2007–12		15	Individual datapoints for concentrator PV systems
Nemet and Husmann (2012)	23		1976–2006	Global		OFEC for PV modules
Zhang et al. (2012)	14	0.9791	2005–10	China	6	OFEC for PV cells
Yu et al. (2011)	19	0.9828	1976–2006	Global	31	OFEC
	13.5	0.993				MFEC including manufacturing scale, silicon price, silver price
Junginger et al. (2010)	20.6	0.992	1976–2006	Global	15	OFEC for PV modules
Bione et al. (2009)	18.4		1975–2007	Global	33	OFEC

BOS, Balance of system; *LCOE*, levelized cost of electricity; *MFEC*, multifactor experience curve; *OFEC*, one-factor experience curve; *PV*, photovoltaic; *TFEC*, two-factor experience curve.

appropriate if (1) LCOE is calculated based on a harmonized or standardized solar irradiance value, and (2) there is empirical data on the development of PV system parameters such as the performance ratio and lifetime. This contrasts strongly with, for example, wind energy, where it has become obvious that the developments of costs per unit of rated capacity do not necessarily reflect the developments of wind energy LCOE (see Chapter 6, of this book and Williams et al., 2017).

As shown in Table 5.1, there is still a large range of learning rate values reported, although the average is rather firm around 20%. Depending on the period of the dataset, the learning rate value determined can be significantly lower or higher compared to this rough average of 20%. Fig. 5.2 shows an analysis based on the most up-to-date dataset for PV modules and analyzes the effect of the time frame of the dataset. Especially apparent from this analysis is the effect of a period in which the price development of PV modules stabilized between 2004 and 2008. As was also shown by van Sark et al. (2010), where the variation in learning rate was analyzed by taking a 10-year moving window over a dataset with a long time frame, the prices at the beginning of the dataset also have profound influence on the estimated learning rate (LR) value.

The data we have gathered here builds on this body of literature. Several reputable sources annually publish reports of the status of the solar PV market, such as Fraunhofer ISE, the International Energy Agency's Photovoltaic Power Systems Programme (IEA PVPS), and the European solar PV industry association SolarPower Europe. PV modules are traded more or less as a global commodity, and factory-gate and average spot prices for PV modules are quite readily available. For complete, installed, PV systems, there is much more geographical variation in prices, especially for small-scale systems. Here, we discuss datasets from two main markets: Germany and the United States, as well as a dataset that describes average prices of PV modules on a global market.

In Table 5.2, we give an overview of the data collection issues applicable to the datasets we have gathered for PVs. As is common when collecting data for experience curves, we have used price data, rather than cost data. Compared to other technologies, the datasets for PV are possibly one of the most comprehensive, as price data as well as cumulative production data is widely available. As will be discussed in Section 5.3.1, the geographical scope of the datasets used, as well as the time frame of the datasets, can substantially affect the derived experience curve parameters. Although PV modules are traded as an international commodity, other components in total installed PV system costs show considerable variation depending on the country of installation. As shown in Fig. 5.3, residential systems in the United States are still quite a bit more expensive compared to those in Germany (Fig. 5.4), since soft costs (including overhead, labor, and procurement) represent over 50% of total system costs in the United States in 2017 (Barbose and Darghouth, 2018).

The long-term dataset for PV modules in Fig. 5.2 shows signs of market dynamics affecting the price trends, resulting in large temporary deviations from the long-term experience curve trend. In Section 5.3.1 we discuss the mechanisms behind these deviations. Short-term datasets can exhibit very different trends compared to this long-term dataset, as is

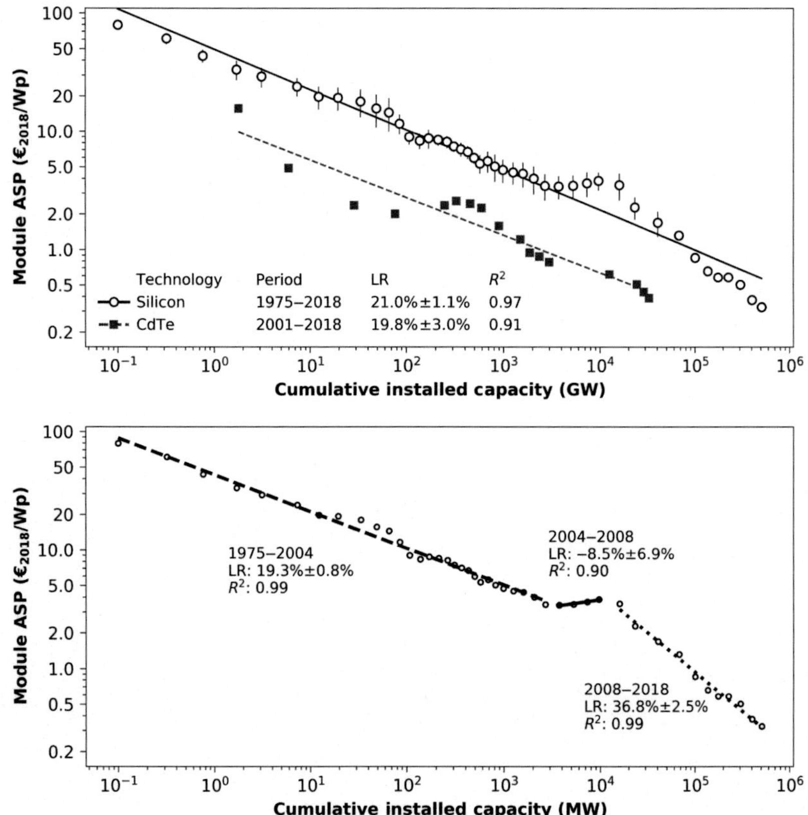

Figure 5.2

Top: Experience curves for ASP of silicon and CdTe-based PV modules. The error bars in the plot for silicon PV modules indicate the standard error for module ASP across the datasets used. Bottom: segmented experience curve for silicon PV modules. *ASP*, Average selling price; *CdTe*, cadmium telluride; *PV*, photovoltaic. *Data from: Louwen et al. (2016a,b); van Sark et al. (2010); Verlinden (2014); IEA PVPS (2017,2018); Fraunhofer ISE (2019), own data collection.*

highlighted in Figs. 5.2 and 5.4. Because of the short time horizon of the dataset in Fig. 5.4, the established learning rate for PV modules is much higher (30%) compared to what is found for the long-term datasets (21%).

5.3 Results

5.3.1 State-of-the-art experience curves for photovoltaic modules, systems, and balance of system components

Figs. 5.2–5.5 show the results of our analyses, with experience curves for PV modules, complete PV systems, and BOS components for different markets. The first figure shows

Table 5.2: General data collection issues for photovoltaics.

Issue	Resolution applied	Applicability
Data is not for cost but for price	Use price data as indicator for costs	
Data not available for desired cost unit		
Data is valid for limited geographical scope	Convert currency	
	Combine with other datasets from various geographical scopes	
Cumulative production figures not available		
Data is in incorrect currency or currency year	Convert currency and correct for inflation and purchasing power parity (PPP) when necessary	
Early cumulative production figures are not clear or available		
Supply/Demand affecting costs significantly	Keep data as is	
Lack of empirical (commercial-scale) data		

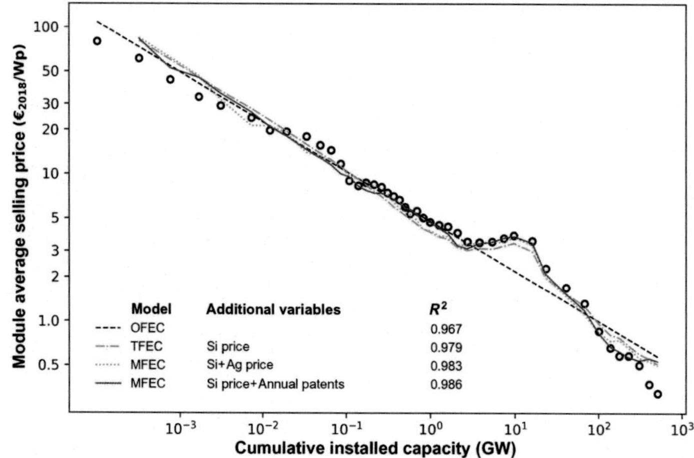

Figure 5.3

Two- and multifactor experience curves for the ASP of PV modules between 1975 and 2018. *ASP*, Average selling price; *PV*, photovoltaic. Source: *Data sources shown in Table 5.3.*

the development of the average selling price of silicon PV modules over the time frame 1975–2018, and CdTe PV modules from 2001 to 2018.

Over this period, with this state-of-the-art dataset, we establish for silicon PV modules a learning rate of 21.0% ± 1.1%, with a coefficient of determination of 0.97 for the fitted model. Within this dataset, prices declined from around 80€/Wp in 1975 to 0.33€/Wp in 2018. From the figure, some obvious developments in the silicon PV module prices can be observed. First is the stabilization (and even small increase) of these PV module prices starting in 2004. As we will discuss later on, and as is shown in Fig. 5.2, this price plateau

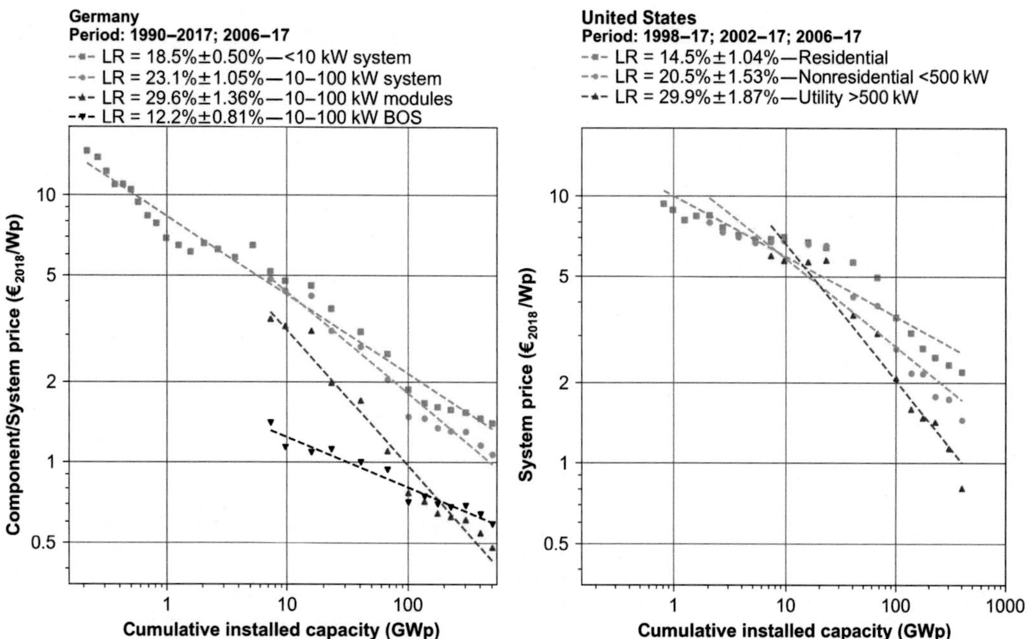

Figure 5.4

Overview of PV system, modules, and BOS components prices and experience curves between 1990 and 2017 for Germany and for residential, nonresidential and utility-scale systems in the United States between 1998 and 2017. *BOS*, Balance of system; *PV*, photovoltaic. *Data from: Fraunhofer ISE (2019); Barbose and Darghouth (2018).*

Table 5.3: Overview of data collected.

Dataset	Source
Cumulative capacity	Louwen et al. (2016a,b), IEA PVPS (2017, 2018), and SolarPower Europe (2019)
Modules ASP	van Sark et al. (2010), Verlinden (2014), and Fraunhofer ISE (2019) Own data collection
BOS components	Fraunhofer ISE (2019)
Systems < 10 kW	Fraunhofer ISE (2019)
Systems 10–100 kW	Fraunhofer ISE (2019)
Systems residential	Barbose and Darghouth (2018)
Systems commercial (scale)	Barbose and Darghouth (2018)
Systems utility	Barbose and Darghouth (2018)
Patents filed for PV technology	EPO PATSTAT (https://data.epo.org/expert-services/index.html)
Silicon price	Yu et al. (2011), PV insights (http://pvinsights.com/)
Silver price	The Perth Mint (https://www.perthmint.com/historical_metal_prices.aspx)

ASP, Average selling price; *PV*, photovoltaic.

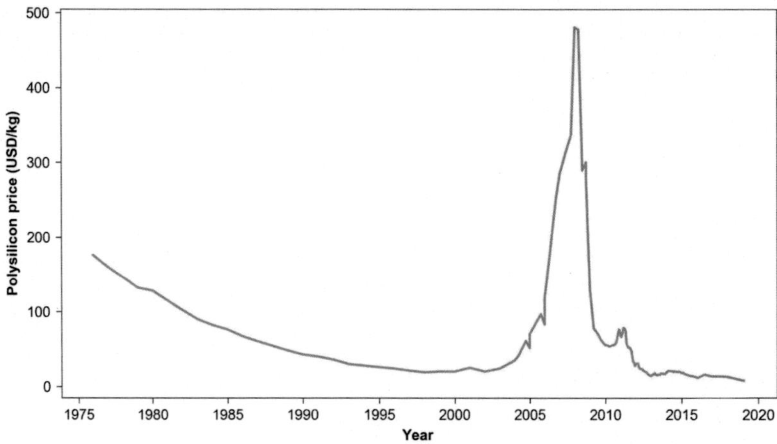

Figure 5.5
Silicon price developments between 1975 and 2018. *Data from: Yu et al. (2011), own data collection.*

coincided with a sharp increase of the price of silicon. From 2009 onward, a steep decline of silicon PV module prices can be observed. From 2012 onward, a period was started when PV module prices dropped substantially below the trend of the experience curve derived for the whole period from 1975 to 2018. Although after 2012 prices for silicon PV modules seemed to stabilize briefly, in 2018 and especially 2017, prices dropped sharply again.

For CdTe PV modules, we observe a learning rate of 19.8% ± 3.0%, and an R^2 of 0.91. Prices for CdTe modules have varied significantly from the long-term experience curve trend, especially the first four datapoints. It is obvious from the graph that CdTe seems to be the cheaper technology compared to silicon-based PV modules, but the deployment of CdTe is much lower, compared to that of silicon PV modules. This is largely due to the fact that CdTe (and other thin film technologies) has historically had much lower module efficiencies, thus requiring a larger installed area for the same rated capacity.

The dataset for silicon PV modules seems to indicate that the regression is segmented, for example, the dataset consists of different segments for which the regression is much more accurate. Therefore we analyzed the position of the breaks in the dataset using an approach documented by Muggeo (2008), who implemented a procedure in *R*, based on the Davies test, to analyze breaks in linear regressions. The test indicates that the dataset can be split up into three segments, with statistically significant breaks occurring in 2004 and between 2007 and 2008. This is shown in Fig. 5.2 (right). The three segments have very different learning rates, and for the period 2004–08 (coinciding with a spike in silicon prices), the learning rate found is negative, meaning that prices of PV modules temporarily increased. Although the period between 1975 and 2004 contains another deviation from the experience curve trend starting from 1982, we did not find a structural break around this year. By segmenting the dataset, and having separate linear regressions for the three segments, a much higher R^2 value is found. Although the segmented linear model describes the price developments more accurately, its application for (long-term) extrapolation of PV module prices is not recommended, given that the sharp price decrease of the last 10 years (LR of 36%) is likely not realistic for price developments over longer time frames.

In Fig. 5.4, we show the development of rooftop PV system prices (data from Germany) for two different PV system size ranges: up to 10 kW, and from 10 to 100 kW, including the contribution of PV modules and BOS components for the latter PV system scale. For the rooftop PV systems up to 10 kW, we observe a learning rate of 18.5%, which is slightly lower compared to the PV module *LR* we show in Fig. 5.2. For the dataset of 10–100 kW PV systems (which covers a much shorter period), we observe some very striking developments. First of all, the contribution of PV modules to the overall PV system price has dropped below that of BOS components starting from 2013 to 2014, due to the sharp decline in PV module prices. Second, PV system prices in this dataset have a high learning rate of 23.5% ± 1.14%, due to the sharp decrease in PV module prices. Third, 10–100 kW systems cost less per Wp compared to smaller PV systems, but this difference seems to have materialized only in the last 10 or so years. Finally, we observe in this (relatively short) dataset that the learning rate for BOS components is much lower at 12.1% ± 0.93% than that for PV modules, which we found to be 21.0% (shown in Fig. 5.2). The high learning rate for PV modules in the dataset shown in Fig. 5.3 confirms the analysis of Fig. 5.2, for example, that the price decline has been very strong in the last 10 years. Given the very different learning rates of BOS versus those of PV modules, it becomes apparent that for extrapolations of PV system prices on a long term, separate experience curves for these components need to be applied. In Section 5.4, we further analyze this. Since the BOS includes all costs not related to the PV modules, it also includes most likely a number of more or less fixed costs for which hardly any learning is observed (e.g., cost of land purchasing and installation labor).

Data from the United States for residential, commercial, and utility-scale systems is also shown in Fig. 5.4 (right). The figure indicates that, in the last 8–9 years, there is strong differentiation in the PV system price for different PV system sizes. Current prices (in 2018EUR) for PV systems are about 0.80€/Wp for large, utility-scale systems, 1.45€/Wp for nonresidential systems, and about 2.20€/Wp for residential PV systems. Learning rates for these three PV system sizes also vary substantially and are 29.9% ± 1.9%, 20.5% ± 1.5%, and 14.5% ± 1.10%, respectively. It must be noted however that the different time frames over which these learning rates were established strongly affects these estimated values. The relatively high prices for residential and nonresidential systems up to 500 kW seem to indicate that especially the residential PV market is not as well-developed in the

United States as compared to Germany. Recent prices for utility-scale systems seem however much more in line with those of well-developed PV markets.

5.3.2 Multifactor learning curves for silicon photovoltaic module prices

Several authors have analyzed MFECs for PV, taking into account additional inputs (aside from cumulative production) such as R&D (Zheng and Kammen, 2014) and prices of input materials such as silicon and silver (Yu et al., 2011). The aim of these studies is generally to gain more insight into the mechanisms behind cost reductions and to separate and quantify the effects of learning-by-doing and learning-by-searching.

Here, we have analyzed the effect of input material prices (silicon and silver) and R&D activities (based on developments of cumulative patent applications in the PV industry) on the price developments of PV modules, by including them in MFECs.

In Fig. 5.3, we show four distinct experience curves for PV modules prices: a one-factor experience curve based on cumulative installations, a two-factor experience curve (TFEC) taking into account the silicon price, a MFEC, which additionally includes the silver price, and a MFEC that includes the *annual* patents as an independent variable. Hence, we performed regressions of the following equations:

$$\log C_{PV} = \log C_{PV1} + b \log n \tag{5.1}$$

$$\log C_{PV} = \log C_{PV1} + b \log n + c \log \mathrm{Si} \tag{5.2}$$

$$\log C_{PV} = \log C_{PV1} + b \log n + c \log \mathrm{Si} + d \log \mathrm{Ag} \tag{5.3}$$

$$\log C_{PV} = \log C_{PV1} + b \log n + c \log \mathrm{Si} + e \log \mathrm{Pat} \tag{5.4}$$

Here, C_{PV} is the cost of the PV modules, given respectively by cumulative capacity n, the silicon price Si, the silver price Ag, and the annual patents Pat. Data for patents was taken from the European Patent Office service PATSTAT.[3] To account for the delay between filing of the patent and its application in production environments, we assume a lag period of 3 years.

Normally in two- or MFECs that take into account R&D, authors use cumulative patents (Kittner et al., 2017) or use the "knowledge stock" (*KS*) as described in Chapter 2. Here we found that there was a strong correlation between cumulative patents and cumulative production. In several related studies (Zheng and Kammen, 2014; Kittner et al., 2017), authors successfully make use of a residual variable to prevent problems with this multicollinearity. Although this approach eliminates the problem of multicollinearity, the results obtained for our dataset are counterintuitive, as we observe a strongly negative

[3] <https://data.epo.org/expert-services/index.html>.

learning rate of ($LR_{Pat} = 1-2^d$) for cumulative patents Pat. More or less similar results were obtained when applying the patent data in the form of the KS, given by

$$KS = (1 - \delta)KS_{t-1} + Pat_{t-x} \tag{5.5}$$

where δ is the depreciation rate, and x is the time lag before the patent knowledge is implemented in the production environment. We found that the same multicollinearity problems arise as with cumulative capacity. When a sufficiently high time-lag and/or depreciation rate is applied, the problem decreases, however, in that case the KS starts to resemble the annual patent data.

The results obtained for the TFEC with silicon price show that including the silicon price as independent variable gives a learning rate of 19.8% for learning-by-doing ($LR_n = 1 - 2^b$), and a "learning rate" for silicon price of -15.1%, for example, for every doubling of the silicon price the price of PV modules increases by 15.1%. Compared to the one-factor model, the fit quality is improved substantially, most notably by the TFEC curve more closely following the price plateau around 2008, which we discussed earlier.

The use of silver in metallization paste for the electrical contacts of crystalline silicon cells contributes about 10% to the overall module price (Louwen et al., 2016a,b). However, as was found before by Yu et al. (2011), we observe a positive learning rate for silver of 16.2% when we include it in our model. Contrary to the price of silicon, the price of silver has been on average increasing during the time frame of our dataset, with some notable short-term spikes in price. At the same time the use of silver in PV cells has decreased substantially. It is likely that these two factors, combined with the overall decrease of the PV module price, lead to the fitting of a negative learning rate for silver. When we include silver price in the model of Eq. (5.4), the quality of the fit hardly increases.

As can be seen in Fig. 5.3, including annual patents as an additional input variable further increases the fit quality compared to the TFEC and has an effect on especially the learning rate for cumulative capacity, which decreases to 15.6%. We find a learning rate of 17.5% for annual patents. In Table 5.4, we give an overview of the different models we fitted and their associated experience curve parameter values.

It is apparent from Fig. 5.3 that none of the models that we have analyzed here are able to follow the very recent price decline of the last two datapoints in our dataset. Although the result of a variety of mechanisms, it is likely that this fast decline in recent years is mainly a result of increased competition in the PV market due to oversupply as a result of declining demand, forcing lower market prices for PV modules. For the near future, it is expected that market prices for PV modules will stabilize, if not slightly rise. It will be interesting to see whether the price developments will return to the trends of the experience curve models.

Table 5.4: Overview of the regression results for the multifactor experience curves for PV modules.

	Eq. (5.1) One-factor model	Eq. (5.2) Two-factor model	Eq. (5.3) Three-factor model	Eq. (5.4) Three-factor model
$\log(C_{PV1})$	1.6961 (0.032)	1.2931 (0.087)	1.3484 (0.081)	1.9411 (0.169)
b	− 0.3393 (0.010)	− 0.3183 (0.010)	− 0.2823 (0.014)	− 0.2451 (0.019)
LR_n	21.0%	19.8%	17.8%	15.6%
c		0.2039 (0.040)	0.2638 (0.041)	0.1945 (0.034)
LR_{Si}		− 15.1%	− 20.1%	− 14.4%
d			− 0.2554 (0.082)	
LR_{Ag}			16.2%	
e				− 0.2791 (0.066)
LR_{Pat}				17.5%
Period	1975–2018	1976–2018	1976–2018	1976–2018
P-Value	0.0000	0.0000	0.0000	0.0000
BIC	− 66.3	− 83.1	− 88.9	− 95.8

The values in parentheses indicate the standard errors of the parameters. The *P*-value reported refers to the probability of the *F*-statistic test for overall significance. Lower values indicate a better fit result. *BIC*, Bayesian Information Criterium.

5.3.3 Analysis and quantification of observed cost reductions

Several studies of the last 10–20 years have offered detailed insights into the mechanisms underlying the cost reductions observed for PV modules and BOS components. Cost reductions in PV module production arise from a variety of mechanisms. When examining the cost breakdown of a PV module, it becomes apparent that there are a limited number of technical parameters that largely determine the cost per Wp of capacity, being module efficiency, silicon price, silicon consumption, nonsilicon material costs (mainly silver), and cell area (Goodrich et al., 2013; Louwen et al., 2016a,b; Kavlak et al., 2018). Experience curve mechanisms such as R&D (learning-by-researching), learning-by-doing, and economies of scale help drive down production cost by improving the set of technical parameters that determine the cost of PV modules (Boston Consulting Group, 1970; Nemet, 2006; Kavlak et al., 2018).

Since the majority of the PV modules installed are based on crystalline silicon, it is obvious that the development of silicon prices should be analyzed if we want to understand the decline in PV module prices. Shown in Fig. 5.6, the price of silicon used in the PV industry has largely declined between 1976 and 2018, although immediately apparent from this figure is the peak in silicon prices in 2008. While the price of polysilicon in 1976 was around 170 USD_{2018}/kg, the current price is only 8.9 USD/kg. In 2008, at the peak, the price of silicon was around 475 USD/kg, quickly dropping in 2010 to just over 50 USD/kg. The price spike in 2008 has been attributed to an increase in silicon demand by the PV sector combined with a silicon production shortage, as a result of delayed investment in

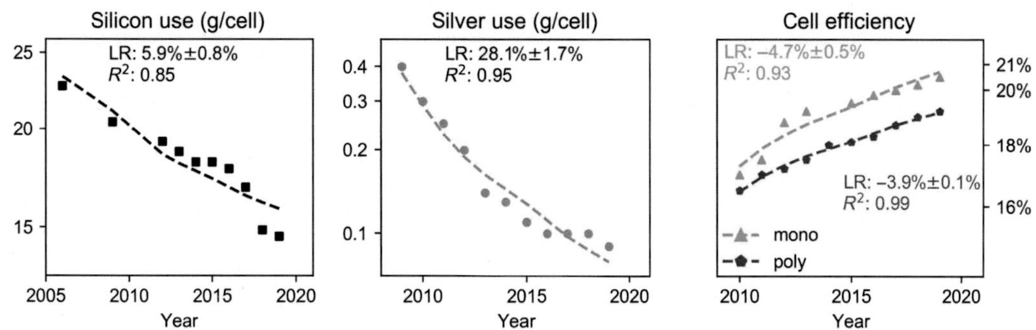

Figure 5.6
Development of silicon use, silver use, and cell efficiency for crystalline silicon solar cells. Note that the *y*-axis shows years, but the experience curves were fitted as a function of cumulative capacity.

silicon feedstock production facilities, but it also preceded the global financial crisis that occurred during that time frame. After the price surge the production capacity for solar grade polysilicon was strongly expanded, resulting in a decline in polysilicon prices to "normal" levels. Even though demand for polysilicon from the PV industry has grown strongly since 2010, the prices have dropped to current levels around 2012−13 and have been relatively stable ever since.

Since silicon has been a major cost component, R&D in the PV industry has tried to decrease the silicon usage in PV modules. The silicon usage (in g/Wp) can decrease in two ways: increases in module efficiency and decrease in the amount of silicon used per PV module. In both ways the PV industry has made significant steps. The development of PV module efficiency is discussed later. Hence, in this section, we discuss the decrease in silicon usage per PV module. The silicon usage has decreased due to advances along the PV production chain. First of all, the thickness of the wafers used for PV cells has decreased substantially, from around 500 μm (Kavlak et al., 2018) in the 1970s to around 160 μm nowadays (ITRPV, 2019). Furthermore, the kerf losses, sawing losses, which occur when slicing the wafers, have also decreased significantly. Fig. 5.6 shows the development of silicon use as a function of these parameters. Indirectly, the usage of silicon also decreases due to the improvement of production yield. The development of silicon use can be described with an experience curve (which was fitted against cumulative capacity) and shows a learning rate for silicon use of 5.9%.

Also shown in Fig. 5.6 are two other technological improvements in PV cell manufacturing: silver use per cell, and the efficiency developments of both mono- and polycrystalline silicon PV cells over the past decade. Silver use was found to have a learning rate of 28.1% over the last 11 years, and current cells have only 0.09 g of silver per cell compared to 0.4 g/cell in 2009.

Concurrent with these improvements in material consumption, cell efficiency of crystalline silicon PV cells has increased steadily. Over the past decade the efficiencies show learning rates of −3.9% and −4.7% for poly and monocrystalline silicon cells, respectively, implying that the power output per unit area of these types of cells increased by 3.9% and 4.7% for every doubling of cumulative capacity. Current cell efficiencies are 19.2% and 20.5%, respectively.

A recent study discusses and quantifies the contribution of what the authors define as low-level and high-level mechanisms, which explain observed cost decline for PV modules between 1980 and 2012 (Kavlak et al., 2018). In that study, low-level mechanisms include cost components of a PV module, such as module efficiency, materials cost, cell size, production plant size, and production yield. High-level mechanisms refer to strategies or activities, which lead to cost decline by affecting the low-level mechanisms, and include mechanisms such as R&D, learning-by-doing, and economies of scale (Kavlak et al., 2018). The authors conclude that of the low-level mechanisms, the largest fraction of the cost decline is due to increases in PV module efficiency (23%) and nonsilicon materials costs (21%). Due to the overall decline in the price of silicon between 1980 and 2012, a cost reduction of 16% can be attributed to the price of silicon, while a decrease of the silicon used per watt peak of PV module capacity explains 14% of the observed cost decline. Although over the whole period of 1980–2012 the contribution of plant size to overall cost reduction is smaller (11%) compared to the mechanism mentioned above, the authors found that upscaling of plant sizes between 2001 and 2012 was the major cause of cost decline in this period, explaining 36% of the cost decline between these years. The typical plant size scaled up from 1 MW in 1980 to 13 MW in 2001, and to 1000 MW in 2012 (Kavlak et al., 2018). Plant upscaling was also found to have a profound effect on cost reductions in the PV industry in a MFEC analysis (Yu et al., 2011).

In terms of high-level mechanisms, Kavlak et al. find that the majority of the cost decline can be attributed to public and private R&D expenditures, while the effect of learning-by-doing and economies of scale were found to be much smaller. This is seemingly in contrast with the results we show in Fig. 5.4, where based on the learning rates established for learning-by-doing and learning-by-searching, these contributions are roughly equal. Furthermore, since the cumulative capacity has grown much stronger compared to the number of published patents, the contribution of learning-by-doing to price reduction in our experience curve models is far greater than that of learning-by-searching. However, Kavlak et al. classify a variety of technological improvements in the PV supply chain, such as cell and module efficiency, silicon utilization, and wafer area as results of R&D, while only improvements directly related to repeated manufacturing operations are considered the result of learning-by-doing. As is mentioned in Junginger et al. (2010), it is generally difficult to accurately separate learning-by-doing from learning-by-searching in MFECs; hence, it is likely that the effect of learning-by-doing in our experience curve models includes the technological progress in areas that Kavlak et al. deemed the result of R&D activities.

5.4 Future outlook for prices of photovoltaic modules, systems, and balance of system components

Using the experience curves shown in this chapter, it is possible to extrapolate the prices of PV modules, systems, and BOS components. Given the experience curve equation and estimated parameters, price trajectories can be estimated when we know the future developments of cumulative PV installations. In this section, we analyze the future costs that we derive from the experience curves, using forecasts of future cumulative PV-installed capacity as inputs. Given the sometimes unexpectedly strong developments of the global PV market, it is inherently difficult to accurately predict the future development of this very dynamic market. We use several scenarios that describe these future developments, as well as an approach that extrapolates the development of the PV market based on historical data. We will compare the extrapolated costs based on different experience curves (e.g., those for complete systems, modules, and BOS components) taking into account the uncertainty in the experience curve parameters that were determined. Finally, we compare these extrapolations with bottom-up cost estimates.

In Table 5.5, we show the forecasted cumulative installations based on a set of modeling studies, for different scenarios, and based on expert elicitations from key market players. Shown in the table is a large range of forecasted cumulative PV capacity from 2025 to 2040. One of the most well-known scenario studies that outline the development of the global energy system is the International Energy Agency's World Energy Outlook (WEO). Published in 2018, the most recent WEO analyses two distinct scenarios that vary significantly in the amount of projected cumulative PV installations. At the low end the new policies scenario shows a global cumulative installation of 1589 GWp in 2030, and 2540 GWp in 2040, while the sustainable development scenario shows installed capacities of 2346 and 4240 GWp, respectively.

The scenarios for 2030 and 2040 that were calculated based on the short-term market outlooks by SolarPower Europe (2019) are much more optimistic in terms of deployment of PV. From a "low" to "high" scenario the forecasted deployment of PV installations ranges from 2200 to 4120 GWp in 2030 and from 4740 to 9835 GWp in 2040.

Even more optimistic is the scenario by Lappeenranta University of Technology (LUT) that is based on an exploration of a global power sector that is 100% based on renewable electricity. This analysis shows a growth of the PV-installed capacity to over 13 TWp in 2040. The associated extrapolated PV module costs are shown in Table 5.4 and range from 0.237 ± 0.006 to 0.391 ± 0.004€$_{2018}$/Wp in 2030 and from 0.188 ± 0.006 to 0.333 ± 0.005€$_{2018}$/Wp in 2040. The extrapolations were calculated by continuing the experience curve trend from 2018 onward, and the intervals represent the effect of the standard error in the derived learning rate.

Table 5.5: Overview of scenario studies and expert forecasts for which have performed experience-curve cost extrapolations for PV modules and systems.

Source	Scenario	Installed capacity (GWp)			Extrapolated costs (2018€/Wp)					
					PV modules			PV systems		
		2025	2030	2040	2025	2030	2040	2025	2030	2040
Greenpeace	Advanced [r]evolution	2000	3725	6678	0.36	0.29	0.24	0.82	0.70	0.61
LUT	100% RES power sector	3513	6980	13805	**0.30**	**0.24**	**0.19**	**0.71**	**0.60**	**0.51**
BNEF	NEO 2018	1353	2144	4527	0.41	0.35	0.27	0.91	0.81	0.67
IEA WEO	New policies	1109	1589	2540	*0.44*	*0.39*	*0.33*	*0.96*	*0.87*	*0.77*
IEA WEO	Sustainable development	1472	2346	4240	0.40	0.34	0.28	0.89	0.79	0.68
SolarPower Europe[a]	Low	1315	2200	4740	0.42	0.35	0.27	0.91	0.80	0.66
SolarPower Europe[a]	Med	1700	3000	6725	0.38	0.32	0.24	0.86	0.74	0.61
SolarPower Europe[a]	High	2180	4120	9835	0.35	0.28	0.21	0.80	0.68	0.55
Average		1830	3263	6636	0.37	0.31	0.24	0.84	0.73	0.61

Bold text indicate the lowest extrapolated prices, cursive text the highest extrapolated prices. *PV*, Photovoltaic.
[a]The values in the row for SolarPower Europe (2019) were obtained by linearly extrapolating the low, med, and high scenarios in the most recent Global Market Outlook.
Data from: Installed capacity scenarios from Jäger-Waldau, A., 2019. Snapshot of photovoltaics—February 2019. Energies 12 (5), 769. Available from: https://doi.org/10.3390/en12050769; SolarPower Europe, 2019. Global Market Outlook for Solar Power 2019–2023. Available from: <http://www.solarpowereurope.org/>.

We also calculated costs based on extrapolation with the derived learning rate but with the current price level as a starting point. This results in significantly lower extrapolations, as current prices are well below the experience curve trend. Future costs based on this approach range from 0.134 ± 0.004 to 0.222 ± 0.002€$_{2018}$/Wp in 2030, and from 0.106 ± 0.004 to 0.189 ± 0.003€$_{2018}$/Wp in 2040. Fig. 5.7 shows the extrapolated cost trends.

It must be noted that current (2018) price levels for PV modules are substantially below the experience curve trend and are likely at a level that is unsustainable for current manufacturing, due to sharp price drops in 2017 and 2018 resulting from the oversupply situation that was discussed earlier. Roughly since October 2018, PV module prices have been more or less stable, and no price decrease is expected for 2019.[4] On the short term, prices of PV modules could even increase by 10%–15% as the PV manufacturing market consolidates.[5] With this in mind the extrapolations of PV prices based on continuing the long-term experience curve trend are recommended as a more conservative estimation of future PV prices.

4 <https://www.pv-magazine.com/features/investors/module-price-index/>.
5 <https://fortune.com/2019/01/24/chinese-made-solar-panels-recover/>.

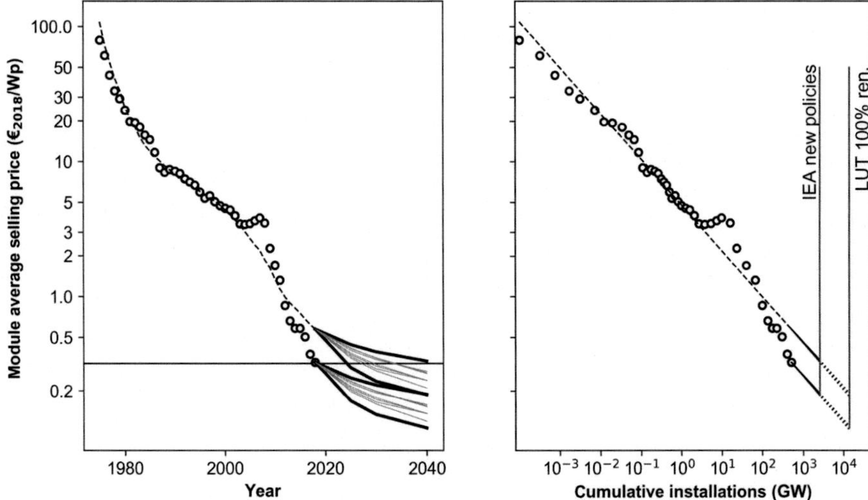

Figure 5.7

Extrapolated PV module costs. Left: Cost developments over time, thick lines show the least and most optimistic scenarios, thin gray lines show the other scenarios. Right: Cost as a function of cumulative installations. Dotted lines show the most optimistic scenario, solid lines show the least optimistic scenarios. *PV*, Photovoltaic.

In Fig. 5.8, we show cost extrapolations for PV systems from 10 to 100 kW, BOS components and modules. Here, we used the derived experience curves for PV modules (*blue lines*) and BOS components (*orange lines*) to calculate the total system costs (*green lines*). Based on this approach, depending on the scenario for future installed capacity developments, PV systems costs range from 0.51 to 0.77€/Wp, based on module prices of 0.19–0.33€/Wp and BOS components prices of 0.32–0.44€/Wp. When we extrapolate the experience curve derived for complete systems, shown in Fig. 5.8 with a dashed black line, we end up with much lower system cost of 0.28–0.53€/Wp. This underlines the issue (discussed in Section 5.3.1) of making cost extrapolations using a single experience curve for a technology that is an aggregate of multiple components.

5.5 Conclusions and recommendations for science, policy, and business

Almost 50 years of PV module data indicates that the price of PV modules decreases with 21% for each doubling of cumulative installed capacity. Although the price trend was quite consistent between 1975 and 2004, we observe significant variations of this overall trend in the period after 2004. First, due to a shortage in silicon supply and a resulting surge in silicon prices, prices stabilized and even increased between 2004 and 2008. From 2008 onward, PV module prices have been declining with a learning rate of 36%. This sharp

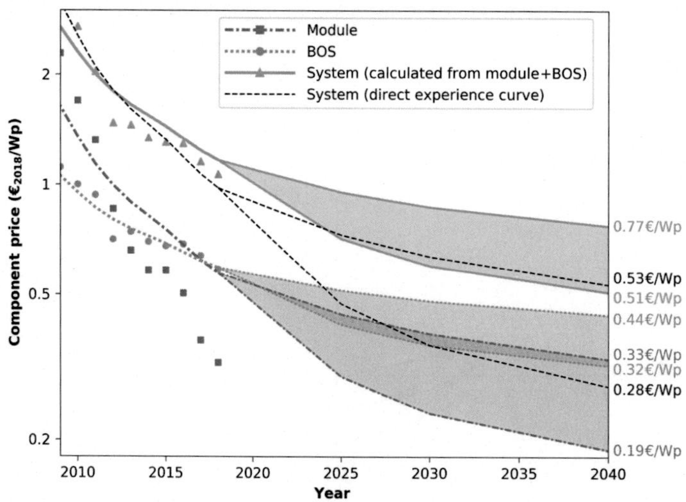

Figure 5.8

Cost extrapolations for complete PV systems (*green lines and datapoints*), calculated based on separate experience curves for PV modules (*blue*) and balance of system components (*orange*). The dashed black line shows the experience curve for complete systems and its extrapolation. The ranges shown by the green, orange, and blue areas are based on the least and most optimistic scenarios from Table 5.5. *PV*, Photovoltaic.

price decline followed a period of polysilicon shortage, and is attributed to an oversupply of polysilicon that resulted from large expansions of the production capacity, resulting in a steep drop of the polysilicon price. More recently, oversupply of PV modules has led to further strong declines in PV module and system prices.

The fast decline in PV module prices of the last 10 years also has its effect on residential PV system prices, which show an overall learning rate of 18.5% between 1990 and 2017, but show a steeper decline in the last 10 years, also observed for larger scale rooftop systems in Germany, and residential, commercial, and utility-scale systems in the United States. The different datasets analyzed here show that the time horizon over which the experience curve is established significantly affects the estimated learning rate. Furthermore, depending on the location for which the price data is obtained, large differences may occur in the observed trends and estimated learning rates, as evidenced by comparing the datasets of Germany with those for the United States.

The contribution of BOS components to overall systems prices has increased to over 50% of installed PV system prices, due to a much lower learning rate compared to that for PV modules. As a result, experience curves for PV modules or for complete PV systems are not recommended to be used for extrapolations of PV system prices, but rather, an approach using separate curves for PV modules and BOS components (especially inverters) is to be used. Since the dataset on BOS components is of a relatively short time frame, further research should focus on expanding this dataset. Furthermore, R&D should focus on achieving cost reductions in BOS components.

Given the already low cost of PV systems, especially for large-scale systems, electricity from PV is already competitive with fossil generation in high irradiance locations and has achieved grid parity for private consumers years ago in a much larger geographical region (Breyer and Gerlach, 2013). Given that there is still room for further cost reductions, it is likely that electricity generation with PV will be cost-competitive in many more locations, even those with relatively low solar irradiance. This means that PV could substantially contribute toward low-GHG-emissions electricity generation, but policy makers should put into place measures that allow for high penetration of intermittent PV in electricity grids, for instance by means of grid reinforcement or incentivizing local electricity storage and demand-side management.

References

Barbose, G., Darghouth, N., 2018. Tracking the Sun—Installed Price Trends for Distributed Photovoltaic Systems in the United States, 2018 Edition. Berkeley, CA. Available from: <https://emp.lbl.gov/tracking-the-sun>.

Bhandari, R., 2018. Riding Through the Experience Curve for Solar Photovoltaics Systems in Germany. Available from: https://doi.org/10.1109/IESC.2018.8439945.

Bione, J., Fraidenraich, N., Vilela, O.C., 2009. Analysis of the Learning Curve of Photovoltaic Technology and the Underlying Significant Factors, pp. 1682–1691. Available from: <https://www.scopus.com/inward/record.uri?eid = 2-s2.0-84873876072&partnerID = 40&md5 = 1028920eb8161e19fde3f90183a79952>.

Boston Consulting Group, 1970. Perspectives on Experience. Boston Consulting Group, Boston, MA.

Breyer, C., Gerlach, A., 2013. Global overview on grid-parity. Prog. Photovoltaics Res. Appl. 21 (1), 121–136. Available from: https://doi.org/10.1002/pip.1254.

Elshurafa, A.M., et al., 2018. Estimating the learning curve of solar PV balance-of-system for over 20 countries: implications and policy recommendations. J. Cleaner Prod. 196, 122–134. Available from: https://doi.org/10.1016/j.jclepro.2018.06.016.

Fraunhofer ISE, 2019. Photovoltaics Report. Fraunhofer ISE, Freiburg, Germany. Available from: <https://www.ise.fraunhofer.de/en/publications/studies/photovoltaics-report.html>.

Gan, P.Y., Li, Z., 2015. Quantitative study on long term global solar photovoltaic market. Renew. Sustain. Energy Rev. 46, 88–99. Available from: https://doi.org/10.1016/j.rser.2015.02.041.

Goodrich, A., et al., 2013. A wafer-based monocrystalline silicon photovoltaics road map: utilizing known technology improvement opportunities for further reductions in manufacturing costs. Sol. Energy Mater. Sol. Cells 114, 110–135. Available from: https://doi.org/10.1016/j.solmat.2013.01.030.

Haysom, J.E., et al., 2013. Concentrated Photovoltaics System Costs and Learning Curve Analysis, pp. 239–243. Available from: https://doi.org/10.1063/1.4822240.

Haysom, J.E., et al., 2015. Learning curve analysis of concentrated photovoltaic systems. Prog. Photovoltaics Res. Appl. 23 (11), 1678–1686. Available from: https://doi.org/10.1002/pip.2567.

Hong, S., Chung, Y., Woo, C., 2015. Scenario analysis for estimating the learning rate of photovoltaic power generation based on learning curve theory in South Korea. Energy 79 (C), 80–89. Available from: https://doi.org/10.1016/j.energy.2014.10.050.

IEA PVPS, 2014. Trends 2014 in Photovoltaic Applications. International Energy Agency (IEA) PVPS Task 1, Report T1-25:2014.

IEA PVPS, 2017. Trends 2017 in Photovoltaic Applications. Available from: <www.iea-pvps.org>.

IEA PVPS, 2018. Trends 2018 in Photovoltaic Applications. Available from: <www.iea-pvps.org>.

ITRPV, 2019. *International Technology Roadmap for Photovoltaic (ITRPV)*. Frankfurt, Germany. Available at: https://itrpv.vdma.org/.

Jäger-Waldau, A., 2019. Snapshot of photovoltaics—February 2019. Energies 12 (5), 769. Available from: https://doi.org/10.3390/en12050769.

Junginger, M., Van Sark, W., Faaij, A., 2010. Technological Learning in the Energy Sector : Lessons for Policy, Industry and Science. Edward Elgar Publishing, Cheltenham.

Kavlak, G., McNerney, J., Trancik, J.E., 2018. Evaluating the causes of cost reduction in photovoltaic modules. Energy Policy 123, 700–710. Available from: https://doi.org/10.1016/j.enpol.2018.08.015.

Kittner, N., Lill, F., Kammen, D.M., 2017. Energy storage deployment and innovation for the clean energy transition. Nat. Energy 2 (9), 17125. Available from: https://doi.org/10.1038/nenergy.2017.125.

de La Tour, A., Glachant, M., Ménière, Y., 2013. Predicting the costs of photovoltaic solar modules in 2020 using experience curve models. Energy 62, 341–348. Available from: https://doi.org/10.1016/j.energy.2013.09.037.

Louwen, A., et al., 2016a. Re-assessment of net energy production and greenhouse gas emissions avoidance after 40 years of photovoltaics development. Nat. Commun. 7, 13728. Available from: https://doi.org/10.1038/ncomms13728.

Louwen, A., et al., 2016b. A cost roadmap for silicon heterojunction solar cells. Sol. Energy Mater. Sol. Cells 147. Available from: https://doi.org/10.1016/j.solmat.2015.12.026.

Muggeo, V.M.R., 2008. Segmented: an R package to fit regression models with broken-line relationships. R News 8 (1), 20–25.

Nemet, G., 2006. Beyond the learning curve: factors influencing cost reductions in photovoltaics. Energy Policy 34, 3218–3232. Available from: https://doi.org/10.1016/j.enpol.2005.06.020.

Nemet, G.F., Husmann, D., 2012. PV learning curves and cost dynamics. Semicond. Semimetals 85–142. Available from: https://doi.org/10.1016/B978-0-12-388419-0.00005-4.

PV Magazine International, 2019. Module Price Index. Available from: <https://www.pv-magazine.com/features/investors/module-price-index/> (accessed 06.06.19.).

Rekinger, M., et al., 2015. Global Market Outlook for Photovoltaics 2015–2019. Brussels, Belgium. Available from: <www.solarpowereurope.org>.

Sampedro, M.R.G., Gonzalez, C.S., 2016. Spanish photovoltaic learning curve. Int. J. Low Carbon Technol. 11 (2), 177–183. Available from: https://doi.org/10.1093/ijlct/ctu026.

SolarPower Europe, 2019. Global Market Outlook for Solar Power 2019–2023. Available from: <http://www.solarpowereurope.org/>.

Trappey, A.J.C., et al., 2016. The determinants of photovoltaic system costs: an evaluation using a hierarchical learning curve model. J. Cleaner Prod. 112, 1709–1716. Available from: https://doi.org/10.1016/j.jclepro.2015.08.095.

van Sark, W., et al., 2010. Photovoltaic solar energy. In: Junginger, M., van Sark, W., Faaij, A. (Eds.), Technological Learning in the Energy Sector: Lessons for Policy, Industry and Science. Edward Elgar Publishing, Cheltenham, pp. 93–114.

Verlinden, P., 2014. Cost Analysis of Current PV Production and Strategy for Future Silicon PV Modules. Sydney, Australia. Available from: <http://www2.pv.unsw.edu.au/videos/Pierre-Verlinden-22January2014/slides/Presentation_PJV_20140122.pdf> (accessed 13.07.15.).

Williams, E., et al., 2017. Wind power costs expected to decrease due to technological progress. Energy Policy 106, 427–435. Available from: https://doi.org/10.1016/J.ENPOL.2017.03.032.

Yu, C.F., Van Sark, W.G.J.H.M., Alsema, E.A., 2011. Unraveling the photovoltaic technology learning curve by incorporation of input price changes and scale effects. Renew. Sustain. Energy Rev. 15 (1), 324–337. Available from: https://doi.org/10.1016/j.rser.2010.09.001.

Zhang, D., et al., 2012. Economical assessment of large-scale photovoltaic power development in China. Energy 40 (1), 370–375. Available from: https://doi.org/10.1016/j.energy.2012.01.053.

Zheng, C., Kammen, D.M., 2014. An innovation-focused roadmap for a sustainable global photovoltaic industry. Energy Policy 67, 159–169. Available from: https://doi.org/10.1016/J.ENPOL.2013.12.006.

CHAPTER 6

Onshore wind energy

Martin Junginger[1], Eric Hittinger[2], Eric Williams[2] and Ryan Wiser[3]

[1]Copernicus Institute of Sustainable Development, Utrecht University, Utrecht, The Netherlands, [2]Rochester Institute of Technology, Rochester, NY, United States, [3]Lawrence Berkeley National Laboratory, Berkeley, CA, United States

Abstract

Onshore wind has shown cost reductions for more than three decades, but the importance of underlying factors has varied over time. While in the past the cost of capacity was mainly used to measure cost reductions, this chapter analyzes the reduction of the levelized cost of electricity (LCOE). Next to lower upfront capital expenditures (Capex), the capacity factor has also increased significantly, going hand in hand with higher hub heights and larger rotor diameters. While Capex and LCOE have also temporarily increased between 2005 and 2011, the overall learning rate for LCOE for data between 1990 and 2017 is 11.4%. Combining this learning rate with anticipated growth in global onshore wind deployment yields a projected LCOE of 3.7$ cents/kWh by 2030, a reduction of approximately 25% from 2018 levels, making it highly competitive with expected prices of new coal and natural gas generation. A recent expert elicitation study of future costs of wind power yielded a range of implicit future learning rates between 14% and 18%. The 11.4% learning rate estimated here yields less bullish prospects for cost reductions; this distinction can be partly explained by the fact that there are a few factors still excluded or partially excluded from the learning-rate analysis (specifically, improvements in project lifetime, operational expenditures, and financing costs).

Chapter outline

Technological Learning in the Transition to a Low-Carbon Energy System.
DOI: https://doi.org/10.1016/B978-0-12-818762-3.00006-6

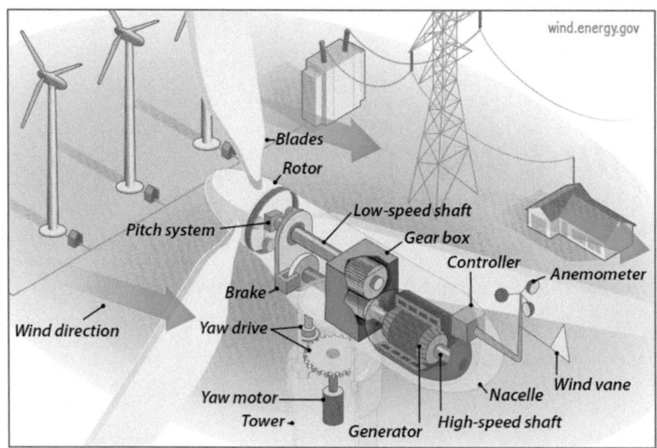

Figure 6.1
The inside of a wind turbine. *Source: Reproduced with permission from EERE (2019).*

6.1 Introduction

Wind turbines are based on the principle of converting the kinetic energy of a wind resource into mechanical work, such as water pumping, or via mechanical work into electricity via a generator (Da Rosa, 2005). Nowadays, wind turbines produced worldwide are almost solely of the electricity-generating type, though mechanical turbines used for water pumping are still of essential use in some areas (Twidell and Weir, 2015). Wind turbines come in a variety of blade designs, but the most common types of wind turbine for electricity generation, both on- and offshore, are horizontal axis turbines with three blades (IRENA, 2018).

Modern-day conventional (horizontal axis) wind turbines are built up from a steel or concrete tower, a yaw system between the tower and the nacelle (the housing for the hub section of the wind turbine) that orientates the wind turbine toward the wind, a drivetrain (gearbox or direct drive generator), a convertor, and the rotor with the blades (Twidell and Weir, 2015). The inner structure of a wind turbine is shown in Fig. 6.1.

The power generated by a wind turbine is related to the quality of the wind, height of the tower (hub height), the rotor diameter, and management of operation and maintenance. In general, wind turbines are able to generate electricity at wind speeds between 3–5 and 25 m/s. The maximum electricity generation is usually achieved from 11 to 25 m/s (IRENA, 2018).

6.2 Market development

Simple devices exploiting the energy available in wind date back thousands of years. The first large wind device for electricity generation was a 12 kW turbine introduced in 1888 in

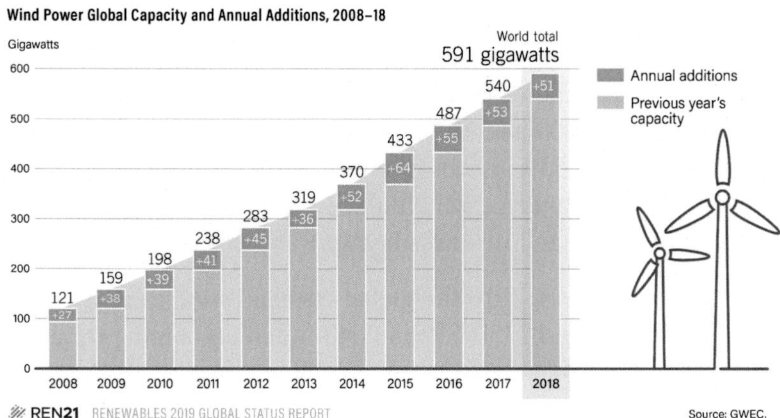

Figure 6.2
Cumulative wind power capacity between 2008 and 2018 (on- and offshore) and annual additions. *Source: REN21, 2019.*

Cleveland, OH, United States. As with many other technologies, renewed interest in wind energy was stimulated by the 1973 oil crisis. While the first developments of wind energy were mainly in the United States (most notably California), market activity shifted to Europe from 1990 onward (Kaldellis and Zafirakis, 2011). More recently, the United States regained a leading position, together with China (REN21, 2019).

In the last 10 years, wind energy has grown from an installed capacity of 121 GW in 2008 to 591 GW in 2018, with about 568.4 GW onshore and the rest operating offshore (REN21, 2019). The current market is dominated by deployment in Asia, and especially China (see Fig. 6.2), being the largest regional market for 9 years straight. China is the largest in terms of both installed capacity and capacity growth. Wind power provides a substantial share of electricity in a growing number of countries. In 2018 wind energy contributed about 14% to the EU's electricity consumption. Higher shares were achieved in at least six EU member states, including Denmark, which met 40.8% of its annual electricity consumption with wind energy (REN21, 2019).

On the supplier side of onshore wind turbines, activities are concentrated at a number of manufacturers in China, the EU, India, and the United States. The largest turbine manufacturers in 2018 were Vestas (Denmark), Goldwind (China), Siemens Gamesa (Germany/Spain), and GE (United States) (GWEC, 2019).

The costs of turbines generally represent 64%–84% of total installed system costs for onshore wind farms. Besides turbines, major cost components include construction and foundation work, grid connection, land, and project costs. Prices of wind turbines were lowest in 2000–04 and rose for some years afterward. The increase occurred because of several factors that are as follows: higher material and labor costs, more demand than supply, and development of larger wind-turbine technologies, leading to higher construction costs (Bolinger and Wiser, 2012). After prices peaked between 2007 and 2010, a downward trend started again. Between 2009 and 2017, turbines with rotor diameters smaller than 95 m showed price declines of 53%, prices of turbines with rotor diameters more than that size declined by 41%. In 2017 average wind turbine prices were less than 1000$/kW in most markets, returning to capital cost levels last seen in 2002 (IRENA, 2018).

In the IEA World Energy Outlook 2018 the global installed wind capacity is expected to increase to 1066 GW in 2030 and 1350 GW in 2040, when accounting for current and announced policy plans. In the alternative scenario where global temperature increase is kept to 2°C, installed capacity increases to 1712 GW in 2030 and 2819 GW in 2040 (also including offshore wind). Together with solar photovoltaics (PV), wind energy is expected to contribute two largest shares to the renewable portfolio (OECD/IEA, 2017). For comparison, IRENA estimates that a total wind capacity of 5445 GW is expected by 2050, of which 4923 GW is onshore and 521 GW offshore (IRENA, 2018).

6.3 Trends in capital expenditures and levelized cost of energy

The cost of installed wind turbines and projects has been declining more or less continuously between the 1980s and 2002 and between 2009 and 2017. In the intermediate time period, there were several factors idiosyncratic to wind power (explained in the previous section) that caused the installed costs of turbines and projects to temporarily increase. Generally, it can be stated that the overall reductions in the levelized cost of wind energy, in the first two decades after the introduction of modern wind turbines in the early 1980s, were driven in large measure by the reduction of installed costs, or upfront capital expenditures (Capex), which was in turn mainly driven by the spectacular increase in size, whereas typical wind turbines capacities were around 10–50 kW in the early 1980s, the largest turbines reached 2 MW in the early 2000s. Due to the economies of scale, use of higher wind speeds at greater height, and other factors, there was a decline in the levelized cost of electricity (LCOE) for more than 400–600$$_{2014}$/MWh in the early 1980s (Wiser et al., 2016) to about 60–80$$_{2014}$/MWh in the early 2000s in the United States (Wiser and Bolinger, 2018); generally consistent trends are observed in many other countries (IRENA, 2019; Wiser et al., 2016).

After 2000 the trend of increasing wind turbine capacities slowed down for onshore wind (however, rapid turbine capacity increases continued offshore, see Chapter 7). While nameplate capacity, hub heights, and blade lengths all grew, growth was slowed, in part, due to the logistical challenges of delivering increasingly large components to sites. A second factor potentially limiting turbine size increase is the visual obstruction of landscapes (including production of noise and casting of moving shadows), which limits the

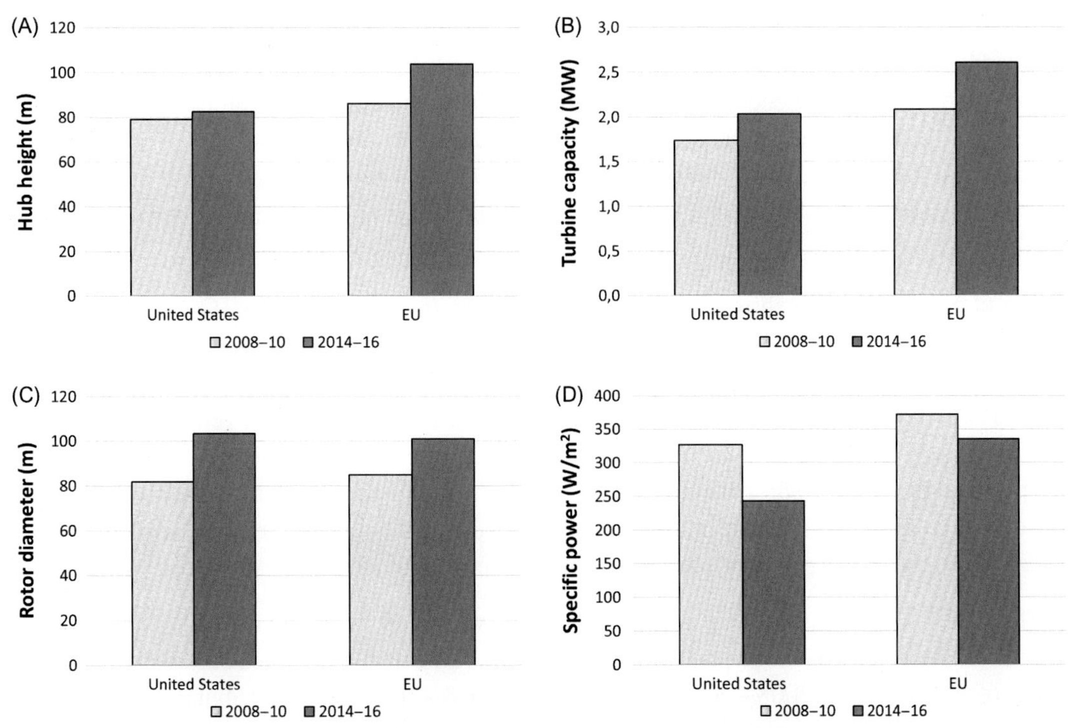

Figure 6.3
Technology trends in EU and the US (weighted averages) for the periods 2008–10 and 2014–16: (A) hub height; (B) turbine nameplate capacity; (C) rotor diameter; (D) turbine specific power. *Source: Derived from Duffy et al. (2019), based on data from IEA Wind Task 26 (The Cost of Wind Energy).*

number of possible sites in densely populated areas and is related to public resistance to wind energy. Nowadays, weighted average installed capacities are approaching 2.5 MW in the United States and between 2.5 and 3.5 MW in many European countries, with hub heights around 80–130 m (see Fig. 6.3). While this is still roughly a doubling compared to the early 2000s, the growth trend has clearly slowed down compared to the 20 × factor in capacity increase observed between 1980 and 2000. Nonetheless, Duffy et al. (2019) find that capital costs have generally fallen between 2006 and 2016 in the EU and the United States, from a range of approximately 1100–2100 to 1200–1600€/MW. Of the countries analyzed the largest cost declines were observed in the United States (27%) to a level in line with EU countries. Globally, average costs reached 1500$/kW in 2018, with costs as low as 1200$/kW in China and India (IRENA, 2019).

Duffy et al. (2019) also find that, especially in the EU, recently installed wind parks are at sites with somewhat lower average wind speeds compared to the past, most likely due to the fact that the best sites are already occupied. Yet at the same time, they find a pronounced increase in capacity factor, from 22%–36% in 2008–10 to 27%–42% in 2014–16. In other words, while new wind projects are often installed at sites with somewhat lower wind speeds, overall capacity factors have increased due to taller towers and longer blades. For comparison, capacity factors were around 18%–22% in the 1980s and the first half of the 1990s (Williams et al., 2017). This trend toward higher capacity factors is not restricted to Europe and the United States but is instead a global phenomenon (IRENA 2019).

Duffy et al. (2019) also show downward pressure on operational costs (Opex), from approximately 45–60 to 40–50 €/kW-year over the 2008–10 to 2014–16 periods. Wiser et al. (2019) conducted a dedicated study on operational cost developments in the United States reaching even further back, finding all-in lifetime Opex reductions from approximately 80$/kW-year for projects built in the late 1990s to roughly 40$/kW-year for projects built in 2018. Turbine operations and maintenance (O&M) costs—inclusive of scheduled and unscheduled maintenance—represent the single largest component of overall Opex and the primary source of cost reductions over the last decade.

Duffy et al. (2019) also investigated the change in weighted average cost of capital (WACC): the period-average real after-tax WACCs fell for all countries studied, from a range of 2.9%–5.6% (2008–10) to 1.2%–4.6% (2014–16). This is attributed to the internationally low cost of finance and to the growing perception of wind energy as a mature technology by finance institutions, thus resulting in lower project-risk premiums. The latter is a spin-off from technological learning and could be considered a form of "financial" learning, which may also be relevant for other technologies, for example, offshore wind (Egli et al., 2018).

Overall, it is clear that many different factors have influenced the LCOE in the past decade. Fig. 6.4 depicts the contribution of capacity factor, Capex, Opex, and WACC to reduction in LCOE between 2008–10 and 2014–16 for the EU (averaged) and the United States.

As can be seen from Fig. 6.4, there are two dominant factors that have driven down LCOE in the past decade: reduced Capex and increased capacity factors, with generally less significant contributions from improvements in WACC and Opex. It must be stressed that these are averages, and that on a country basis, the contributions of each factor may vary, and the overall LCOE reduction may also be significantly different than the average 35€/MWh shown for both the EU and the United States. Also, this graph does not take into account country-specific tax regimes. When these are factored in, Duffy et al. (2019) find LCOE as low as 34€/MWh for Denmark and as high as 68€/MWh in Ireland over the 2014–16 period. Using even more recent data, IRENA (2019) reports a global average wind LCOE of 54$/MWh in 2018, whereas Wiser and Bolinger (2019) report the average in the United States in 2018 as 38$/MWh.

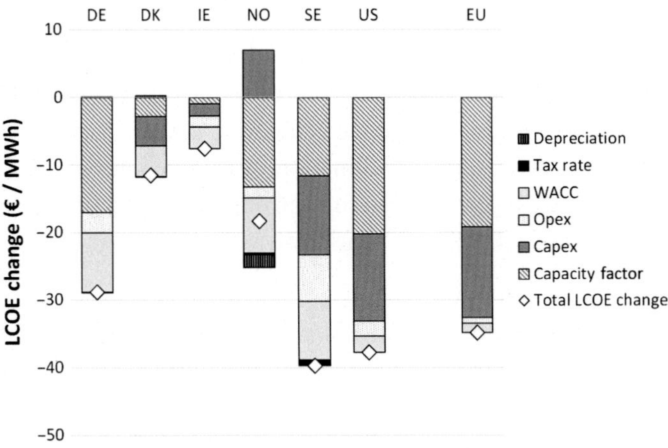

Figure 6.4

Contribution of input variables to changes in average national LCOEs between 2008–10 and 2014–16 periods for various EU member states, the EU as an average and the United States. *DE*, Germany; *DK*, Denmark; *IE*, Ireland; *NO*, Norway; *SE*, Sweden. *Source: Duffy et al. (2019).*

From the earlier statement, it can be concluded that the overall LCOE of onshore wind energy has declined substantially and that the most important drivers for the cost decline were the reduction in Capex and increasing capacity factor. Thus any tools/experience curves solely focusing on Capex as a representation for the cost of wind energy are increasingly missing important contributions from other variables, discussed in more detail in the next section.

6.4 Experience curves for onshore wind energy

Together with solar photovoltaics, onshore wind energy is likely the most extensively studied technology for experience curve analyses. Since the mid-1990s, dozens of studies have been published, covering technological progress curves for onshore wind energy from many different angles. The scope of the studies varies in terms of time series analyzed, geographical scope (e.g., national or global analyses), the use of one-factor (capacity) learning curves or multifactor learning curves (typically also taking R&D expenditures into account), analyzing either the costs of wind turbines, installed wind parks, or (in few cases) the LCOE. This chapter does not aim to provide a comprehensive overview of these studies; for this, we refer to Junginger et al. (2009) for older studies and two meta-analyses (Lindman and Söderholm, 2012; Rubin et al., 2015) for more recent studies. However, this body of work finds a wide range of learning rates: Williams et al (2017) summarize the literature, showing a full range from −5% to 35%, with most learning rates reported between 5% and 10%. This wide variation arises from three main factors: (1) differences in start and end dates, (2) which country's data (or global data) are used, and (3) what type of experience curve model is used. The dates of the studies reviewed in Williams et al. (2017) range from 1980 to 2010, with many focusing on the 1990–2000 period. Countries studied include Denmark, Germany, Spain, the United Kingdom, and the United States, with some aggregating to a global scale via some combination of these countries. The most common model used is a single-factor experience curve focusing on capital cost and installed capacity, but multifactor models, or those that focus on production or LCOE, are also represented.

There are three primary considerations to evaluate when constructing a technological progress model for wind power: the temporal scope, the geographical scope, and the structure of the model. Earlier studies where date ranges for data were constrained resulted in estimated learning rates far from the consensus values. It is thus important to choose a data set with the longest possible coverage (and hence the greatest number of cumulative doublings of capacity) in order to derive meaningful trends least affected by temporary outlier effects. Fig. 6.5 illustrates the reasons why using data from short-time series is tricky: using data from 1982 to 2016 gives a learning rate of around 6%. If one were to instead use recent data from 2009 onward only, one would find learning rates between 24% and 26%, much higher than for, for example, PV, and it would seem unrealistic compared

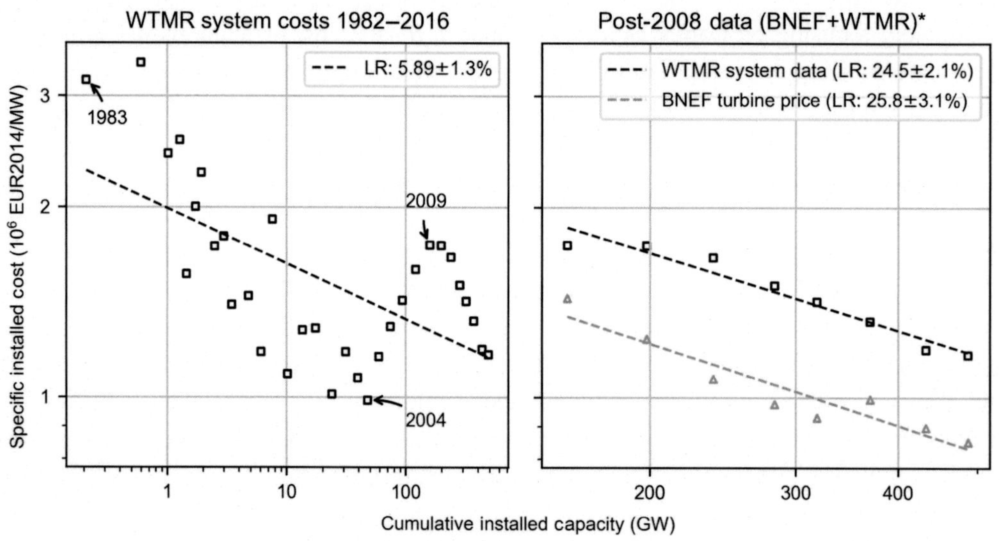

Figure 6.5
Overview of onshore system and turbine price data and experience curves. (Left) Whole dataset from WTMR (Wiser and Bolinger, 2017). (Right) Post-2008 data from Wiser and Bolinger (2017) and BNEF wind turbine price index (BNEF, 2017). *Note that the curves on the right are only shown as an example of short term datasets, and the learning rates shown are not recommended to be used in any modeling activity.

to the long-term trend. On the other hand, using data between 2004 and 2009 would yield negative learning rates—an even more unlikely rate of long-term learning given the data before and after this time period.

Similarly, it is logical that a global scope best fits this technological system: modern wind turbines and wind parks have been deployed globally since the 1980s, with 15 wind turbine manufacturers typically covering more than 90% of the global market at any given time (GWEC, 2019), all using the same basic technological concept (the pitch controlled, three-blade horizontal axis turbine). It therefore seems appropriate to assume that technological learning of onshore wind energy can be best measured using global deployment.

The third point, the modeling approach used, particularly the explanatory variables included, warrants additional discussion. The first issue is whether to rely on a single dependent variable (typically installed capacity) or attempt to disentangle different factors driving technological learning and associated reductions in production cost using a multifactor learning curve approach. Several studies have been published separating cost reductions into learning-by-doing and learning-by-research (Miketa and Schrattenholzer, 2004; Klaassen et al., 2005; Jamasb and Koehler, 2007; Söderholm and Sundqvist, 2007; Ek and Söderholm, 2010). Learning-by-doing rates often vary from 1% to 17%, whereas learning-by-research varies from 5% to 27%. However, these studies often suffer from limited data availability and are typically not able to include the impact of private R&D expenditures. This analysis therefore focuses on single-factor learning curves.

The vast majority of studies use installed capacity as the independent variable and the installed cost of wind turbines or entire wind parks as the dependent variable. This has been the primary focus of the historical literature for several reasons: data availability (cost/prices of wind turbines and wind parks have been collected for many countries and world regions), consistency with experience curves for other renewable energy technologies (e.g., for PV, almost all existing experience curves are also based on upfront costs, most often just for PV modules), and the fact that, between the mid-1980s and 2004, there was a continuous decline of turbine and wind park upfront costs, as shown in Fig. 6.5, typically yielding learning rates between 10% and 20%. The latter trend changed fundamentally with the observed increases in upfront costs between 2004 and 2009 (see Fig. 6.5). While upfront costs have declined between 2009 and 2019 and have reached 2004 levels again, it is hard to argue that upfront costs consistently followed the learning curve observed before 2004. Partly, this can be explained by factors such as increased demand for wind turbines and increased raw material prices between 2004 and 2009 (see the previous section), yet the fact that capital costs in 2017 have only fallen back to those in 2004 requires further explanation.

So, given the strong fluctuations observed in the past 15 years, is the long-term trend with learning rates of about 5.8% (Figs. 6.5) to 6.5% (Fig. 6.6A) for onshore wind reliable? Is it

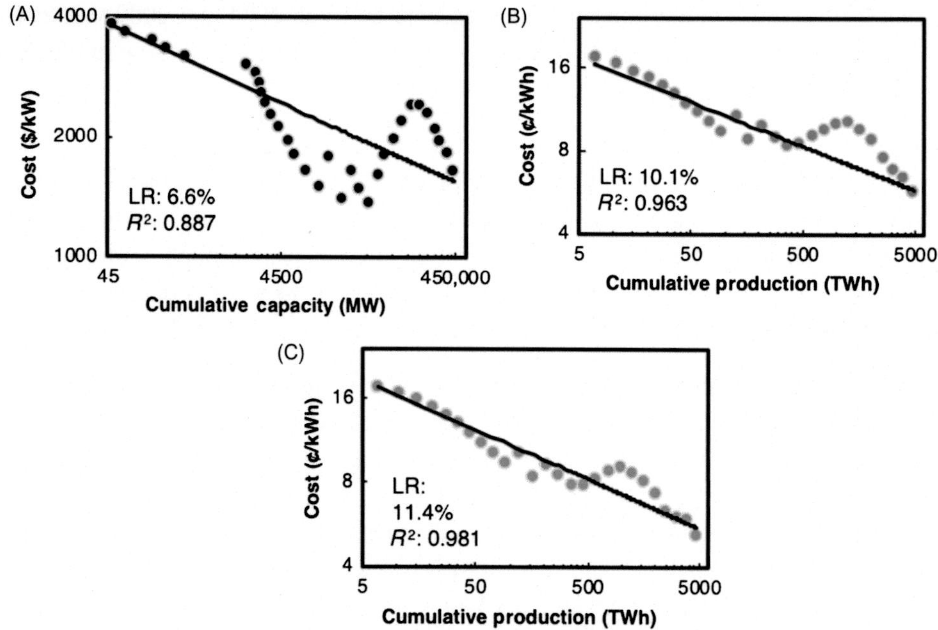

Figure 6.6
Global experience curves for wind power costs in three different models, in log–log scale, with LR and R^2 value reported. Model A considers only power capacity of plants, model B tracks total energy generated (accounting for capacity factor improvements), and model C removes the effect of changes in quality of wind sites. The data for (A) covers years 1984–2017; for models (B) and (C), the range is 1990–2017 (due to lack of data on capacity factors before that period). Costs are shown in real $\$_{2017}$. *LR*, Learning rate. *Source: Adapted from Williams et al. (2017).*

the best model for estimating future cost developments? To answer these questions, we refer to Section 6.3, where some of the underlying factors of the levelized cost of energy reductions observed in the 2006–16 period are discussed. As shown in that section, during this period, many wind parks were placed at sites with somewhat lower wind speeds (presumably because the sites with higher wind speeds had already been occupied). The geographical potential for wind energy in the world is limited, and as a general rule, sites with lower wind speeds will ultimately deliver higher LCOE. While this effect in itself has little to do with learning, the fact that this trend has been paired with a strong increase in capacity factors is very much linked to learning. However, this effect cannot be measured using the traditional method of using upfront cost of capacity as the dependent variable. As a related argument, it is clear that the primary aim of wind plant operators (and, indirectly, wind turbine manufacturers) is not to build *capacity* at lowest cost but to produce *electricity* at the lowest possible cost. This may not necessarily be at the lowest upfront cost of capacity, as higher per-MW capital costs might allow for improvement in other factors that affect the primary design variable (LCOE): capacity factor of the plant, operational costs, and financing costs. Perhaps most importantly, the sizable increases in tower heights and the longer blades that are hallmarks of modern wind turbines have yielded sizable increases in the capacity factors of wind plants, helping to contribute to lower LCOE.

Given these considerations, a small but growing literature has emphasized the need to develop learning curves based on the levelized cost of wind energy. Whereas recent analyses suggest historical global learning rates of 6%–9%, when considering only the upfront costs of land-based wind (IRENA, 2018; Wiser et al., 2016), LCOE-based learning rates have been shown to be higher, typically ranging from slightly less than 10% to nearly 20% given the greater number of cost-reduction drivers considered (Wiser et al., 2016; Williams et al., 2017; IRENA, 2018).

Focusing on the approach used in Williams et al. (2017), this study develops three different experience curve models, using global installed capacity/electricity produced as independent variables and US data on the cost of land-based wind (as proxy for global costs) as the dependent variable:

1. Capacity model—upfront capital cost ($/W) versus total capacity (W)
2. Generation model—LCOE ($/kWh) versus cumulative generation (kWh)
3. Wind quality–adjusted generation model—levelized cost assuming same wind site quality ($/kWh) versus cumulative generation (kWh)

This method attempts to account for all factors that are related to learning while controlling for changes in those factors that affect LCOE but are unrelated to learning. This includes changes in reference wind quality for individual years, and the average capacity factors for each year for new wind parks. Global capacity and cumulative generation are global totals, given lack of global data, US prices, capacity factors, and wind site quality were used as proxies. A discount rate of 7% and a lifetime of 20 years are assumed for all calculations of LCOE. For the detailed method and data sources, we refer to the underlying publication in Williams et al. (2017), and the focus here on the results, shown in Fig. 6.6.

As seen in Fig. 6.6A, the experience curve using the upfront installed cost of capacity as the independent variable (model A) shows a pattern similar to Fig. 6.5. Sensitivity analysis revealed that the capacity-based learning rate is very sensitive to the chosen starting and end years. The R^2 of the fitted trend improved successively for models B and C, and the average learning rate increased about 10%–11%. The C model has the highest R^2 value, 0.981, with an average learning rate of 11.4% and confidence interval of 10.9%–12.0%. Models B and C were shown to be much less sensitive than model A to starting and end point years for the data series. Thus taking into account the additional factors driving overall LCOE reductions led to an improved experience curve and a significant narrowing of the range in learning rates.

Three factors that Williams et al. (2017) did not account for were cost reductions in Opex post 2006, changes in project lifetime, and the cost of finance. Wiser et al. (2019) find that when plotting historical Opex against cumulative global installed capacity, a 9% Opex learning rate is reached. That work also concluded that expected project design lives have increased from 20 years to 25–30 years; this lowers LCOE by about 4%–6% with a 7% discount rate. Thus the learning rate of about 10% cited earlier could be even (slightly) higher; the same would also be true if the effect of technological maturity on the cost of finance was considered.

6.5 Data collection and methodological issues

An overview of the general data collection issues applicable to wind energy is given in Table 6.4. There are no publicly available data for the production cost of wind turbines or wind parks, and so the use of market prices is very common for onshore wind. Price trends ought to follow costs over the long term, but there can easily be deviations over shorter time periods. For example, prices were inflated in the period of 2004–08 when wind demand clearly exceeded supply, and a number of other exogenous influences impacted prices. Care is thus needed using price as a proxy for underlying production costs. Also, as argued in the previous section, the upfront cost of capacity is increasingly becoming an inadequate metric to capture the overall technological learning of onshore wind—elements such as higher capacity factors, lower O&M costs, longer design lives, and lower interest rates (as a consequence of increased trust in onshore wind technology) are not reflected by this metric. By taking into account (some of) these factors in an experience curve for electricity, a more accurate model to project future LCOE developments can be achieved. Further research should therefore focus on this formulation.

Table 6.4: General data collection issues for onshore wind power.

Issue	Resolution	Applicability
Production cost data is unavailable	Price data are used	fx1
Data not directly reported/available for desired cost unit (LCOE)	Estimating LCOE from upfront capacity cost, capacity factor, and other variables	fx1
Data only available for limited geographical scope, while technology is deployed globally	Different datasets combined and compared	fx1
Data are in different currencies or currency year	Convert currency and correct for inflation	fx1
Early cumulative production figures are not agreed upon or available		
Supply/Demand affecting costs significantly	Use long-term trends; try to control and correct for exogenous price drivers, or exclude periods in which prices are driven by demand	fx1

LCOE, Levelized cost of electricity.

Finally, there is a geographical component to LCOE for onshore wind, that is, there are sites with low and high production costs, depending in part on average wind speeds, accessibility, and possibilities to connect to the grid, etc. Typically, the low-cost sites would be expected to be occupied first, with higher cost sites used later on. This effect will, to some extent, cancel out learning effects—yet the previous sections have shown wind turbine designs can be optimized for such sites, leading, thus far, to continued reductions in average LCOE. If and how long this trend will continue is subject to uncertainty.

6.6 Discussion, conclusion, and future outlook

Based on the findings presented earlier, it has become clear that using upfront capital cost as the basis for experience curves for wind energy may result in increasingly inaccurate estimates regarding the past (and future) cost reduction of electricity from onshore wind parks. Unless improvements in, for example, capacity factor and O&M costs are estimated exogenously, using capacity-only experience curves may result in a serious underestimation of the potential future LCOE reductions that may be achieved.

Based on the Opex learning rate derived by Wiser et al. (2019), a further \$5–\$8/kW-year (12%–18%) Opex reduction from 2018 to 2040 is projected. When compared with the broader literature, these findings suggest that continued Opex reductions may contribute 10% or more of the expected reductions in land-based wind LCOE. Moreover, these estimates may understate the importance of Opex owing to the multiplicative effects through which operational advancements influence not only O&M costs but also component reliability, performance, and plant-level availability—thereby affecting levelized costs though Opex reduction and by enhancing annual energy production and plant lifetimes. Beyond Opex, further improvements in project performance, project life, and the cost of finance can also be expected (Wiser et al., 2016), confirming the inadequacy of learning rates based solely on upfront costs.

Using global wind adoption rates from the 2018 IEA New Policies Scenario (IEA, 2018), combined with a 11.4% learning rate for the LCOE of wind power, future wind-power prices are projected to 2030. Historical data follow the same method as model C on Fig. 6.6. The projected data in Fig. 6.7 start with an estimated wind power cost of 4.6 \$cents/kWh in 2018 and apply an 11.4% learning rate (see Fig. 6.6) to the installed capacity estimates from the 2018 IEA New Policies Scenario. The 2018 levelized cost of 4.6\$cents/kWh is an average of the 5.4\$cent/kWh estimate from IRENA (2019) and the 3.8\$cent/kWh estimate from Wiser and Bolinger (2019).

In this scenario the price of wind power falls to 3.7\$cents/kWh by 2030, making it highly competitive with expected prices of new coal and natural gas generation. This constitutes a 50% reduction in LCOE between 2011 and 2030, which is quite a bit higher than

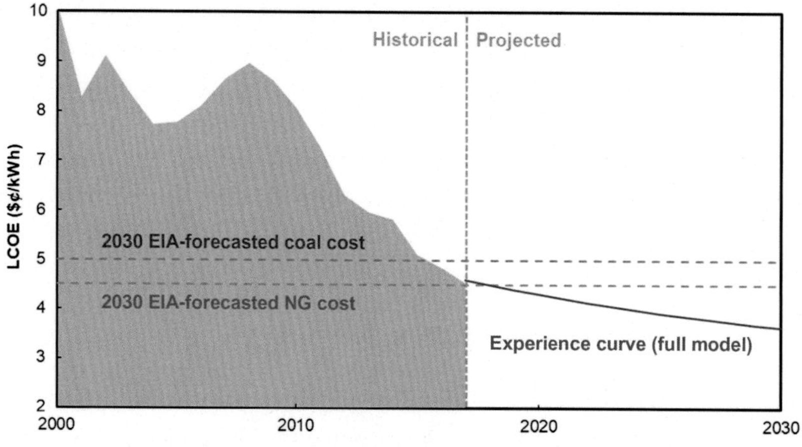

Figure 6.7
Historical and projected onshore wind generation costs (global estimate), with wind generation costs falling to 3.7cents/kWh by 2030.

the 20%–30% reduction described in the IEA report Past and Future Cost of Wind Energy (Lantz et al., 2012). It represents a reduction in LCOE of approximately 25% from 2018 to 2030. Another relevant comparison is an expert elicitation study of future costs of wind power, which yielded a range of implicit future learning rates between 14% and 18% (Wiser et al., 2016). While there is overlap in results from learning curves and expert elicitation, the 10%–12% learning rate estimated here yields less bullish prospects for cost reductions. This distinction can be partly explained by the fact that there are a few factors still excluded or partially excluded from the learning rate analysis (specifically, improvements in project lifetime, Opex costs, and financing costs).

Considering all of these factors together and leveraging the best available current research, the long-term learning rate for the LCOE from onshore wind power is estimated to be at least 10%–12%, with higher estimates possible if additional cost-reducing drivers are considered. However, history has shown that external effects can temporarily shift LCOE far from the long-term trend, so considerations of supply and demand, labor and material costs, or changes in finance structure are relevant to translating the underlying learning effects to actual market prices.

References

BNEF, 2017. Wind Turbine Price Index. Bloomberg New Energy Finance. Available from: <https://about.bnef.com/blog/2h-2017-wind-turbine-price-index/>.

Bolinger, M., Wiser, R., 2012. Understanding wind turbine price trends in the U.S. over the past decade. Energy Policy 42, 628–641. Available from: https://doi.org/10.1016/j.enpol.2011.12.036.

Da Rosa, A.V., 2005. Fundamentals of Renewable Energy Processes. Elsevier Academic Press, Burlington, MA.

Duffy, A., Hand, M., Wiser, R., Lantz, E., Dalla Riva, A., Berkhout, V., et al., 2019. Land-Based Wind Energy Cost Trends in Selected Countries: Germany, Denmark, Ireland, Norway, Sweden and the United States. Submitted to Applied Energy March 2019. (Under review).

EERE, 2019. The Inside of a Wind Turbine. Office of Energy Efficiency and Renewable Energy. Available from: <https://www.energy.gov/eere/wind/inside-wind-turbine-0>.

Egli, F., Steffen, B., Schmidt, T.S., 2018. Nat. Energy 3, 1084–1092. Available from: https://doi.org/10.1038/s41560-018-0277-y.

Ek, K., Söderholm, P., 2010. Technology learning in the presence of public R&D: the case of European wind power. Ecol. Econ. 69, 2356–2362.

GWEC, 2019. 1 in 5 Wind Turbines Installed by Vestas in 2018, According to New Market Intelligence Report. Global Wind Energy Council. Available from: <https://gwec.net/gwec-1-in-5-wind-turbines-are-installed-by-vestas-according-to-new-market-intelligence-report/>.

IEA (2018), World Energy Outlook, 2018, International Energy Agency. Available from: https://www.iea.org/weo/

IRENA, 2018. Renewable Power Generation Costs in 2017. International Renewable Energy Agency (IRENA), Abu Dhabi. Retrieved from: <https://www.irena.org/-/media/Files/IRENA/Agency/Publication/2018/Jan/IRENA_2017_Power_Costs_2018.pdf>.

IRENA, 2019. Renewable Power Generation Costs in 2018. International Renewable Energy Agency (IRENA), Abu Dhabi. Retrieved from: <https://www.irena.org/publications/2019/May/Renewable-power-generation-costs-in-2018>.

Jamasb, T., Koehler, J., 2007. Learning curves for energy technology: a critical assessment. In: Grubb, M., Jamasb, T., Pollitt, M.G. (Eds.), Delivering a Low Carbon Electricity System: Technologies, Economics and Policy. Cambridge University Press, Cambridge.

Junginger, M., Lako, P., Neij, L., Engels, W., Milborrow, D., 2009. Chapter 6: Onshore wind energy. In: Junginger, M., van Sark, W., Faaij, A. (Eds.), Technological Learning in the Energy Sector. Lessons for Policy, Industry and Science. Edward Elgar Publishing, Cheltenham, pp. 65–78. ISBN 978 2 84844 834 6.

Kaldellis, J.K., Zafirakis, D., 2011. The wind energy (r)evolution: a short review of a long history. Renew. Energy 36, 1887–1901.

Klaassen, G., Miketa, A., Larsen, K., Sundqvist, T., 2005. The impact of R&D on innovation for wind energy in Denmark, Germany and the United Kingdom. Ecol. Econ. 54 (2-3), 227–240.

Lantz, E., Wiser, R., Hand, M., 2012. The Past and Future Cost of Wind Energy. National Renewable Energy Laboratory, Golden, CO, Report No. NREL/TP-6A20-53510.

Lindman, Å., Söderholm, P., 2012. Wind power learning rates: a conceptual review and meta analysis. Energy Econ. 34 (3), 754–761.

Miketa, A., Schrattenholzer, L., 2004. Experiments with a methodology to model the role of R&D expenditures in energy technology learning processes; first results. Energy Policy 32, 1679–1692.

OECD/IEA, 2017. World Energy Outlook 2017. International Energy Agency, Paris, France, Online tool viewed April 4, 2018 on <https://www.iea.org/weo/>; Executive summary retrieved from <https://www.iea.org/Textbase/npsum/weo2017SUM.pdf>.

REN21, 2019. Global Status Report 2019. REN21 Secretariat, Paris. Available from: <https://www.ren21.net/gsr-2019>.

Rubin, E.S., Azevedo, I.M., Jaramillo, P., Yeh, S., 2015. A review of learning rates for electricity supply technologies. Energy Policy 86, 198–218.

Söderholm, P., Sundqvist, T., 2007. Empirical challenges in the use of learning curves for assessing the economic prospects of renewable energy technologies. Renew. Energy 32, 2559–2578.

Twidell, J., Weir, T., 2015. Renewable Energy Resources, third ed. Routledge, New York.

Williams, E., Hittinger, E., Carvalho, R., Williams, R., 2017. Wind power costs expected to decrease due to technological progress. Energy Policy 106, 427–435.

Wiser, R., Bolinger, M., 2017. 2016 Wind Technologies Market Report. US Department of Energy Office of Energy Efficiency and Renewable Energy, Washington, DC.

Wiser, R., Bolinger, M., 2018. 2017 Wind Technologies Market Report. Lawrence Berkeley National Laboratory, Berkeley, CA. Available from: <https://emp.lbl.gov/sites/default/files/2017_wind_technologies_market_report.pdf>.

Wiser, R., Bolinger, M., 2019. 2018 Wind Technologies Market Report. US Department of Energy Office of Energy Efficiency and Renewable Energy, Washington, DC.

Wiser, R., Jenni, K., Seel, J., Baker, E., Hand, M., Lantz, E., et al., 2016. Expert elicitation survey on future wind energy costs. Nature Energy 1, Article no. 16135 (2016). Available from: https://doi.org/10.1038/nenergy.2016.135.

Wiser, R., Bolinger, M., Lantz, E., 2019. Benchmarking Wind Power Operating Costs in the United States: Results From a Survey of Wind Industry Experts. Renewable Energy Focus.

Further reading

Junginger, M., Faaij, A., Turkenburg, W.C., 2005. Global experience curves for wind farms. Energy Policy 33, 133–150.

CHAPTER 7

Offshore wind energy

Martin Junginger[1], Atse Louwen[1,2], Nilo Gomez Tuya[1], David de Jager[3], Ernst van Zuijlen[4] and Michael Taylor[5]

[1]Copernicus Institute of Sustainable Development, Utrecht University, Utrecht, The Netherlands, [2]Institute for Renewable Energy, Eurac Research, Bolzano, Italy, [3]GROW Foundation (Growth Through Research, Development & Demonstration in Offshore Wind), Utrecht, The Netherlands, [4]WindWerk BV, Utrecht, The Netherlands, [5]International Renewable Energy Agency, Bonn, Germany

Abstract

Offshore wind farms have been deployed since 1991, mainly in the North sea and Baltic sea but in recent years also in the United States and East Asia. Learning effects in offshore wind parks have been masked by various other developments over the past two decades, leading first to an increase in capital expenditures (Capex) and average levelized cost of electricity (LCOE) between 2003 and 2012, followed by a steep decline between 2015 and 2018. Based on changes in Capex, capacity factor, weighted average cost of capital (WACC), and operational expenditures (Opex), the LCOE increased from 120€/MWh in 2000 to 190€/MWh in 2015 and then decreased to about 100€/MWh at the end of 2018, with average projections for 2021 reaching 70€/MWh. Especially the increase in capacity factor has been a major driver in reducing the LCOE. Given the strong fluctuations in the past and many factors influencing the LCOE of offshore wind projects, it was not possible to derive meaningful one-factor experience curves and learning rates that would allow extrapolation for the future cost projections. Multifactor learning curves approach taking into account raw material costs, location-specific properties, and soft factors such as developments in WACC show more promise, but more deployment of offshore wind is needed to demonstrate whether such models can provide more accurate cost trend forecasts for the coming years.

Chapter Outline

Technological Learning in the Transition to a Low-Carbon Energy System.
DOI: https://doi.org/10.1016/B978-0-12-818762-3.00007-8

7.1 Introduction

The first offshore wind park was installed almost 30 years ago in Denmark in 1991. The very first such wind farm in Vindeby, Denmark, comprised 11 turbines with a power of 450 kW each, resulting in a total capacity of 4.95 MW (Henderson, 2015). Since then, offshore wind farm deployment has grown exponentially, initially in the North and Baltic seas, but in recent years in new markets outside Europe. Putting wind turbines offshore has several advantages compared to onshore parks, such as a vast techno-economical potential; higher, more constant, and often "smoother" wind speeds; and in many cases less resistance from local stakeholders (depending on distance to shore, given visual impacts are minimal to none when parks are sited far offshore). Also, the possibility to transport turbines by ships rather than trucks has allowed for an impressive increase in turbine size, allowing for higher electricity production per turbine amortizing installation costs across greater output. On the other hand, offshore wind installation costs are higher, requiring dedicated equipment, and the environment is harsher, also making operation and maintenance more challenging.

Europe has been leading the development of offshore wind farms for the past three decades. In 2018 Europe connected 409 new offshore wind turbines to the grid across 18 projects. This brought 2649 MW of net additional capacity. Europe now has a total installed offshore wind capacity of 18,499 MW, as shown in Fig. 7.1 (WindEurope, 2019). The United Kingdom has the largest amount of installed offshore wind capacity in Europe, representing 43% of all installations, followed by Germany with 34%, Denmark with 8% (despite no additional capacity in 2017), the Netherlands 7%, and Belgium 6% (WindEurope, 2019). Combined, the top five countries cover 98% of all grid-connected offshore turbines in

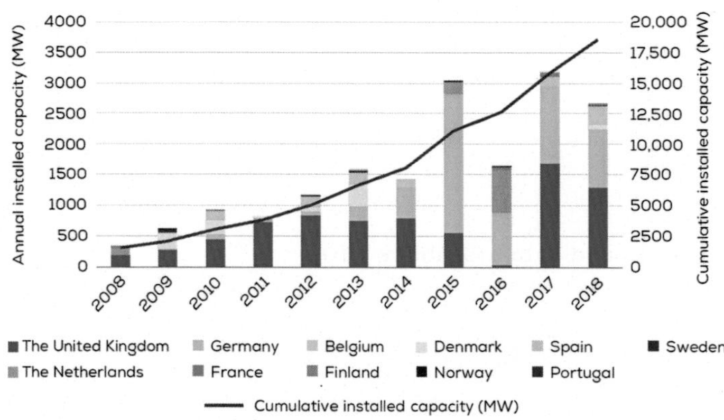

Figure 7.1
Offshore wind energy power installed in Europe in the period 2008–18. *Source: Reproduced with permission from WindEurope (2019).*

Europe and 80% globally. Outside Europe, a number of relatively small offshore projects have been developed in the United States and Asia (China, Japan, Vietnam), and projects are planned in Korea and Taiwan. In addition, China installed about 5 GW of offshore capacity in 2017 and 2018, bringing the global total to around 24 GW by the end of 2018 (IRENA, 2019). In this chapter, for historical developments, we focus mainly on the North and Baltic seas, as the vast majority of deployment (and most likely technological learning) has taken place in this region.

There are important differences between onshore and offshore wind farms. A clear difference with onshore park is the need for a more elaborate foundation, usually a monopile (in shallow water) or a tripod or lattice construction for waters up to 40 m deep (Fig. 7.2). The turbines are interconnected with an array of power cables that deliver electricity to an offshore substation that converts the electricity to a high-voltage current to minimize losses during transport to shore over the export power cable. A second substation connects the park to the conventional high-voltage grid.

7.2 Methodological issues and data availability

An overview of the general data collection issues applicable to wind energy is given in Table 7.1. For both on- and offshore wind, large sets of price data are available, but especially the upfront investment cost shows a large increase between approximately 2003 and 2012, making it difficult to devise meaningful experience curves that can be used to determine learning rates that reflect actual technological progress. Market dynamics, market maturity, technology evolution, and raw material prices have influenced the price

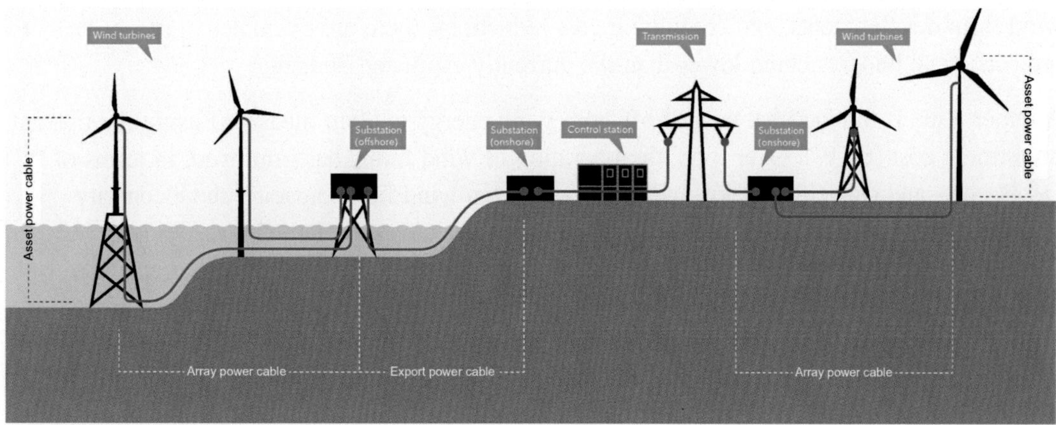

Figure 7.2
Typical layout of an offshore and onshore wind farm. *Source: Reproduced with permission from DNV GL (2015).*

Table 7.1: General data collection issues for offshore wind power.

Issue	Resolution	Applicability
Historical data is not for cost	Price data is used	☑
Data not available for desired cost unit (LCOE)	Capex and capacity factors are known for most projects, WACC, and O&M usually need to be estimated	☑
Data is valid for limited geographical scope (five countries, covering North Sea and Baltic Sea)	Different datasets combined and compared	☑

Capex, Capital expenditures; *LCOE*, levelized cost of electricity; *WACC*, weighted average cost of capital.

developments of wind turbines (Wiser and Bolinger, 2017). By taking into account these factors in a multifactor experience curve, a more accurate model to project future wind turbine, and on- and offshore wind farm prices, could possibly be established, as shown in Section 7.3.1.

Data for total installed costs, capacity factors, O&M, and costs of capital is required to calculate the levelized cost of electricity (LCOE) of offshore wind projects, with assumptions used in the absence of project level data. For many recent offshore tenders, only the complete, estimated costs for the wind farms as a whole are often available, and the price per unit of electricity may or may not reflect and LCOE equivalent depending on the contract terms. An additional complication is that the comparability of data between countries is often compromised due to different boundary conditions. For instance, costs reported for recent Dutch tendered wind farms do not include the grid connection to shore, while in the United Kingdom, the results do. In this chapter, all capital expenditures (Capex) include the cost of grid connection. An additional complication is that given the longer lead times of offshore wind farm developments, and continuing cost reductions, there are examples in recent years of projects' final budgets being lower than the currently estimated budgets.

Another issue is the fact that cost of offshore wind energy is often measured using the upfront investment cost (€/kW). Over time, the operation of wind farms has improved, in terms of O&M costs and capacity factors, and thus for a given wind farm capacity, the electricity generated has increased, and the operating costs have decreased relatively, resulting in lowered costs of electricity produced (LCOE). Furthermore, offshore wind farms are characterized by higher capital costs but in general have much higher capacity factors compared to most onshore wind farms. The increase of capacity factors is also the result of other improvements that might actually increase the costs of capacity (such as application of higher towers) yet still result in lower LCOE. In this chapter, all LCOE presented have been calculated as much as possible with project-specific data on Capex (including grid connection), capacity factor, and lifetime. The weighted average cost of capital (WACC) has been based on either project or (if not available) country-specific data, whereas operational expenditures (Opex) have been estimated based on general literature estimates. For more details, see Gomez Tuya (2019).

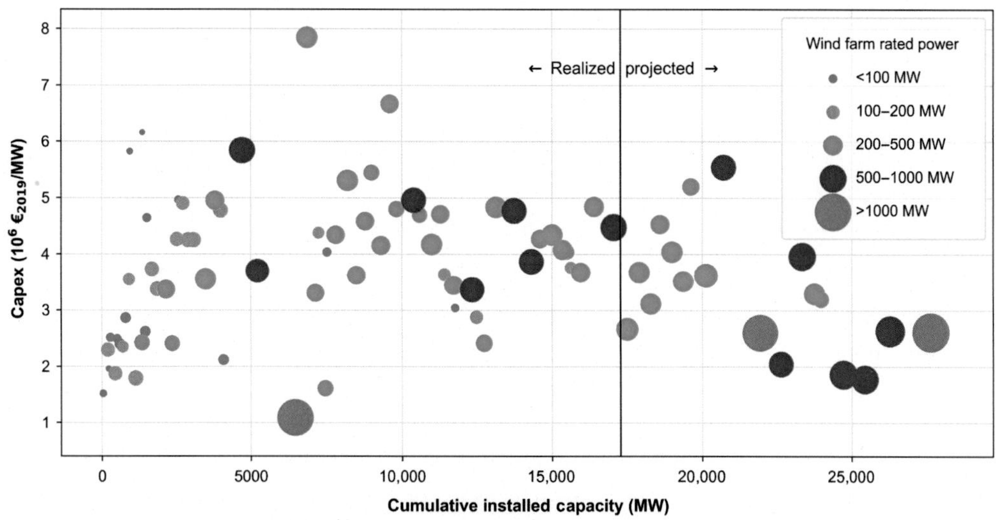

Figure 7.3
Development of global offshore wind park specific capital expenditures (Capex) between 2001 and 2018 (realized) and 2019–22 (projected). *Source: Based on own data collection by Gomez Tuya and Taylor.*

As shown in the previous research (Junginger et al., 2010; Voormolen et al., 2016), there can be large country-specific differences between wind system prices, due to differences in market maturity, regulatory requirements, government support schemes, and structural factors (e.g., labor and material costs). While recent offshore wind farm data (see Fig. 7.8) indicates that recent prices are becoming similar for four major European markets, further research should establish whether these price differences are still occurring for, for example, onshore wind farms.

7.3 Results

7.3.1 Experience curve analyses

For offshore wind farms, data was collected on all wind farms installed offshore in Europe between 2001 and 2018 and announced wind parks of 2019–22. The resulting specific Capex cost (including grid connection costs) is plotted against cumulative installed offshore capacity in Fig. 7.3.

As can be seen on the left side of Fig. 7.3, the general Capex trend increased between 2001 and about 2012. Especially around 2012, a wide divergence of Capex was found, with prices ranging from as low as 1.2 up to 8 $M€_{2019}$/MW. Many reasons can be attributed to this increase in Capex, which will be discussed in more detail in the next section. After

2013 a general declining trend can be observed, especially with some projected wind farms expecting to reach Capex under 2 M€/MW. Nevertheless, no meaningful experience curve could be fitted for the entire data series.

For the development of LCOE the picture looks very similar (see Fig. 7.4A–D)—based on the estimated LCOE of individual wind parks, no clear experience curve could be established, and only after correcting for a variety of factors, the following clear(er) trends could be found:

- Analog to onshore wind (see Chapter 6), using LCOE for experience curve analyses captures more factors than Capex-based analyses—especially the increase in capacity factor (see also the next section). Also, ultimately wind parks are designed to achieve lowest cost of electricity rather than lowest upfront investment cost.
- As shown in Fig. 7.4A, LCOE varied significantly for different countries: the United Kingdom saw a doubling of LCOE from 120 to 240€/MWh, and subsequently a sharp decline to 70–90€/MWh (projected). In contrast, in Denmark, with one exception, LCOE has always remained under 100€/MWh. This is likely due to differences in policy support schemes and other factors (see also next section).
- Typically, in experience curves, no difference between countries is made, and in Fig. 7.4B, all annual LCOE are averaged. Also, next to the one-factor experience curve, additional explanatory variables are introduced, such as the changes in WACC, water depth, and steel prices. However, meaningful experience curves can be derived.
- An additional factor that makes the application of the experience curve concept difficult is the small number of individual wind parks built, an issue earlier discussed by Junginger (2005). This means that in years with only one data point (especially in early phases of the technology), these can determine the fit of the experience curve to a large extent. In Fig. 7.4C an attempt is made to correct for this by only considering years with at least two data points.
- As a last step, in Fig. 7.4D, only offshore wind parks are taken into account with a size of 250 MW or larger, at least 20 km from shore, and in at least 20 m water depth, in order to ensure that wind parks can reasonably be compared with each other. This effectively limits the dataset to the period of 2014–21. As can be seen in Fig. 7.4D, this yields an experience curve with learning rates between 26.8% and 31.2%. Especially the inclusion of water depth as additional explanatory variable increases the quality of the fit.

Fig. 7.4D does show both single and multifactor learning curves with high quality of fit. Yet, we caution against the use this learning rate for any projections: especially Fig. 7.4D only describes a trend over a very limited time period with only two cumulative doublings of electricity generation (compared to more than nine doublings in Fig. 7.4B). It is highly unlikely that this steep decline in LCOE can be continued in the coming years, and also

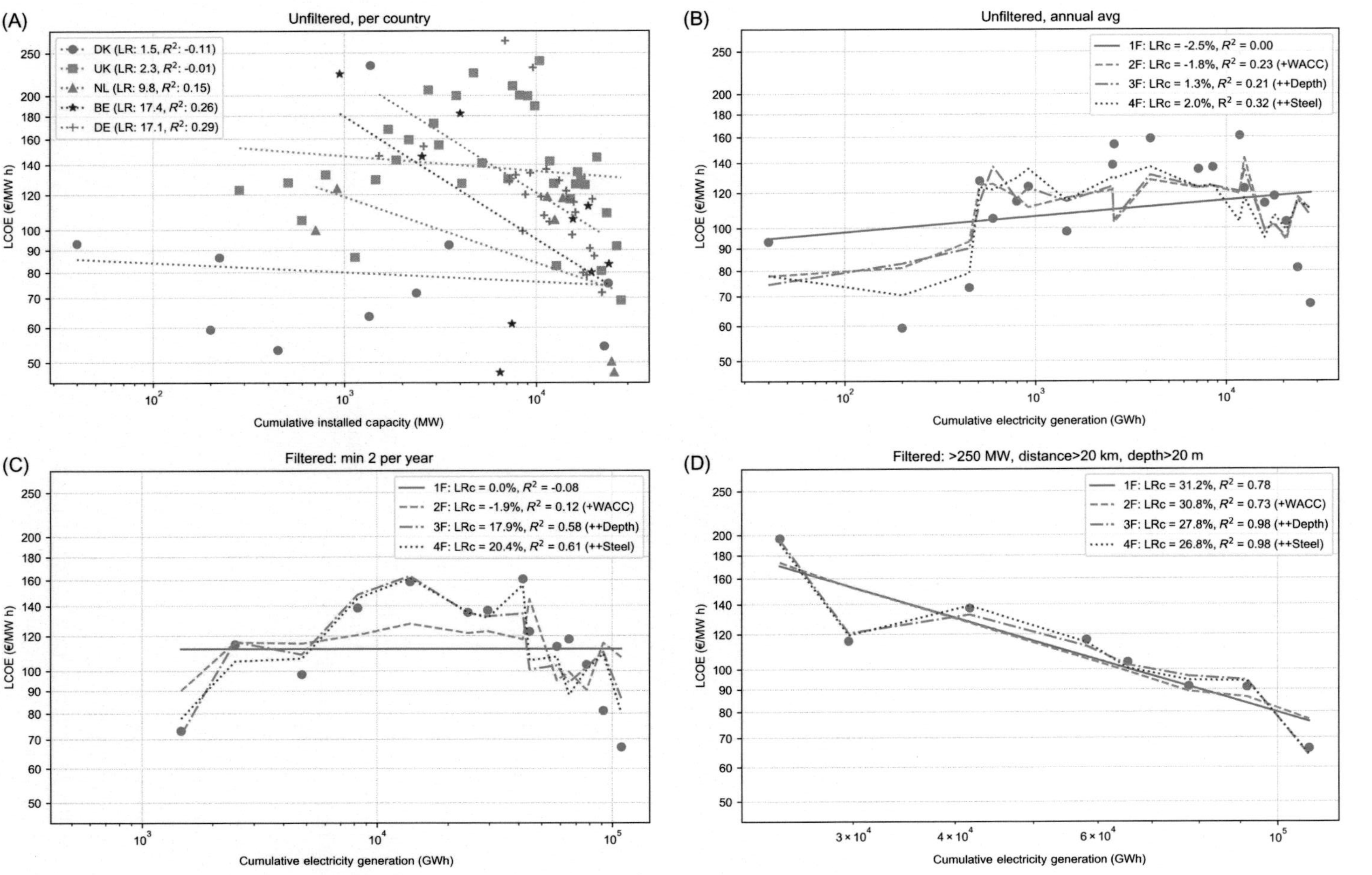

Figure 7.4

(A) Levelized cost of electricity (LCOE) for individual offshore wind farm in Denmark, the United Kingdom, the Netherlands, Belgium, and Germany; (B) annual averages, with additional explanatory variables weighted average cost of capital (WACC), water depth, and steel prices; (C) only annual data with at least two data points; (D) only considering wind parks of at least 250 MW, at least 20 km from shore, and a water depth of at least 20 m. *Source: Based on own data collection.*

when comparing it to the LCOE, learning rates found for onshore wind parks of 10%–12% (see Chapter 6).

7.3.2 Main drivers behind cost and price changes

When offshore wind parks were first installed in the early 1990s, it seemed to be a reasonable idea to use the same technology for offshore wind farm as the one used in onshore with slight adjustments. However, the needs for further offshore R&D and expertise were underestimated. Wüstemeyer et al. (2015) point out that competition for onshore products in the 2000s was high, and that most companies manufacturing components for wind farms were reluctant to divert R&D to specific offshore turbines, but this strategy failed. One example is the 160 MW Horns Rev 1 wind farm in Denmark with 80 Vestas V80 turbines that were adapted for offshore usage (Richardson, 2010). Two years after the commissioning, all wind turbines had to be removed for refurbishment, maintenance, and replacement works due to eminent transformer and generator problems (Sweet, 2008). Companies soon realized that offshore wind farms needed new turbine designs, specifically adapted to the harsh marine environment and not simply an adaptation of existing onshore turbines. However, optimizing products for offshore usage meant at the same time making them inefficient for onshore wind power, since additional features, such as an extended corrosion resistance, are unnecessary cost drivers (Wüstemeyer et al., 2015). Nevertheless, in the past decade, most manufacturers have set up separate R&D lines for offshore wind farms, and thus dedicated R&D has certainly added to LCOE reductions.

Since the establishment of the first offshore wind parks in the 1990s, the Capex and the price trend in general have been increasing, together with the size and rated power of the turbines, the water depth, and the distance from shore, as shown in Fig. 7.5.

This led in the period of 2000–15 to an increase of the Capex from around 1.5 M€/MW in 2000 to 4.0 M€/MW in 2010. A large number of factors have been mentioned in literature, which may all partly have contributed to this:

- Between 2001 and 2015, offshore wind parks have continuously been placed further offshore and in deeper water. Obviously, with increasing water depth to about 25 m in 2018, the cost of foundations increases. The same goes for the average cost of grid connection and installation costs with increasing distance to shore (the average distance of farms built in in 2018 was about 40 km). However, even when correcting for these factors, it can only partially explain the increase in Capex (Voormolen et al., 2016).
- A second factor is the increase of raw material prices such as steel (factor 2 between 2004 and 2009) and copper (factor 3–5 from 2004 to 2009–14). Yet again, these factors contributed to the overall Capex increase but cannot fully explain the increase.

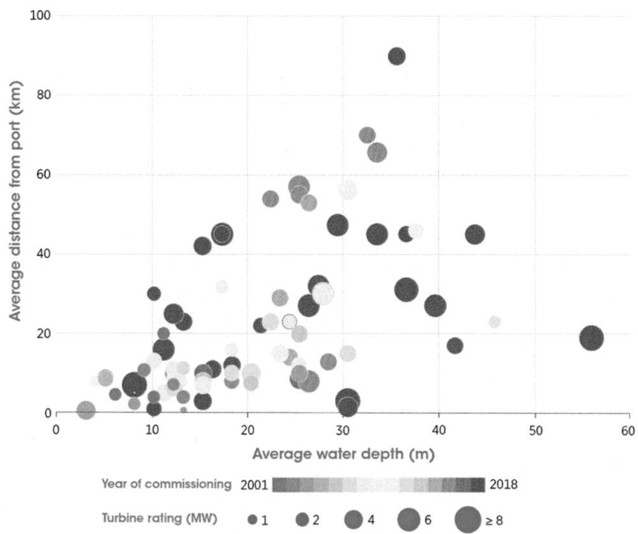

Figure 7.5
Average distance from port and water depth in commissioned offshore wind projects, 2001–18. *Source: Reproduced with permission from IRENA (2019)*

- Voormolen et al. (2016) find that also competition between turbine suppliers and other market actors was limited. For example, Siemens supplied turbines for 23 of the 45 projects surveyed by Voormolen et al., with a capacity of 5.2 GW (62% of total capacity assessed). In 2014 Siemens had a market share of 86% (EWEA, 2015). Under such monopolistic conditions, it is unlikely that market prices reflect production costs.
- Voormolen et al. (2016) also reported that in the United Kingdom, environmental permits specified the turbine size several years before the project was actually realized, requiring the project owners to use outdated technology at the point of construction.

Probably, a combination of these and other factors is responsible for the increase in Capex between 2000 and 2015. However, from 2015 onward, decreasing Capex can be observed. This is most likely again due to a number of factors, including technology improvements, greater experience among project developers, greater economies of scale in supply chains, as well as competition among suppliers, benefits from multiple wind farms in specified zones.

The improvements in offshore wind turbine technologies have been impressive. The first turbines installed had a diameter around 65 m and a capacity of 2 MW, while in 2018 the average rated capacity of new installed turbines was 6.8 MW (Fig. 7.6), and an average rotor diameter was 160 m. Since 2014 the average rated capacity of newly installed wind turbines has grown at an annual rate of 16% (WindEurope, 2019). For the commercialization of a 10 MW turbine (featuring a 164 m rotor diameter) has been

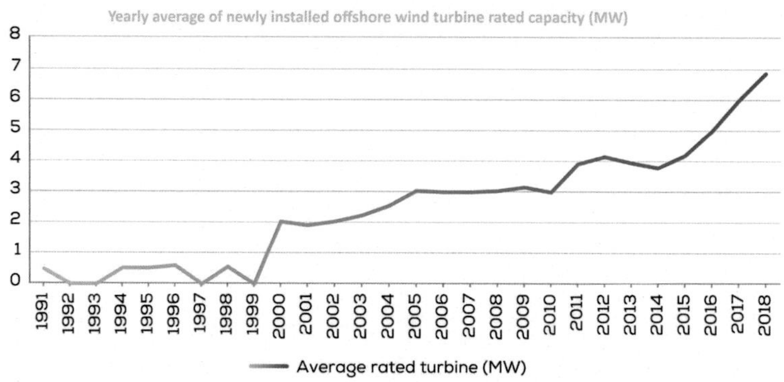

Figure 7.6
Yearly average rated capacity of newly installed offshore wind turbines (MW). *Source: Reproduced with permission from WindEurope (2019).*

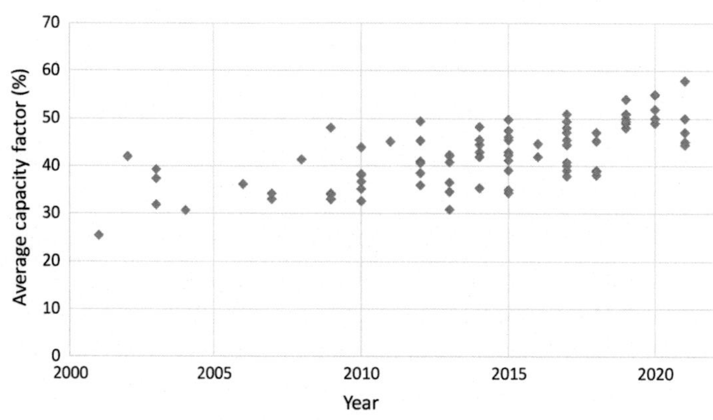

Figure 7.7
Capacity factor trend for European offshore wind farms; values from 2001 till 2021 (2019–21 projected). *Source: Based on (Wiser et al., 2016a) with permission from Springer Nature.*

announced by Vestas for the beginning of 2021, and a 12 MW turbine by Haliade-X is under development for 2022 (IRENA, 2018).

But the changes in Capex only tell part of the story: other factors also have significantly influenced the overall LCOE. Perhaps even more importantly, the average capacity factor has increased continuously over the years. While in the early 2000s, capacity factors of around 30% were common, by 2018, capacity factors had increased to 45% (with a range between 40% and 50% covering most projects) and are projected to increase further to 50% and more by 2022, as shown in Fig. 7.7. Similarly to the development for onshore wind, this impressive increase in electricity production is not reflected in the Capex but has played a major role in lowering LCOE in recent years.

But cost reductions appear to be an even more complex set of factors. In a report published by Vree and Verkaik (2017) the cost structures of three specific offshore wind farm projects were analyzed in detail. In this report, three offshore wind farms are discussed, which (in 2019) have not yet been built. The analyses reveal that a multitude of different factors not only led to overall significant reductions of LCOE, improved wind turbine and foundation design, increased experience in installation, and reduced steel prices (all impacting upfront Capex), but also increased availability (from 94% to 96%). Next to these technical improvements, also "soft" costs have been reduced, such as insurance cost, financing cost, return on equity, and permit and policy improvements. The impact of individual factors, however, varied between the three parks assessed. The report concludes that strong tender pressure (and competition), low raw material prices, low interest rates and equity costs, good policies in Denmark and the Netherlands (government as "project developer": risk reduction for the investors) resulted in LCOE reduction from 150–200 to 40–80€/MWh (depending on site condition, tax regulations, inclusion of exclusion of grid costs).
The range of 40–80€/MWh does not reflect current prices, but bids are made some years back for projects to be operational as of 2020 and beyond. Overall, based on some case studies, we estimate that about roughly 60% of the cost reduction can be attributed to institutional innovation, whereas 40% to "hardcore" technological innovation.

The cost of financing, and more specifically the WACC, has also likely played a role in past cost increases between 2003 and 2012. Gomez Tuya (2019) found an increasing trend in WACC between 2004 and 2012 in the United Kingdom, Denmark, Germany, and the Netherlands, but all sharply declined between 2014 and 2015. Interestingly, the WACC fluctuated much less in Denmark (always between 7.5% and 9% between 2001 and 2018) than in the United Kingdom, which displayed structurally higher WACC, ranging between 9% (between 2015 and 2017) and almost 14% (in 2009 and 2012). This is part of the explanation for the large differences observed between the United Kingdom and Denmark as shown in Fig. 7.4A. Possibly, the sharp decline in WACC in recent years is (partly) due to a perception by equity providers that offshore wind is less risky than it used to be. If so, then technological learning had an indirect impact on the LCOE, but this is speculation, and incorporating such effects in experience curves remains a challenge.

7.3.3 Future outlook

Given the ambitious Paris climate goal, offshore wind is expected to continue to grow strongly over the coming years. Next to the main growth area in the European Union, also the United States and China are expected to invest increasingly in offshore wind parks. For the near future (2030), it is expected that global installed capacity will reach 130–140 GW (IEA, 2018), compared to less than 20 GW installed at the end of 2018. For 2050, EWEA expects a growth in Europe alone of 400 GW, and so globally installed capacity could increase up to well above 500 GW. Gernaat et al. (2014) show that integrated assessment models assume significant deployment of offshore wind, especially in scenarios with climate policies (e.g., to reach a 2°C target), showing deployment between 200 and 1500 GW by 2050. For the baseline scenario in the IMAGE TIMER model, 850 GW alone is deployed, with highest deployment in the rest of Asia (20% of global offshore capacity), Europe (15%), South America (14%), and China (12%). According to this projection, only 1% of the offshore technical potential will have been developed by 2050, increasing to 3% by 2100 (Gernaat et al., 2014).

Given this expected massive increase in installed capacity, it could be both expected and societal interest that the cost of offshore wind energy is further reduced. However, as discussed in the previous sections, it is very difficult to identify robust cost trends, given the fact that a large number variables influence the LCOE, and not all of them are linked to technological learning. Therefore based on the trends identified in this chapter, no estimates are made given future cost reduction prospects.

For comparison, elicitation surveys by several experts have been done before predicting the future development of wind costs. This method is widely in use and offers a close estimation of the future expectations as shown in Wiser et al. (2016a,b).

In Fig. 7.8 the baseline of 2014 and the middle point in 2030 were determined by consulting experts on the expected average LCOE and by requesting details on five core input components of LCOE: total upfront Capex to build the project (€/kW); levelized total annual Opex over the project design life (€/kW-year); average annual energy output

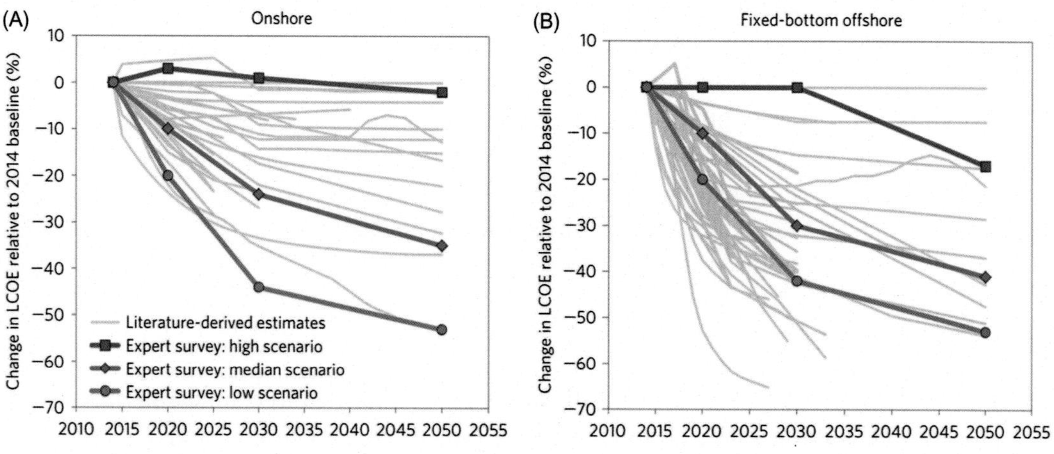

Figure 7.8
Estimated change in levelized cost of electricity (LCOE) for onshore wind parks (A) and fixed-bottom offshore wind parks (B); comparing expert survey results with other forecasts (Wiser et al., 2016a,b).

(capacity factor, %); project design life considered by investors (years); and costs of financing, in terms of the after-tax, nominal WACC % (Wiser et al., 2016a). As a 2015 starting point, an LCOE of 170 US$/MWh was taken, and for 2030, most experts expected a drop of LCOE to about 120 US$/MWh by 2030. This survey was taken at (or shortly after) the peak of LCOE in UK and German projects. But based on the latest bids, LCOE are expected to decline to 50–90€/MWh already by 2021. This elucidates again the difficulty to make sound projections for offshore LCOE.

For the future, we see several trends. First of all, with recent capacity additions outside Europe, the scenarios used in integrated assessment models are starting to materialize. In these regions the lowest cost sites are still largely unoccupied. In the North and Baltic seas, however, new sites will have to move ever further from shore, which may cause higher LCOE. Then again, we also observe that new technological concepts are emerging for installation and operation, such as hydraulic drivetrains, airborne wind, off-grid wind energy with H_2 production, and automated O&M—some expected before and some after 2030. These may bring down the LCOE. In the North Sea the intense use means that combination with interconnectors is likely to occur, energy islands with O&M stations are envisaged, etc. It is difficult to say how the interaction between depletion of geographic potential on the one hand and technological learning and new concepts on the other will ultimately impact LCOE, but they will typically be site specific and thus difficult to assess with experience curves alone.

One factor that has hampered the determination of actual LCOE is the lack of reliable data on WACC for individual projects but also of the actual electricity production and development of operation and maintenance cost of offshore wind farms. Publication of such data would also aid the more accurate assessment of current and future LCOE of offshore wind.

7.3.4 Conclusions and recommendations for science, policy, and business

Almost three decades ago, offshore wind started as an experiment. The Vindeby was dismantled in 2016/17 after 25 year of operation (it was designed in 1991 for 20 years max), and so by now, a truly new generation of offshore wind farms is expected to power a large share of the renewable energy needs in the coming years. In order to reach a 100% renewable electricity system in the European Union, offshore wind will be essential (Zappa et al., 2018). Offshore wind is the backbone of the decarbonization strategy for many European countries, with plans to produce green hydrogen for transport and industry use. As such, the recently achieved reductions in LCOE are promising. However, due care should be taken when assuming further LCOE reductions and extrapolating recent trends. The geographic potential of offshore wind is ultimately limited, and locations further offshore will require floating foundations and longer cables for grid connections.

Also reductions in WACC achieved recently may just as well be canceled if interest rates increase again. Last but not least, in this chapter, only the production cost of electricity has been assessed, but not how electricity prices will develop. With increasing penetration of offshore wind and other intermittent renewables, prices may decrease as well (both yearly averages and during specific times of high wind speeds), and so profit margins may become smaller, and policy support may have to be maintained for longer periods. Then again, with increasing availability of low-cost electricity storage technologies (see chapter 8), and the increasing demand for green hydrogen may enable the development of new business models.

Given the strong fluctuations in the past and many factors influencing the LCOE of offshore wind projects, it was not possible to derive meaningful one-factor experience curves and learning rates that would allow extrapolation for future cost projections. Multifactor learning curves approach taking into account raw material costs, location-specific properties, and soft factors, such as developments in WACC, show more promise, but more deployment of offshore wind is needed to demonstrate whether such models can provide more accurate cost trend forecasts for the coming years. Ultimately, experience curves will likely remain more applicable to modular technologies (e.g., photovoltaics, batteries) or at least more standardized supply chains and homogeneous site conditions such as onshore wind.

References

DNV GL, 2015. Project certification of wind power plants. Service specification. In: DNVGL-SE-0190, Edition December 2015. Available from: <http://www.dnvgl.com>.

EWEA (2015). The European Offshore Wind Industry Key Trends and Statistics 2014. The European Wind Energy Association. Availabel at: https://www.ewea.org/fileadmin/files/library/publications/statistics/EWEA-European-Offshore-Statistics-2014.pdf

Gernaat, D.E.H.J., van Vuuren, D.P., van Vliet, J., Sullivan, P., Arent, D.J., 2014. Global long-term cost dynamics of offshore wind electricity generation. Energy 76, 663–672.

Gomez Tuya, N., May 2019. Offshore Wind Energy Cost Trends and Learning Curves (M.Sc. thesis). Utrecht University.

Henderson, A.R., 2015. Offshore wind in Europe. Walking the tighttrope to success. Refocus 3 (2), 14–17. Available from: https://doi.org/10.1016/S1471-0846(02)80021-X.

International Energy Agency, 2018. World energy outlook series offshore energy outlook. Retrieved from: <https://webstore.iea.org/download/direct/1034?fileName = WEO_2018_Special_Report_Offshore_Energy_Outlook.pdf>.

IRENA, 2018. Offshore innovation widens renewable energy options. In: Opportunities, Challenges and the Vital Role of International Co-Operation to Spur the Global Energy Transformation. International Renewable Energy Agency. Available from: <https://www.irena.org/-/media/Files/IRENA/Agency/Publication/2018/Sep/IRENA_offshore_wind_brief_G7_2018.pdf>.

IRENA, 2019. Renewable Power Generation Costs in 2018. Available from: <https://www.irena.org/-/media/Files/IRENA/Agency/Publication/2019/May/IRENA_Renewable-Power-Generations-Costs-in-2018.pdf>.

Junginger, M. (2005) Learning in Renewable Energy Technology Development (Ph.D. thesis). Copernicus Institute, Utrecht University, p. 216. Available at: https://dspace.library.uu.nl/bitstream/handle/1874/7123/full.pdf

Junginger, M., van Sark, W., Faaij, A., 2010. Technological Learning in the Energy Sector: Lessons for Policy, Industry and Science. Edward Elgar Publishing, Cheltenham, p. 332.

Richardson, P., 2010. Relating Onshore Wind Turbine Reliability to Offshore Application. Offshore, Conroe, TX. Retrieved from: <http://etheses.dur.ac.uk/1408/>.

Sweet, W., 2008. Danish wind turbines take unfortunate turn. IEEE Spectr. Available from: https://doi.org/10.1109/mspec.2004.1353791.

Voormolen, J.A., Junginger, H.M., van Sark, W.G.J.H.M., 2016. Unravelling historical cost developments of offshore wind energy in Europe. Energy Policy 88, 435–444. Available from: https://doi.org/10.1016/j.enpol.2015.10.047.

Vree, B., Verkaik, N., 2017. TKI Wind op Zee. Offshore Wind Cost Reduction Progress Assessment. Version 3.0.

WindEurope, February 2019. Offshore Wind in Europe Key Trends and Statistics 2018. Wind Europe. Available from: <https://windeurope.org/wp-content/uploads/files/about-wind/statistics/WindEurope-Annual-Offshore-Statistics-2018.pdf>.

Wiser, R., Bolinger, M., 2017. 2016 Wind Technologies Market Report. US Department of Energy Office of Energy Efficiency and Renewable Energy, Washington, DC.

Wiser, R., Jenni, K., Seel, J., Baker, E., Hand, M., Lantz, E., et al., 2016a. Expert elicitation survey on future wind energy costs. Nat. Energy 1 (10). Available from: https://doi.org/10.1038/nenergy.2016.135.

Wiser, R., Jenni, K., Seel, J., Baker, E., Hand, M., Lantz, E., et al., June 2016b. Forecasting wind energy costs & cost drivers. IEA Wind. Retrieved from: <http://eta-publications.lbl.gov/sites/default/files/lbnl-1005717.pdf%0Ahttps://2018-moodle.dkit.ie/pluginfile.php/387913/mod_resource/content/1/DCCAE> website_2017 Reference Prices for <REFIT.pdf%0Ahttp://www.dccae.gov.ie/energy/SiteCollectionDocuments/Rene>.

Wüstemeyer, C., Madlener, R., Bunn, D.W., 2015. A stakeholder analysis of divergent supply-chain trends for the European onshore and offshore wind installations. Energy Policy 80, 36–44. Available from: https://doi.org/10.1016/j.enpol.2015.01.017.

Zappa, W., Junginger, M., van den Broek, M., 2018. Is a 100% renewable European power system feasible by 2050? Appl. Energy 233–234 (2019), 1027–1050.

CHAPTER 8

Grid-scale energy storage

Noah Kittner[1,2,3,4], Oliver Schmidt[5,6], Iain Staffell[6] and Daniel M. Kammen[7,8,9]

[1]Group for Sustainability and Technology, ETH Zurich, Zürich, Switzerland, [2]Department of Environmental Sciences and Engineering, Gillings School of Global Public Health, University of North Carolina at Chapel Hill, Chapel Hill, NC, United States, [3]Department of City and Regional Planning, University of North Carolina at Chapel Hill, Chapel Hill, NC, United States, [4]Environment, Ecology, and Energy Program, University of North Carolina at Chapel Hill, Chapel Hill, NC, United States, [5]Grantham Institute for Climate Change and the Environment, Imperial College London, London, United Kingdom, [6]Centre for Environmental Policy, Imperial College London, London, United Kingdom, [7]Energy and Resources Group, UC Berkeley, Berkeley, CA, United States, [8]Renewable and Appropriate Energy Laboratory, UC Berkeley, Berkeley, CA, United States, [9]Goldman School of Public Policy, UC Berkeley, Berkeley, CA, United States

Abstract

Grid-scale storage technologies have emerged as critical components of a decarbonized power system. Recent developments in emerging technologies, ranging from mechanical energy storage to electrochemical batteries and thermal storage, play an important role for the deployment of low-carbon electricity options, such as solar photovoltaic and wind electricity. This chapter details the types of technological learning models to evaluate the experience rates (ERs) for key grid-scale storage technologies, including lithium-ion and lead-acid batteries, pumped hydro storage, and electrolysis and fuel cells. It updates the state of the literature to determine learning rates of these and other grid-scale storage technologies. We discuss methodological issues in determining ERs for grid-scale storage systems, which often provide multiple applications and services on the grid. In addition, the chapter highlights future outlooks and new areas for research, including topics related to learning-by-doing, learning-by-searching, and manufacturing localization to derive further insights. Rapid cost reductions in lithium-ion batteries have the potential to disrupt electricity and transportation sectors, creating further complementarities and innovation cycles. More rigorous data collection for grid-scale storage systems on cost indicators that incorporate multiple services and applications provided by storage, life cycle greenhouse gas emissions from storage options, and materials availability of emerging battery chemistries could inform better policies to enable low-carbon power systems.

Chapter Outline

Technological Learning in the Transition to a Low-Carbon Energy System.
DOI: https://doi.org/10.1016/B978-0-12-818762-3.00008-X

8.1 Introduction

Grid-scale energy storage has the potential to transform the electric grid to a flexible adaptive system that can easily accommodate intermittent and variable renewable energy, and bank and redistribute energy from both stationary power plants and from electric vehicles (EVs). Grid-scale energy storage technologies provide the means to turn the power system into a dynamic market of distributed producers and consumers, indeed, "prosumers" of energy.

Electricity can be stored through the conversion of different types of energy—for example, mechanical energy in the form of pumped hydropower or flywheels, electrochemical energy for batteries, electrical energy storage in capacitors, chemical energy in the form of hydrogen, and thermal energy such as pumped heat or ice cooling devices. Flywheels that use mechanical storage take electric currents and use them to spin a disk, which can store electricity in the rotational inertia of the disk.

On the main grid, pumped hydro storage has provided electricity storage for decades; however, new options are emerging. Storage systems operate at different scales—including those that enable load balancing for mini-grid systems, which includes those that are isolated and those that can interact with the large-scale utility grid.

Technological learning that leads to cost reduction and performance improvements for these storage technologies could enable reliable electricity supply with intermittent renewable sources that are directly competitive with fossil fuel-based electricity. Technological learning curves may reduce the uncertainty level of future capital costs and technology applications.

The market for a diverse variety of grid-scale storage solutions is rapidly growing with increasing technology options. For electrochemical applications, lithium-ion batteries have dominated the battery conversation for the past 5 years; however, there is increased attention to nonlithium battery storage applications including flow batteries, fuel cells, compressed air energy storage, supercapacitors, and flywheels. Globally, lithium-ion batteries have attracted the most attention due to their multiple applications at the grid-scale and rapid cost declines for consumer products and EVs. Pumped hydro storage maintains the largest existing market share of grid-connected energy storage. While certain lithium-ion batteries have the most attention to date in terms of market deployment for both vehicles and grid-scale applications, numerous opportunities remain for newcomer grid-scale mechanical, thermal, or electrochemical storage solutions.

The economic value of storage technologies also varies across application, technology, and, ultimately, through battery chemistry or physical performance. Grid-economics and alternative remuneration schemes for energy storage on the grid provide multiple revenue streams for grid-scale storage owners; opportunities where previous electricity generation technologies may not be able to compete (Stephan et al., 2016; Davies et al., 2019).

Pumped hydro storage historically has the most installed capacity of any energy storage capacity on the grid with nearly 184 GW of installed nameplate capacity (US DOE Global Energy Storage Database, 2019). The basic concept utilizes gravity and potential energy to pump stored water in a reservoir up from a low elevation to a higher elevation. Pumped hydro storage has opportunities for expansion, especially as an option to retrofit existing large-scale hydropower plants and turn them into storage, which has been one option under consideration in places with a large hydropower dependency such as Laos and Switzerland (Schmitt et al., 2019).

Lithium-ion batteries are available today and are a promising electrochemical storage technology for their dual applications on the grid and for EVs offering a wide range of energy densities, operating temperature ranges, and scales for deployment. Key components of lithium-ion batteries include positive and negative electrodes and an electrolyte. Graphite-based electrodes are the most popular; however, new materials and battery chemistries have experimented with different positive electrodes such as lithium-phosphate or manganese-based cells. Typically, lithium-ion cells are distinct from the actual battery and are formed in cylindrical, flat, or pin shapes. The cells are contained in packs. Current research and development includes the increase of cycle life, power density, and safety concerns to reduce flammability risks.

In 2010 the total volume of lithium-ion batteries was 20 GWh largely owing to portable electronics. Since then, production has been growing annually by 26% reaching a total market size of 120 GWh in 2017 (Avicenne Energy, 2018). While the electronics market gradually slowed down, production of lithium-ion batteries continued to increase, primarily due to the growing demand from EVs. Overall, the market share of lithium-ion batteries for EVs and stationary storage increased from about 5% early this decade to more than 60% in 2017, surpassing the sales for electronics (Fig. 8.1). Still, volumetric energy densities of 600−650 Wh/L in cylindrical cells have been reported (Choi and Aurbach, 2016).

As of 2017, global capacity of electrochemical system storage reached about 1.6 GW, and lithium-ion batteries are the main type used, accounting for about 1.3 GW or 81%, in terms of power capacity in 2017 (Fig. 8.1). Deployment of residential lithium-ion batteries behind-the-meter was estimated at around 600−650 MWh (or about 200 MW) in 2016 (Schmidt et al., 2017; Sekine and Goldie-Scot, 2017), which is substantial, considering that it represents almost 20% of the total lithium-ion battery capacity installed for system

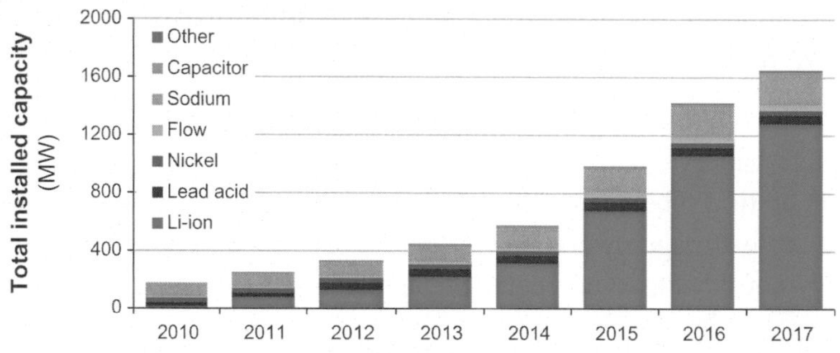

Figure 8.1
Global cumulative installed capacity of electrochemical grid-scale storage (Tsiropoulos et al., 2018).

storage. Bloomberg New Energy Finance reports additional behind-the-meter storage capacity of 650 MWh in commercial and industrial sectors (Sekine and Goldie-Scot, 2017). Dramatic increases are expected in the coming years, with a number of state and federal mandates, and large utility-scale projects expected to result in the deployment of multiples of the 2017 capacity.

Flow batteries offer potential advantages to lithium-ion technologies at the grid scale. Flow batteries are formed of two electrochemical cells that can be separated by a membrane where ion exchange occurs. Often times, the separation of the liquids in an electrolyte mean that one could build a larger battery that scales with the volume of the liquid and area of the membrane. This allows for distinct advantages at the grid scale compared to lithium-ion batteries that are optimized for transportation applications. For instance, the typical flow battery design allows for a decoupling of the power density and energy capacity, which means that batteries can increase their duration. Compared to lithium-ion batteries, flow batteries maintain separate electrolytes and electrodes, which decouples their energy—power ratio and offers a variety of new material chemistries for next-generation batteries and grid-scale storage. Unlike most chemical batteries whose performance degrades after a few thousand cycles, flow batteries can maintain their charge-discharge characteristics for over 100,000 cycles over 20-year lifespans. In addition, flow batteries contain much fewer risks for explosion or fire than their lithium-based counterparts. Conventionally, aqueous redox-flow batteries have dominated the research discussion; however, now alternative materials have emerged ranging from organic metal-free flow batteries based on quinones to vanadium- and zinc-based options. Emerging chemical flow batteries also range across aqueous and nonaqueous solutions (Dunn et al., 2011; Larcher and Tarascon, 2015).

Fuel cells have struggled to achieve commercial success due to the lack of materials science advances in catalyst or membrane technologies, combined with a lack of suitable infrastructure to deploy in a hydrogen-based economy. Nevertheless, with excess and abundant intermittent renewable electricity stemming from utility-scale solar and wind farms, fuel cells are becoming an increasingly viable option to reelectrify hydrogen for power or transport applications.

Compressed air energy storage offers new seasonal and long-duration opportunities for high power and utility-scale energy storage. However, the affordability and availability of compressed air storage varies geographically, thus significantly limiting its potential. Compressed-air-energy storage often uses natural gas as a fuel to combust in the pressurized air and expand with the compressed air to generate electricity. Natural gas expands the capacity and efficiency of operating a compressed-air-storage facility (Succar and Williams, 2011). There are also options to use compressed-air-energy storage without natural gas inputs; however, these projects remain at the demonstration phase. There are two existing commercial compressed air energy storage plants totaling only 400 MW installed capacity—located in underground caverns mined from salt in Germany (290 MW) and the United States (110 MW). Recent studies demonstrate achievable storage in the range of \$0.42–\$4.71/kWh in saline aquifers (Mouli-Castillo et al., 2019).

Flywheels, mechanical energy storage devices using the rotational energy in a spinning disk, also have the potential for rapid performance improvements as technologies gain access to commercial markets. Flywheels are a type of mechanical storage that store rotational energy proportional to the square of their rotational speed. Major applications include frequency regulation and voltage control of power output as a source of torque. Flywheels can be used as spinning reserves. Larger flywheels are also increasing in the duration of their storage, making them another promising grid-scale storage option. The majority of profitable revenue streams for flywheels rely on providing frequency control tasks; however, new economically viable applications are emerging (Diaz-Gonzalez et al., 2015). Mini-grid studies highlight the ability for flywheels to integrate hybrid photovoltaic (PV) mini-grids at a low-cost and reduce the overall system costs by providing stability and storage services. As distributed energy resource architectures change, grid-scale flywheels operating in larger power system networks could gain tractability. Lower cost material advances for flywheel technologies also enable potentially greater learning similar to wind turbines due to the largely mechanical nature of the energy storage compared to electrochemical options.

Supercapacitors can provide short bursts of power on the grid. Physically, they exploit the difference in electrical potential across an electric field and can provide fast-responding bursts of power. Supercapacitors can also be used in vehicle applications and are notably interesting for micro-grid or distribution system level applications that may require higher levels of voltage control than other types of battery storage.

The different technologies noted here have different properties and can provide electricity over seconds, hours, or even weeks. All of these applications fulfill different roles in a power grid with high penetrations of intermittent renewable electricity. For instance, supercapacitors and flywheels that may provide short duration bursts of power into the grid may work in conjunction with longer duration flow batteries that may provide hours of storage. Pumped hydropower storage or hydrogen gas in a power-to-gas facility could provide seasonal storage over long durations in case there are supply–demand mismatches in renewable dependent power systems. It is likely that the different roles of the technology will coevolve as policies and applications create new niches. Grid-scale storage could provide complementary or enabling capabilities to generation sources such as solar and wind. Combined flow-battery and solar-PV systems could generate "baseload" electricity. In addition, the types of storage needed at different timescales may vary, and just because one distribution feeder has a flow battery installation would not preclude the growing technical and economic viability of supercapacitors or flywheels in the same system.

Taken together, the variety of emerging energy storage technologies for grid-scale applications has created a newly competitive ecosystem for clean energy systems to reduce costs, gain experience in manufacturing and deployment, and increase innovations to improve their CO_2 emissions, use fewer rare earth materials, and become safer for human health and the environment.

8.2 Methodological issues and data availability for technological learning

Technological learning and experience curves offer improved analytics and more generalized theories of technological change. Various types of quantitative models have been proposed to quantify and investigate the rates of technology adoption, investment in R&D, innovative cluster effects and technological spillovers, and policies that encourage emergent technological progress. All of these technological change agents have remained active for battery storage technologies. A variety of tools exist to examine technological learning for grid-scale storage technologies in further detail and are summarized here.

Traditional experience curves are based on the idea of "learning-by-doing" and relate the deployment and cumulative production of a storage technology with cost reductions. The one-factor experience curve model is appealing because the idea that firms learn from experience in the past seems intuitive, and by reducing the complex process of innovation into a single parameter, the model is simplified (Gross et al., 2013). They also have been described as the most objective method to project future cost of technologies (Farmer and Lafond, 2015). The underlying reasons for cost reduction as a result of learning-by-doing in manufacturing are identified as spreading overhead cost over larger volumes, reducing inventory cost, cutting labor cost with process improvements, achieving greater division of labor, and improving efficiency through greater familiarity with the process (Abernathy and Kenneth, 1974).

One-factor experience curves focus on relating the unit cost of a technology to its cumulative installed capacity. In the case of storage, this would relate to the amount of electricity stored. The typical learning rate is a useful metric because one can understand that for each doubling of installed cumulative capacity, the associated percentage cost should be reduced. The one-factor experience curve provides a theoretical framework to evaluate cost reductions systematically using a log-linear relationship, dating back to "Wright's law" in the manufacturing sector (Wright, 1936; Rubin et al., 2015). Conceptually simple, a one-factor approach allows for broad technological comparison. This can be achieved in the electricity storage sector as well and encompasses many of the factors related to the cost trajectory of an emerging technology.

However, one-factor curves lack causation and accountancy to the various cost reducing factors. They show how cost may reduce over time but provide no explanation for the underlying reasons beyond its relationship to cumulative output (Junginger et al., 2008). Additional cost-reducing factors are R&D expenditures (learning-by-searching; Cohen and Levinthal, 1989); improvement of product characteristics via user feedback (learning-by-using; Kahouli-Brahmi, 2008); and network relationships between research laboratories, industry, end-users and political decision-makers that can lead to spillover effects (learning-by-interacting) (Kahouli-Brahmi, 2008). Some authors suggest that experience curves largely reflect economies of scale (Hall and Howell, 1985) and may underestimate rapid innovations and materials science advances that change design or standardization of the technology or related changes in inputs for materials or labor. This weakness in the one-factor model has been explored in the development of solar and wind studies and also applied to lithium-ion batteries (Qiu and Anadon, 2012; Nemet, 2006; Zheng and Kammen, 2014; Kittner et al., 2017). These alternative models view cost reductions "beyond the learning curve" and attribute industry structure, technical barriers, and investment in R&D as better or additional indicators driving the cost reduction of critical low-carbon technologies. Efforts to incorporate further two-factor and multifactor experience curve models for a diverse set of energy storage technologies are underway.

Two-factor experience curves typically incorporate a proxy for innovation or R&D investment into the experience curve models. For lithium-ion energy storage, two-factor models have more closely aligned with current projections of battery storage development compared to one-factor experience curves (Kittner et al., 2017). Data availability and model complexity remains one of the foremost challenges to implement two-factor learning rates into a large-scale energy system optimization model that may help assist capacity-expansion efforts.

Therefore many of the major techno-economic energy system optimization models treat technological learning and experience rates (ERs) as exogenous to the model. However, when experience curves become endogenous to the model, there are better synergies and

Table 8.1: General data collection issues for electricity storage technologies.

Issue	Resolution	Applicability
Data is not for cost but for price	Use price data as indicator for costs	☑
Data not available for desired cost unit	Convert data to desired unit if possible Use available data as a proxy	☑
Data is valid for limited geographical scope	Price data assumed to reflect global marketplace Capacity data scaled to global market if applicable	☑
Cumulative production figures not available		
Data is in incorrect currency or currency year	Convert currency and correct for inflation and PPP	☑
Early cumulative production figures are not clear or available		
Supply/Demand affecting costs significantly	Use data as is but recommend tracking and updating	☑
Lack of empirical (commercial scale) data		

PPP, Power purchase parity.

opportunities to understand how R&D investment and/or future deployment policies could help lower overall system costs for electric grid operations in the long term. This could also significantly aid deep decarbonization efforts across the power sector and related industries (Rubin et al., 2015).

An overview of general data collection issues applicable to grid-scale storage technologies is summarized in Table 8.1. One main issue using current experience curve datasets for the electricity storage technologies presented here remains that the number of data-points, as well as the number of doublings of cumulative capacities, is very limited. This is due to the nascence of emerging battery chemistries and alternative technologies. The number of data-points for lithium-based systems is especially low. However, current data stay close to the fitted experience curve, resulting in a low error in the established learning rate. For flow battery systems, however, the data-points represent only about three doublings of cumulative capacity, and the technology is still very much on the cusp of commercialization (Schmidt et al., 2017). Variations in reported prices from the fitted experience curve result in a high error for the established learning rate. The data for utility-scale lithium battery systems reflects the exceptionally fast price decline of lithium batteries, which was observed in 2017. However, one of the main challenges to modeling data based on prices is the inability to fully capture knowledge spillover effects across similar technologies or storage applications. Spillover effects between different storage types are not considered here in detail. It is likely that there are spillover effects for lithium-ion batteries, in terms of cells, pack components, and power electronics across applications such as consumer electronics, EVs, and stationary systems.

Particularly important for battery storage technologies remains the functional unit considered for an experience curve analysis. Given that batteries and other energy storage technologies serve multiple applications, there may be strengths and weaknesses when characterizing costs. For instance, a capital cost expressed in terms of $/kW would certainly weigh the power density application of the storage device more heavily than the duration of storage, which could be a factor better represented by studying the cost in terms of $/kWh or levelized cost of stored electricity (Schmidt et al., 2019a). At the same time, grid-scale storage engineering may continue to measure advances in power or energy density in a volumetric way that considers the $/kg or $/L for critical components such as the electrolyte or the electrode. New metrics exploring the levelized cost of energy storage capture the unit cost of storing energy, subject to the system not charging, or discharging power beyond its rated capacity at any point in time (Comello and Reichelstein, 2019). Yet these metrics require site- and technology-specific data that may not be easy to implement when considering experience curves, since different dimensions of battery performance and economically viable applications can change with new innovations. Most often, experience curve datasets report the declining cost of lithium-ion storage in terms of $/kWh. This remains a useful metric, but as an increasing number of grid-scale applications related to frequency response, voltage control, storage capacity, and ancillary services become economically viable, the challenge in measuring progress related to cost reductions becomes less straightforward than technologies that generate electricity only.

As is common when analyzing experience curves, we use price data as a proxy to reflect all cost input factors (R&D, sales expense, advertising, overhead, etc.), which makes the analysis vulnerable toward pricing policies (Abernathy and Kenneth, 1974). As discussed in Chapter 2, there are several stages in the market deployment of technologies with specific dynamics between the cost and price of a technology. High data variance can lead to significant variations of ERs across studies and data sets. Depending on the spread of the data, it is possible to calculate different learning rates by changing the start and end point of the analysis and the inclusion or exclusion of outliers (Junginger et al., 2010). In particular, when price data is used, a period of at least 10 years' worth of historical data or two orders of magnitude of cumulative output should be available for price trends to be reliably reflective of cost trends (Gross et al., 2013; Junginger et al., 2010), which is rarely the case for many novel electricity-storage technologies.

Geographies and temporal scope of the data matter too. A majority of studies in the United States and Europe highlight dramatic cost reductions for lithium-ion batteries. Tropical regions may utilize alternative battery technologies or storage mechanisms based on technology and materials availability due to the potential for overheating existing technologies. Sodium–sulfur batteries require an operating temperature at nearly 300°C, and lithium-ion battery performance may decline in hot and humid environments. That may cause added stress and cost for these batteries. In addition, more geographically constrained storage options—ranging from compressed air energy storage to pumped hydro storage—require suitable site selection to enable cost-effective deployment. For instance, a recent study identifies the range of storage costs when siting compressed air storage in saline aquifers for the United Kingdom (Mouli-Castillo et al., 2019). This raises a fundamental question whether geographically specific technologies such as compressed air storage can be assessed using learning curves due to their low levels of standardization and site-specific siting requirements that heavily influence the cost of deployment. In mountainous mini-grid or island regions where diesel is often used, there could be synergistic advantages of using energy storage technologies to reduce the overall cost of a mini-grid or improve the mini-grid's energetic performance which also could impact the learning curve (Kittner et al., 2016). Further studies to explore cost reductions across a variety of countries, scales, and geographies would better indicate global progress as often experience curves omit soft-costs or other project development costs that would potentially change based on region.

Experience curves are incapable of predicting step-change innovations or accounting for product changes that might improve performance for the same costs (Abernathy and Kenneth, 1974; Nemet, 2006). It has been argued that radical product changes constitute new products that exhibit new experience rates (Junginger et al., 2010). Moreover, in situations with significant product changes, other indicators than the specific investment costs may be more appropriate to reflect learning outputs, such as product functionality or the levelized cost of electricity for a power-generation technology (Watanabe et al., 2009; Wiesenthal et al., 2012). For example, lithium-ion batteries may experience significant performance improvements with different material compositions in anode or cathode or the development of solid electrolytes.

The idea of experience improvement at a constant rate is also critiqued. Some argue that costs reduction is stronger during the R&D phase due to radical discontinuity (Staffell and Green, 2013; Ferioli et al., 2009). Others argue that learning might be stronger in the commercial phase due to competition (Söderholm and Sundqvist, 2007). ERs can inform near-term forecasts and longer term strategies, better than alternative methods, and provide a standard basis for comparison across technologies. Following the logic that relative cost shares of components with high rates decrease over time, a reduction of the aggregated experience rates for products over time appears feasible (Ferioli et al., 2009). This can be represented in energy systems models with "kinked" (piece-wise linear) curves or ERs that depreciate with time (Kouvaritakis et al., 2000; Seebregts et al., 1998; Epple et al., 1991). Table 8.2 highlights components of electric energy storage technologies and their contribution to further cost reductions. As electrodes currently comprise a large portion of the overall cell cost, new materials innovations for electrodes could achieve further cost reductions. Wiring and interconnections could also provide cost reductions, but perhaps not at the same potential scale as these costs have already moved into mature phases of the learning curve related to other

Table 8.2: Components of electricity storage technologies and indicative cost contributions (Schmidt et al., 2017).

Technology scope	Indicative contribution	Reported technologies
Cell	19%	
Electrodes	46%	–
Electrolyte	14%	*18,650 cell costs for EV packs reported at*
Separators	15%	*145 US$/kWh* (Cobb 2015)
Current Collectors	19%	
Terminals	4%	
Cell container	2%	
Battery (consumer electronics)	*no data*	
Power electronics	*no data*	Lithium-ion (electronics)
Housing	*no data*	
Module	*Included in pack*	
Thermal conductors	9%	Lead-acid (multiple)
Cell group interconnectors	0%	
State-of-charge regulator	85%	
Terminals	1%	
Provision for gas release	2%	
Module enclosure	3%	
Pack	11%	
Wiring, interconnections and connectors	21%	Lithium-ion (EV)
Housing	15%	Nickel–metal hydride (HEV)
Temperature control	7%	Electrolysis (utility)
Power electronics	24%	Fuel cells (residential)
Battery Management System	33%	
Exworks System	35%	
Inverter	*45%*[93]	–
Container	*45%*[93]	
SCADA/controller	*10%*[93]	
System	35%	
Transport	–	Lithium-ion (residential, utility)
Installation	-	Lead-acid (residential)
Commissioning	-	Redox-flow (utility)
		Sodium–sulfur (utility)
		Pumped hydro (utility)
	100%	

EV, Electric vehicle; *HEV*, hybrid electric vehicle.

electrical equipment. Identifying synergistic cost-reduction opportunities between hardware and soft costs could be important for further learning curve research.

Finally, an important distinction between products that require extensive on-site construction and those mass-produced in centralized factories must be made, due to the often highly specific, custom-built nature of the former resulting in lower ERs (Junginger et al., 2008).

8.3 Results

8.3.1 One-factor learning curves

Prices for storage technologies differ by scope, application, and size. Here we review most recent one-factor experience curves for grid-scale storage technologies. The results for electricity storage experience curves are differentiated along two main dimensions, application category, and technology scope. Application category covers portable (electronics), transport (hybrid EV—HEV, and EV) and stationary (residential, utility); technology scope covers cell, battery, module, pack, ex works system, and system level.

Fig. 8.2 shows decreasing product prices as per energy capacity with increasing cumulative installed nominal energy capacities for most electricity storage technologies. Pumped hydro (system), lead acid (module), alkaline electrolysis (pack), and lithium ion for consumer electronics (battery) and EVs (pack) exhibit current prices below 200€/kWh with above 100 GWh cumulative installed capacity. The relatively low ERs below 5% of the first two are contrasted by 17% for electrolysis (pack) and 30% and 22% for lithium-ion batteries and packs, respectively. Technologies between 1 and 100 GWh cumulative installed capacity, such as nickel–metal hydride (pack), utility-scale lithium-ion (system) or sodium–sulfur (system), show current prices between 200 and 600€/kWh and ERs of 11% and 16%. Those below 1GWh, such as residential lithium ion (system), lead acid (system), redox flow (system), and fuel cells (pack), cost more than 800€/kWh with ERs between 13% and 16%.

The price and cumulative capacity data used for electricity storage technologies come from peer-reviewed literature, research and industry reports, news items, energy storage databases, and interviews with manufacturers. In the literature, learning (based on manufacturing cost) and ERs (based on product price) are sometimes used interchangeably. The sources in the referenced literature were therefore double-checked to ensure the use of actual product price data.

The geographic scope of the data is global. Where cumulative deployment data is available on company or country level, the data is scaled to global level. Regarding price data, it is assumed that the global marketplace ensures that these are globally applicable (Wiesenthal et al., 2012) and those technologies where prices are more likely to vary by geography are highlighted. Regardless of geographic applicability, it can be assumed that identified ERs are applicable globally.

Technology scope for electricity storage technologies is differentiated into cell, battery, module, pack, exworks system, and system level. While exworks system refers to the factory-gate price of complete electricity storage systems, system level includes the cost for transportation, installation, and commissioning if applicable. Additional information on the cost components included at each level can be found in Table 8.2.

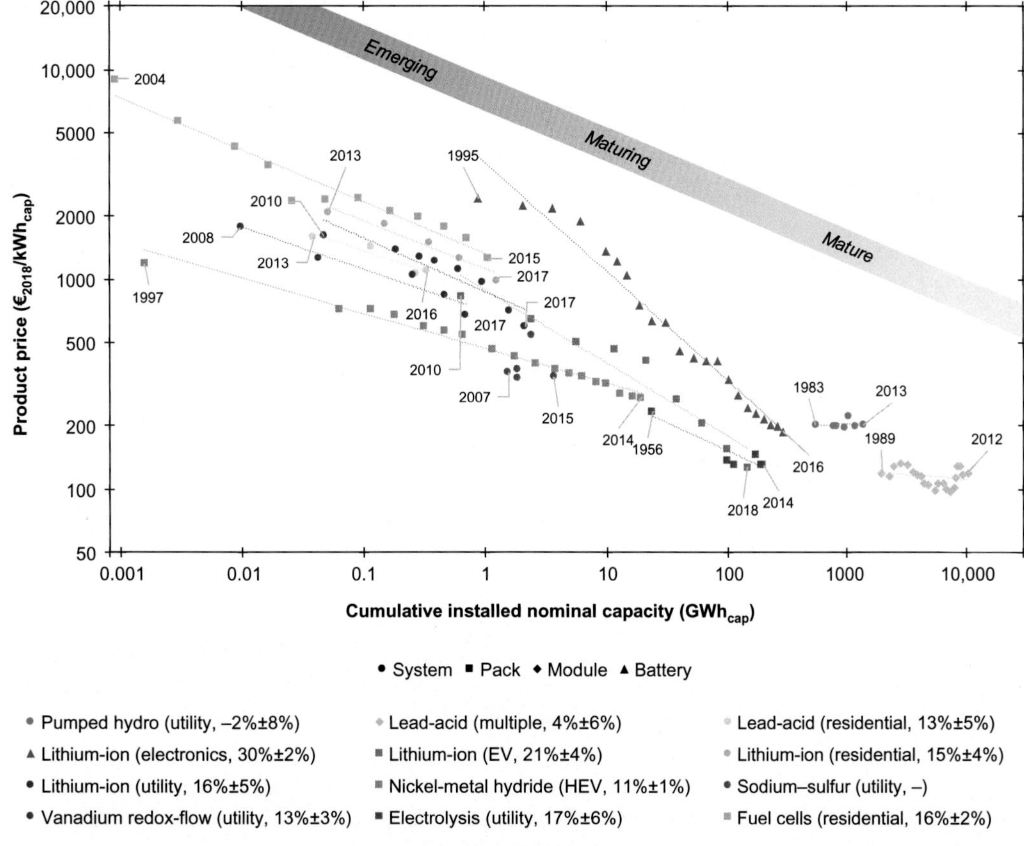

Figure 8.2
Experience curves for electricity storage technologies. Results show product prices per nominal energy capacity. Dotted lines represent the resulting experience curves based on linear regression of the data. Top legend indicates technology scope, and bottom legend denotes technology (including application and experience rate with uncertainty). Experience rate uncertainty is quantified as its 95% standard error confidence interval. Gray bars indicate overarching trend in cost reduction relative to technology maturity. Fuel cell and electrolysis must be considered in combination to form an electricity storage technology. kWh_{cap} is the nominal energy storage capacity. *Source: Updated from Schmidt et al. (2017).*

Experience rate uncertainty is determined using the 95% standard error-based confidence interval (CI). While this is relatively small ($< \pm 5\%$) for most emerging and maturing technologies, most mature technologies (pumped hydro, lead-acid modules, alkaline electrolysis) exhibit high ER uncertainty ($> \pm 5\%$) and are not significantly different from zero ($P > .05$). This is the result of the relatively short data series in terms of doublings of cumulative capacity. Ideally, a dataset for experience curves should cover two magnitudes of cumulative capacity deployment, in order to be significant (Junginger et al., 2010). This is only the case for fuel cells, nickel–metal hydride batteries, and consumer electronics and EV lithium-ion batteries.

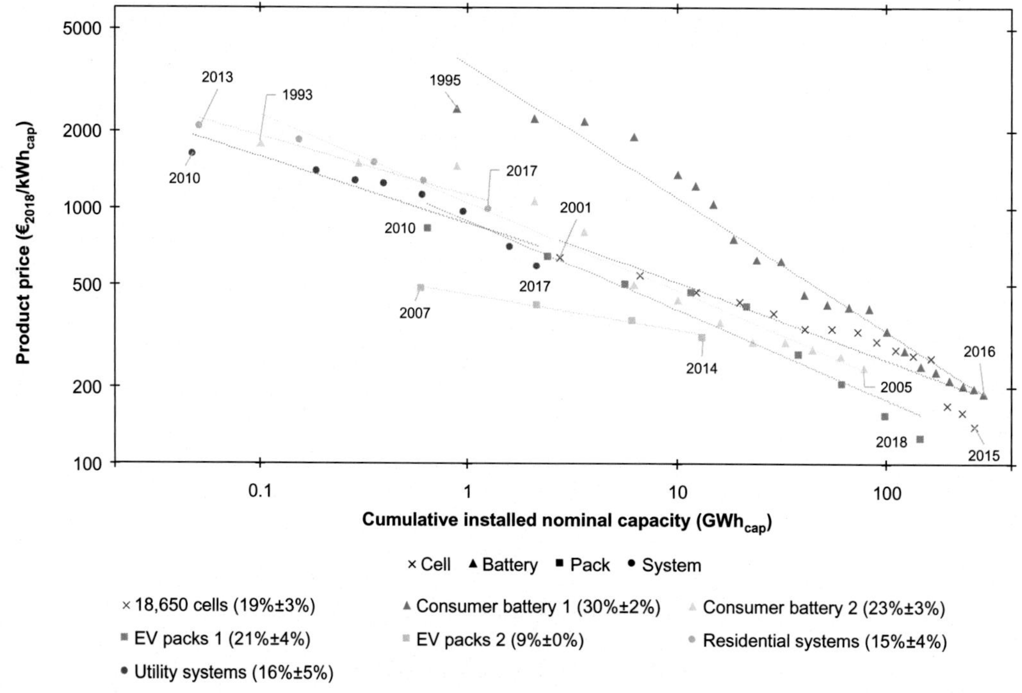

Figure 8.3

Experience curves for lithium-ion technologies (energy terms). Results are shown for product prices per nominal energy capacity. Dotted lines represent the resulting experience curves based on linear regression of the data. Top legend indicates technology scope, and bottom legend denotes technology (including experience rate with uncertainty). Experience rate uncertainty is quantified as its 95% standard error confidence interval. *Source: Updated from Schmidt et al. (2017).*

Electricity storage technologies with insufficient data are excluded, but these may still hold promise in the future. For sodium–sulfur, no feasible ER could be determined from the compiled data (displayed in Fig. 8.2 for reference).

In addition, it can be observed that ERs for lithium-ion technologies decrease with increasing technology scope (Fig. 8.3). Higher ERs for cells and batteries than for packs and systems imply that cost reductions are likely driven by experience in cell manufacturing rather than other components required in packs and systems. Stronger cost reduction for consumer electronics batteries compared to cylindrical cells could reflect the ongoing shift from cylindrical (e.g., 18,650 dimension) to more cost-competitive prismatic and laminate cells used for consumer electronics batteries (Pillot, 2014). Strong cost reduction for cylindrical cells between 2013 and 2015 might be the result of increased demand in EV packs, partly driven by Tesla (Pillot, 2014), which enhanced the experience curve effect and moved the technology down along the experience curve.

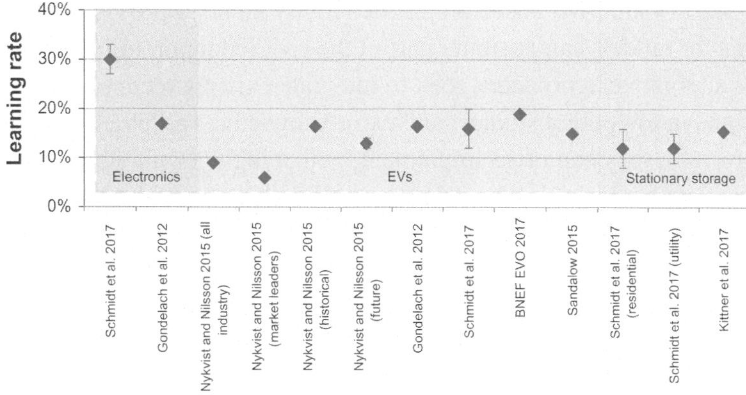

Figure 8.4
Experience and learning rates of lithium-ion batteries across different applications (Tsiropoulos et al., 2018).

Furthermore, these models are not conducted in isolation. Fig. 8.4 shows different learning rates that have reported for lithium-ion batteries based on a variety of methods and data sources including the incorporation of two-factor models (Kittner et al., 2017; industry reports BNEF EVO 2017; and across applications ranging from stationary to vehicle and consumer applications). Though imperfect, this chart begins to also uncover the challenges related to knowledge spillovers and data uncertainty from a lack of cross-industry, cross-sectoral knowledge transfer. The period of analysis, the technology boundaries, and the metrics used (e.g., cost or price, annual or cumulative production) offer possible explanations as to why the values range. For lithium-ion batteries, Schmidt et al. (2017) note that learning rates tend to decrease with increasing technology scope. Learning rates of inverters, a key component of stationary storage systems, are reported at 19% (± 1%) (IRENA, 2016; Fraunhofer ISE, 2015; Schmidt et al., 2017).

8.3.2 Multifactor learning curves

Although conventional one-factor experience curves retain a good level of explanation from 2010 to 2015, recent years of grid-scale storage experience have generally overestimated prices when focused only on economies of scale. In a similar vein to Nemet (2006), multifactor experience curve approaches can integrate the knowledge and innovation acquired from technical improvements, investment in R&D, or alternative industry aspects that could influence cost reductions. For instance, when substituting patent activity for cumulative battery production in a study on ERs for lithium-ion batteries, Kittner et al. (2017) find that the learning rate could nearly double when considering patents only. This is explained by a learning rate of approximately 15% based on cumulative production (over the timescale 1991–2015) and 31% based on patent activity alone (a proxy for R&D efforts). Therefore two-factor models can attribute part of the cost reduction to innovation. While imperfect, these alternative approaches seek to integrate existing technological innovation system theory, innovation policy studies, and extra knowledge spillovers into learning curve models. In these cases, they can often implement, with a high correlation, the critical factors related to cost reductions. From a materials perspective, Kittner et al. (2017) also found that lithium and cobalt prices had weak correlations with the price reduction due to highly diversified materials composition and resilient design features of grid-scale lithium-ion batteries that are not subject to wild price changes due to the lithium and cobalt market.

Multifactor learning curves have the potential to identify key characteristics related to the cost reduction of storage technologies and highlight differences across technologies and applications. For instance, materials chemistry may play a significantly larger role in determining the ER of fuel cells compared to lithium-ion batteries due to the high materials cost of a catalyst for fuel-cell production. Recent assessments do not find cobalt and lithium to significantly limit the cost-reduction potential for existing lithium-ion cells (Kittner et al., 2017; Ciez and Whitacre, 2016). However, for nonlithium storage solutions, there could be raw material bottlenecks as discussed later in the chapter.

The application of multifactor learning curves remains a key component of the innovation literature that is currently evolving. Recent studies are trying to incorporate two-factor and multifactor experience curves endogenously to energy systems optimization modeling tools. This presents a methodological challenge and source of uncertainty given the numerous challenges facing data collection, validation, and verification.

8.4 Future outlook

Using the derived experience curves, future prices for electricity storage based on increased cumulative capacity can be projected (Fig. 8.5), and the feasibility of these projections tested against indicative cost floors is defined by raw material and production costs.

When projecting the experience curves forward to 1 TWh cumulative capacity, the categorization of electricity storage technologies along product prices and cumulative installed capacities can be refined into cost-reduction trajectories for the three-application categories. Prices for stationary systems reduce to a narrow range between 150 and 330€/kWh, and for battery packs reduce to between 80 and 150€/kWh, regardless of technology. This implies that the only technology that manages to bring most capacity to market is likely to be the most cost competitive. Prices for portable batteries reduce to 100€/kWh.

The shaded regions in Fig. 8.5 are visual guides indicating the cost-reduction trajectory for each application category (at a particular technology scope). These narrow to the above-mentioned price ranges. For fuel cells and electrolyzers, prices are only reported on pack

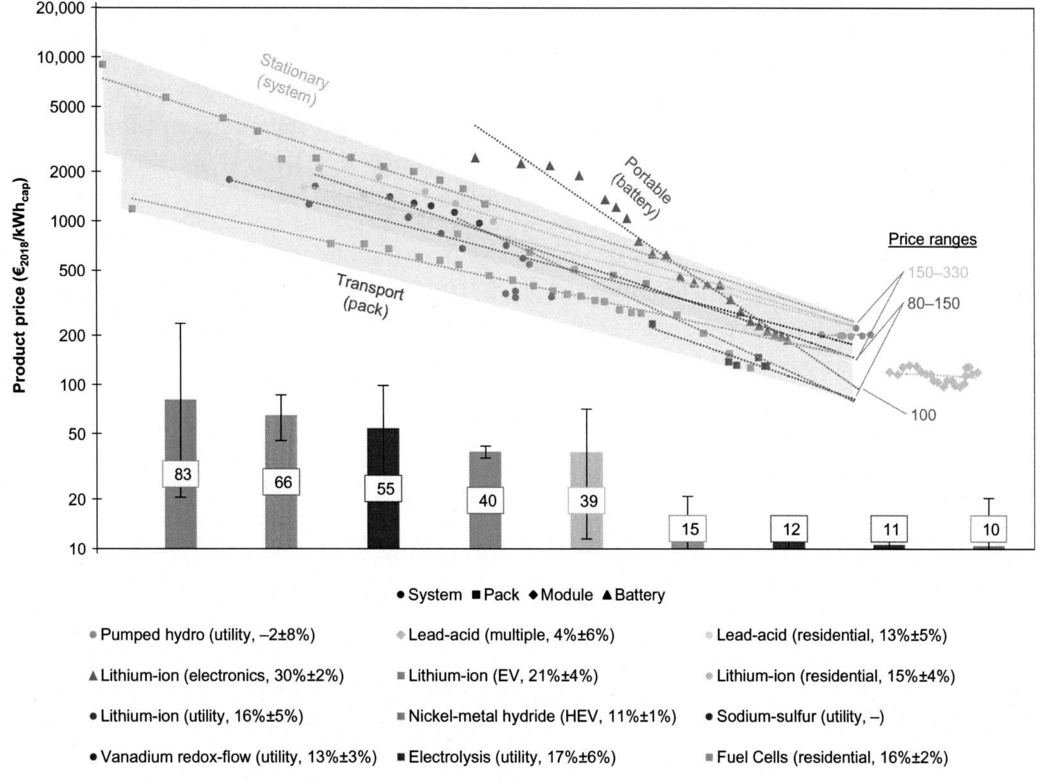

Figure 8.5

Future cost of electricity storage technologies at 1 TWh cumulative capacity. Experience curves (*dotted lines*) are projected forward to analyze product prices at future amounts of cumulative capacity. The bars show the raw material cost of the technologies per system for pumped hydro and per pack for all other technologies. These are calculated by multiplying material inventories from the literature with commodity prices of the past 10 years. Error bars account for variations in each technology's material inventory and commodity prices. Top legend indicates technology scope and bottom legend denotes technology (including application and experience rate with uncertainty). kWh_{cap} is the nominal energy storage capacity. *Source: Updated from Schmidt et al. (2017).*

level. The combination that could be used for stationary storage would cost 330€/kWh at pack-level (electrolysis: 80€/kWh, fuel cell: 250€/kWh), setting the upper bound of the range for stationary system. However, at system level, this combination would cost more, implying a higher upper bound. Pumped hydro systems and lead-acid modules are beyond 1 TWh cumulative installed capacity but cost 200€/kWh (pumped hydro) and 130€/kWh, respectively, which is well within the ranges identified for stationary storage systems and transport packs.

Due to the empirical rather than analytical nature of experience curves, extrapolations are subject to uncertainty of the derived ERs and uncertainty associated with unforeseeable future changes (technology breakthroughs, knowledge spillovers, commodity price shifts) (Junginger et al., 2010; Gross et al., 2013). When accounting for uncertainty of the underlying price and capacity data, the resulting price range at 1 TWh is 90–440€/kWh (systems), 70–160€/kWh (packs), and 95–105€/kWh (batteries).

Experience curve studies should include cost floors in extrapolated forecasts to avoid excessively low cost estimates (Junginger et al., 2010; Gross et al., 2013). Raw material costs for each technology are calculated by multiplying material inventories from the literature with commodity prices of the past 10 years (Fig. 8.5). The average raw material cost across all technologies is significantly below their ER projection. Production and other costs are typically below 20% (Argonne National Laboratory, 2015; James et al., 2014) of final system price for electrochemical, or between 50% and 80% (IRENA, 2016) for mechanical storage technologies, that are technologically mature. This confirms that the identified cost reduction potentials to 80–330€/kWh are feasible without limiting materials availability. Even if this would be the case, material requirements could potentially be further reduced per kW or kWh of energy-storage capacity.

However, it should be acknowledged that despite using price ranges of the past 10 years, there is still high uncertainty on the development of commodity prices. On the one hand, there could be raw material and other input bottlenecks as storage takes off for particular technologies such as fuel cells and increasing commodity prices, while on the other hand, the competition will potentially attract new producers of raw materials and other inputs, depressing commodity prices, and spurring innovation.

To map future cost reductions onto time, the market diffusion process of electricity storage technologies is modeled with the archetypal sigmoid function (*S*-curve) that has been observed for the deployment of several technologies (Rogers 1995; Schilling and Esmundo, 2009).

It is found that 1 TWh cumulative capacity could be installed for most new technology types within 5–20 years (Table 8.3). By 2030 stationary systems may cost between 200 and 440€/kWh, with pumped hydro and an electrolysis–fuel cell combination as minimum and maximum values, respectively. When accounting for ER uncertainty, the price range expands to 150–520€/kWh (min: utility-scale lithium-ion, max: electrolysis–fuel cell). The price range for transport applications in 2030 is 50–190€/kWh (40–200€/kWh with uncertainty). Lithium-ion EV pack prices may fall to 50€/kWh by 2030 due to the high ER of 21% combined with the high demand if 15 million EVs are sold annually by 2030 (MacDonald, 2016). This equals more than 700 GWh annual capacity, compared to 50 GWh for utility storage. Lower demand projections combined with a lower ER for nickel–metal hydride HEV battery packs, means prices could be reduced only to 190€/kWh. Lithium-ion

Table 8.3: Future cost of electricity storage technologies relative to time.

$€_{2018}/kWh_{cap}$	2020	2025	2030	2035	2040	2045	2050
Pumped hydro (system, utility)	205 ± 0	205 ± 1	206 ± 4	207 ± 8	208 ± 13	209 ± 18	209 ± 21
Lead-acid (module, multiple)	109 ± 0	109 ± 0	109 ± 0	109 ± 1	109 ± 1	108 ± 1	108 ± 2
Lead-acid (system, residential)	669 ± 147	416 ± 178	292 ± 173	243 ± 166	220 ± 161	206 ± 158	196 ± 155
Lithium-ion (battery, electronics)	149 ± 3	114 ± 5	94 ± 5	81 ± 6	72 ± 6	65 ± 6	60 ± 6
Lithium-ion (pack, EV)	121 ± 7	79 ± 11	51 ± 12	35 ± 11	27 ± 10	23 ± 10	21 ± 9
Lithium-ion (system, residential)	794 ± 62	465 ± 102	308 ± 102	248 ± 97	221 ± 94	204 ± 92	193 ± 90
Lithium-ion (system, utility)	474 ± 63	302 ± 85	214 ± 85	169 ± 81	146 ± 78	133 ± 76	124 ± 74
Nickel–metal hydride (pack, HEV)	250 ± 3	218 ± 4	191 ± 6	172 ± 6	159 ± 7	149 ± 7	143 ± 7
Vanadium redox-flow (system, utility)	451 ± 65	310 ± 78	236 ± 78	197 ± 76	176 ± 73	163 ± 72	154 ± 71
Electrolysis (pack, utility)	125 ± 1	118 ± 3	103 ± 9	88 ± 13	78 ± 15	72 ± 17	67 ± 17
Fuel cell (pack, residential)	900 ± 50	518 ± 72	334 ± 68	266 ± 64	235 ± 61	216 ± 58	203 ± 57

Cost projections based on experience rates and S-curve type market growth assumptions for consumer electronics, HEVs, electric vehicles, residential storage and utility-scale storage. Hundred percent market share assumed for each technology in their application category (e.g., electronics, EV, HEV, residential, utility). Uncertainty based on experience rate and growth rate uncertainties. Fuel cell and electrolysis must be considered in combination to form an electricity storage technology.
EV, Electric vehicle; *HEV*, hybrid electric vehicle.
Source: Updated from Schmidt et al. (2017).

batteries for consumer electronics would be at 95€/kWh by 2030 (90–100€/kWh with uncertainty).

The identified price range of 200–440€/kWh for stationary systems by 2030 lies within other projections (100–450€/kWh). However, individual products such as the lithium-ion based *Tesla Powerwall 2* were at an estimated retail price of 360€/kWh already by 2017 (Tesla Motors, 2016). A possible explanation could be synergistic learning effects for an electricity storage technology across applications due to shared components, cross-over techniques, or knowledge spillovers, leading to cost reductions not considered in this analysis (Kahouli-Brahmi, 2008). This pricing level could also reflect deliberate underpricing as part of a pricing strategy in newly commercialized products. In contrast, the cost projections in this study are based on the assumption of 100% market share for each technology in their respective application, which yields optimistic trajectories, and would support the projections at the upper end of the literature.

The range of 50–190€/kWh for transport packs is at the lower end of similar projections (50–540€/kWh), but supported by recent industry announcements of lithium-ion cells reaching 70€/kWh as early as 2022 (Cobb, 2015). Since higher estimates come from expert interviews versus lower from ER projections, the difference could be based on the latter placing more emphasis on future capacity additions, which would be significant if transportation is electrified. Conversely, increasingly competitive markets have driven strong price reductions since 2014, which could overestimate the underlying production cost reductions and distort the derived ERs (The Boston Consulting Group, 1970).

It should also be noted that the price projection for lithium ion battery packs beyond 2030 approaches the raw material cost floors identified in Fig. 8.5. This means that if these projections were to come true, significant reductions in commodity prices, improvements in energy density, or changes in commodity composition of lithium-ion batteries must be achieved. The latter two developments are within that timeline in current lithium-ion innovation roadmaps (Global EV Outlook 2018, 2018; Thielmann et al., 2015).

The future outlook for a variety of grid-scale storage technologies and applications provide a rapidly emerging field that requires innovative methods and analytical tools to understand. Major innovation theories related to learning-by-doing, learning-by-searching, economies of scale, and manufacturing localization are all required to consider how grid-scale storage technologies can become key features of a deeply decarbonized power system running on intermittent renewable electricity.

8.5 Conclusions and recommendations for science, policy, and business

Academia, policymakers, and industry can all contribute to the development and encouragement of the use of technological learning models to understand cost reduction and performance improvement related to grid-scale storage technologies and their associated innovations. Because of the wide range of grid-scale solutions that incorporate electrochemical batteries all the way to compressed air and pumped hydro-based storage, experience curve models need to consider a variety of indicators related to cost—beyond simply €/kWh or €/kW. However, the largest obstacle to achieving new indicators related to technological learning in grid-scale storage remains quality data availability.

A major theme revolving uncertainty in technological learning for grid-scale storage is the actual lack of experience and data to make quality forecasts. The rapid pace of materials science advances on the battery chemistry front especially introduces new challenges that have not been faced by energy innovations such as hydropower dams or natural gas–combined cycle plants. Therefore vigilance and increased call for transparency and public access of data remains the key to validating new learning curve models. Furthermore, by encouraging public–private partnerships to share data, there are new opportunities for understanding an innovation ecosystem. The key challenges for grid-scale storage remain quantifying and comparing technologies on a fair basis when batteries may perform multiple applications and functions on the grid. They may provide difficult-to-quantify services such as deferred investment in infrastructure when implemented on the grid and also provide electricity that may be "expensive" from a generation standpoint, but from a systems' approach could lower overall costs. Therefore technological learning studies that incorporate alternative indices related to the life cycle greenhouse gas emissions from storage options, materials availability of emerging battery chemistries, and cost indicators that incorporate multiple services and applications provided by storage can begin to inform policy and research investment.

One particular concern remains that a spate of recent studies (Schmidt et al., 2019b; Fares and Webber, 2017; Hittinger and Azevedo, 2015) demonstrates that under current grid conditions, implementing existing lithium-ion and other battery storage options at the grid-scale, could increase the overall carbon emissions on the grid, due to economically efficient electricity market bidding strategies. However, as highlighted by Louwen et al. (2016) for energy return on investment and CO_2 emission impacts of PV modules, there could be experience-based improvements in materials intensity and CO_2 emission impacts of different battery storage technologies. This can also be coupled with research that quantifies the "energy stored on energy invested." Further research that explores further technological learning along nontraditional indices including materials recycling and footprints, cycle life, roundtrip battery efficiency, and net energy ratios have been underdeveloped related to cost metrics.

From an economic perspective, cost remains one of the key indicators. New data for nonlithium-based grid-scale storage options needs to be transparent and available as new chemistries, flywheels, and fuel cells emerge. Learning curves in the case of grid-scale storage can inform public policy and R&D investments. Low-cost and innovative storage technologies are critical to achieve low-cost, deep decarbonization of the electricity system. Experience curves inform policymakers and industry to work together to develop research-based roadmaps and pathways toward building experience through new technologies. One-factor and multifactor learning curves can both inform society on the cost to deploy new technologies and the expected return on investment and inform strategies to balance resources to effectively gain deployment and research-based experience to drive down the cost of new technologies.

Research development, demonstration, and deployment measures have decreased in the United States over the past decade (Margolis and Kammen, 1999; Nemet and Kammen, 2007; Kittner et al., 2017). Globally, this remains concerning, even as deployment and investment in behind-the-meter storage increases in Germany and as spillover effects from consumer battery industries in South Korea and China develop. However, there are key reasons to believe, based on the data presented in this chapter, that energy-storage systems are already cost competitive and will outcompete traditional electricity-generation technologies in the next 5 years. Therefore better policies to understand this evolution and manage the transition to enable "baseload" renewable electricity systems require the methodological development of experience curves. More data and studies are needed across a variety of technologies and geographies to increase model accuracy and validation.

Research is largely concentrated in Europe and the United States, whereas grid-scale storage manufacturers are typically located in China and South Korea. The main markets for deployment of electricity grid-scale storage technologies are expected to occur most significantly in China, South Korea, and South and Southeast Asia. Therefore in rapidly growing regions, where high levels of investment are moving toward new technology deployment, there is a need for further information sharing, collaboration, and studies across markets to improve the understanding of technological learning and experience curves.

Fundamental theoretical and applied research that integrates an interdisciplinary approach to economic change, materials science advances, physical grid engineering, and human behavior is necessary to advance grid-scale storage to the level of technological maturity rivaling solar panels and wind turbines. The future of storage is exciting and therefore should invite strategic efforts to share information resources, invest in critical R&D expenditures necessary to promote innovation, and coevolved deployment activities to gain experience and understanding in low-carbon energy systems.

References

Abernathy, W.J., Kenneth, W., 1974. Limits of the learning curve. Harv. Bus. Rev. 109–119.

Argonne National Laboratory, 2015. BatPac Version 3.0. Excel Spreadsheet, 2015. Argonne National Laboratory. Available from: <http://www.cse.anl.gov/batpac/about.html>.

Avicenne Energy, 2018. The Rechargeable Battery Market and Main Trends 2017–2025: 100 Pages Update. Avicenne Energy.

Choi, J.W., Aurbach, D., 2016. Promise and reality of post-lithium-ion batteries with high energy densities. Nat. Rev. Mater. 1 (4), 16013.

Ciez, R.E., Whitacre, J.F., 2016. The cost of lithium is unlikely to upend the price of Li-ion storage systems. J. Power Sources 320, 310–313.

Cobb, J., 2015. GM Says Li-Ion Battery Cells Down to $145/KWh and Still Falling. Available from: <http://www.hybridcars.com/gm-ev-battery-cells-down-to-145kwh-and-still-falling/>.

Cohen, W., Levinthal, D., 1989. Innovation and learning: the two faces of R&D. Econ. J. 99, 569–596. Available from: https://doi.org/10.1111/ecoj.12199.

Comello, S., Reichelstein, S., 2019. The emergence of cost effective battery storage. Nat. Commun. 10 (1), 2038.

Davies, D.M., Verde, M.G., Mnyshenko, O., Chen, Y.R., Rajeev, R., Meng, Y.S., et al., 2019. Combined economic and technological evaluation of battery energy storage for grid applications. Nat. Energy 4 (1), 42.

Díaz-González, F., Hau, M., Sumper, A., Gomis-Bellmunt, O., 2015. Coordinated operation of wind turbines and flywheel storage for primary frequency control support. Int. J. Elec. Power 68, 313–326.

Dunn, B., Kamath, H., Tarascon, J.M., 2011. Electrical energy storage for the grid: a battery of choices. Science 334 (6058), 928–935.

Epple, D., Argote, L., Devadas, R., 1991. Organizational learning curves: a method for investigating intra-planet transfer of knowledge acquired through learning by doing. Organ. Sci. 2 (1), 58–70. Available from: https://doi.org/10.1287/orsc.2.1.58.

Fares, R.L., Webber, M.E., 2017. The impacts of storing solar energy in the home to reduce reliance on the utility. Nat. Energy 2 (2), 17001.

Farmer, J.D., Lafond, F., 2015. How predictable is technological progress? Res. Policy 45, 647–665. Available from: https://doi.org/10.2139/ssrn.2566810.

Ferioli, F., Schoots, K., van der Zwaan, B.C.C., 2009. Use and limitations of learning curves for energy technology policy: a component-learning hypothesis. Energy Policy 37 (7), 2525–2535. Available from: https://doi.org/10.1016/j.enpol.2008.10.043.

Fraunhofer ISE, 2015. Current and Future Cost of Photovoltaics Current and Future Cost of Photovoltaics. Freiburg.

Gross, R., Heptonstall, P., Greenacre, P., Candelise, C., Jones, F., Castillo, A.C., 2013. Presenting the Future – An Assessment of Future Costs Estimation Methodologies in the Electricity Generation Sector. UK Energy Research Center, London.

Global EV Outlook 2018, 2018. Paris. Available from: <www.iea.org/t&c/>.

Hall, G., Howell, S., 1985. The experience curve from the economist's perspective. Strateg. Manage. J. 6 (3), 197–212.

Hittinger, E.S., Azevedo, I.M., 2015. Bulk energy storage increases United States electricity system emissions. Environ. Sci. Technol. 49 (5), 3203–3210.

IRENA, 2016. The Power to Change: Solar and Wind Cost Reduction Potential to 2025. International Renewable Energy Agency (IRENA).

James, B.D., Moton, J.M., Colella, W.G., 2014. Mass Production Cost Estimation of Direct H2 PEM Fuel Cell Systems for Transportation Applications: 2013 Update. Strategic Analysis.

Junginger, M., Lako, P., Lensink, S., Weiss, M., 2008. Technological learning in the energy sector. Netherlands Programme on Scientific Assessment and Policy Analysis Climate Change (WAB) no. April: 1–190. Available from: <http://igitur-archive.library.uu.nl/chem/2009-0306-201752/UUindex.html>.

Junginger, M., van Sark, W., Faaij, A., 2010. In: Andre Junginger, M., Van Sark, W., Faaij, A. (Eds.), Technological Learning in the Energy Sector: Lessons for Policy, Industry and Science. Edward Elgar Publishing, Cheltenham. Available from: <http://www.scopus.com/inward/record.url?eid = 2-s2.0-84882003478&partnerID = 40&md5 = 4dab43b29c78b889f8f2b5e9126f3a71>.

Kahouli-Brahmi, S., 2008. Technological learning in energy-environment-economy modelling: a survey. Energy Policy 36, 138–162. Available from: https://doi.org/10.1016/j.enpol.2007.09.001.

Kittner, N., Gheewala, S.H., Kammen, D.M., 2016. Energy return on investment (EROI) of mini-hydro and solar PV systems designed for a mini-grid. Renew. Energy 99, 410–419.

Kittner, N., Lill, F., Kammen, D.M., 2017. Energy storage deployment and innovation for the clean energy transition. Nat. Energy 2 (9), 17125.

Kouvaritakis, N., Soria, A., Isoard, S., 2000. Modelling energy technology dynamics: methodology for adaptive expectations models with learning by doing and learning by searching. Int. J. Global Energy Issues 14 (1–4), 12.

Larcher, D., Tarascon, J.M., 2015. Towards greener and more sustainable batteries for electrical energy storage. Nat. Chem. 7 (1), 19.

Louwen, A., Van Sark, W.G., Faaij, A.P., Schropp, R.E., 2016. Re-assessment of net energy production and greenhouse gas emissions avoidance after 40 years of photovoltaics development. Nat. Commun. 7, 13728.

MacDonald, J., 2016. Electric Vehicles to Be 35% of Global New Car Sales by 2040. Bloomberg New Energy Finance. Available from: <http://about.bnef.com/press-releases/electric-vehicles-to-be-35-of-global-new-car-sales-by-2040/>.

Margolis, R., Kammen, D.M., 1999. Underinvestment: the energy technology and R&D policy challenge. Science 285, 690–692.

Mouli-Castillo, J., Wilkinson, M., Mignard, D., McDermott, C., Haszeldine, R.S., Shipton, Z.K., 2019. Inter-seasonal compressed-air energy storage using saline aquifers. Nat. Energy 4 (2), 131.

Nemet, G.F., 2006. Beyond the learning curve: factors influencing cost reductions in photovoltaics. Energy Policy 34 (17), 3218–3232. Available from: https://doi.org/10.1016/j.enpol.2005.06.020.

Nemet, G.F., Kammen, D.M., 2007. U.S. energy research and development: declining investment, increasing need, and the feasibility of expansion. Energy Policy 35 (1), 746–755.

Pillot, C., 2014. The rechargeable battery market and main trends 2013–2025. In: 31st International Battery Seminar & Exhibit.

Qiu, Y., Anadon, L.D., 2012. The price of wind power in China during its expansion: technology adoption, learning-by-doing, economies of scale, and manufacturing localization. Energy Econ. 34 (3), 772–785.

Rogers, E., 1995. Diffusion of Innovations, fourth ed. Free Press, New York, London, Toronto.

Rubin, E.S., Azevedo, I.M., Jaramillo, P., Yeh, S., 2015. A review of learning rates for electricity supply technologies. Energy Policy 86, 198–218.

Schmidt, O., Hawkes, A., Gambhir, A., Staffell, I., 2017. The future cost of electrical energy storage based on experience curves. Nat. Energy 2, 17110. Available from: https://doi.org/10.1038/nenergy.2017.110.

Schmidt, O., Melchior, S., Hawkes, A., Staffell, I., 2019a. Projecting the future levelized cost of electricity storage technologies. Joule 3 (1), 81–100.

Schmidt, T., Beuse, M., Zhang, X., Steffen, B., Schneider, S.F., Pena-Bello, A., et al., 2019b. Additional emissions and cost from storing electricity in stationary battery systems. Environ. Sci. Technol. 53 (7), 3379–3390.

Schilling, M.A., Esmundo, M., 2009. Technology S-curves in renewable energy alternatives: analysis and implications for industry and government. Energy Policy 37 (5), 1767–1781.

Schmitt, R.J.P., Kittner, N., Kondolf, G.M., Kammen, D.M., 2019. Deploy diverse renewables to save tropical rivers. Nature 569, 330–332.

Seebregts, A.J., Kram, T., Scaeffer, G.J., Stoffer, A., 1998. Endogenous Technological Learning: Experiments With MARKAL. Petten.

Sekine, Y., Goldie-Scot, L., 2017. 2017 Global Energy Storage Forecast. Bloomberg New Energy Finance, London.

Söderholm, P., Sundqvist, T., 2007. Empirical challenges in the use of learning curves for assessing the economic prospects of renewable energy technologies. Renew. Energy 32 (15), 2559–2578. Available from: https://doi.org/10.1016/j.renene.2006.12.007.

Staffell, I., Green, R., 2013. The cost of domestic fuel cell micro-CHP systems. Int. J. Hydrogen Energy 38 (2), 1088–1102. Available from: https://doi.org/10.1016/j.ijhydene.2012.10.090.

Stephan, A., Battke, B., Beuse, M.D., Clausdeinken, J.H., Schmidt, T.S., 2016. Limiting the public cost of stationary battery deployment by combining applications. Nat. Energy 1 (7), 16079.

Succar, S., Williams, R.H., 2011. Compressed air energy storage. In: Large Energy Storage Systems Handbook. CRC Press, pp. 111–152.

Tesla Motors, 2016. Powerwall, 2. Available from: <https://www.tesla.com/powerwall>.

The Boston Consulting Group, 1970. Perspectives on Experience. Boston Consulting Group, Boston, MA. Available from: <https://doi.org/10.1016/0305-0483(83)90010-5>.

Thielmann, A., Sauer, A., Wietschel, M., 2015. Gesamt-Roadmap Energiespeicher Für Die Elektromobilität 2030. Available from: <https://www.isi.fraunhofer.de/content/dam/isi/dokumente/cct/lib/GRM-ESEM.pdf>.

Tsiropoulos, I., Tarvydas, D., Lebedeva, N., 2018. Li-ion Batteries for Mobility and Stationary Storage Applications – Scenarios for Costs and Market Growth, EUR 29440 EN. Publications Office of the European Union, Luxembourg. 978-92-79-97254-6, JRC113360. Available from: http://doi.org/10.2760/87175.

US DOE Global Energy Storage Database. 2019. https://www.energystorageexchange.org/ (Last accessed 20.08.19).

Watanabe, C., Moriyama, K., Shin, J.H., 2009. Functionality development dynamism in a diffusion trajectory: a case of Japan's mobile phones development. Technol. Forecasting Soc. Change 76 (6), 737–753. Available from: https://doi.org/10.1016/j.techfore.2008.06.001.

Wiesenthal, T., Dowling, P.P., Morbee, J., Thiel, C., Schade, B., Russ, P., et al., 2012. Technology Learning Curves for Energy Policy Support. JRC Scientific and Policy Reports. European Commission - Joint Research Centre, Brussel. Available from: https://doi.org/10.2790/59345.

Wright, T.P., 1936. Factors affecting the cost of airplanes. J. Aeronaut. Sci. 3 (4), 122–128.

Zheng, C., Kammen, D.M., 2014. An innovation-focused roadmap for a sustainable global photovoltaic industry. Energy Policy 67, 159–169.

Further reading

Anderson, N., Rosser, J., 2016. Berenberg Thematics : Battery Adoption at the Tipping Point. Berenberg Bank, Hamburg.

Germany Trade & Invest, 2016. Batteries for Stationary Energy Storage in Germany: Market Status & Outlook. Germany Trade & Invest, Berlin.

Frith, J., 2018. 2018 Lithium-Ion Battery Price Survey. Bloomberg New Energy Finance, London.

Hocking, M., Kan, J., Young, P., Terry, C., Begleiter, D., 2016. F.I.T.T. for Investors: Welcome to the Lithium-Ion Age. Deutsche Bank – Market Research, Sydney.

Liebreich, M., 2013. Global Trends in Clean Energy Investment Clean Energy Ministerial. Bloomberg New Energy Finance, Delhi. Available from: <http://www.cleanenergyministerial.org/Portals/2/pdfs/BNEF_presentation_CEM4.pdf>.

Matteson, S., Williams, E., 2015. Learning dependent subsidies for lithium-ion electric vehicle batteries. Technol. Forecasting Soc. Change 92, 322–331. Available from: https://doi.org/10.1016/j.techfore.2014.12.007.

Nykvist, B., Nilsson, M., 2015. Rapidly falling costs of battery packs for electric vehicles. Nat. Clim. Change 5 (4), 329–332. Available from: https://doi.org/10.1038/nclimate2564.

Pitt, A., Buckland, R., Antonio, P.D., Lorenzen, H., Edwards, R., 2015. Citi GPS: Global Perspectives & Solutions – Investment Themes in 2015. Citigroup. Available from: <http://www.qualenergia.it/sites/default/files/articolo-doc/VO2E.pdf>.

International Renewable Energy Agency, 2012. Renewable Energy Technologies: Cost Analysis Series – Hydropower. International Renewable Energy Agency. Available from: <http://www.irena.org/documentdownloads/publications/re_technologies_cost_analysis-hydropower.pdf>.

Sandia National Laboratories, 2016. DOE Global Energy Storage Database 2016. Available from: <http://www.energystorageexchange.org/>.

Tepper, M., 2016. Solarstromspeicher-Preismonitor Deutschland 2016 [German]. Bundesverband Solarwirtschaft e.V. und Intersolar Europe, Berlin. Available from: <http://www.solarwirtschaft.de/fileadmin/media/pdf/speicherpreismon_1hj2013.pdf>.

CHAPTER 9

Electric vehicles

Noah Kittner[1,2,3,4], Ioannis Tsiropoulos[5,*], Dalius Tarvydas[5,*], Oliver Schmidt[6,7], Iain Staffell[7] and Daniel M. Kammen[8,9,10]

[1]Group for Sustainability and Technology, ETH Zurich, Zürich, Switzerland, [2]Department of Environmental Sciences and Engineering, Gillings School of Global Public Health, University of North Carolina at Chapel Hill, Chapel Hill, NC, United States, [3]Department of City and Regional Planning, University of North Carolina at Chapel Hill, Chapel Hill, NC, United States, [4]Environment, Ecology, and Energy Program, University of North Carolina at Chapel Hill, Chapel Hill, NC, United States, [5]Joint Research Centre, European Commission, Petten, The Netherlands, [6]Grantham Institute for Climate Change and the Environment, Imperial College London, London, United Kingdom, [7]Centre for Environmental Policy, Imperial College London, London, United Kingdom, [8]Energy and Resources Group, UC Berkeley, Berkeley, CA, United States, [9]Renewable and Appropriate Energy Laboratory, UC Berkeley, Berkeley CA, United States, [10]Goldman School of Public Policy, UC Berkeley, Berkeley, CA, United States

Abstract

Electric vehicles will play a dominant role in the transition to a low-carbon transportation system. As we track and forecast this evolution, learning rates help to quantify the historical rate and pace of change for emerging transportation options, the interaction of technology-specific and system-wide changes, and the economics of different policy options. In this chapter, we review the leading issues related to determining learning rates for electric vehicles and the potential scale-up for battery electric vehicles worldwide. Globally, electric vehicle deployment has increased rapidly over the past decade. Continued growth over the coming decade remains critical to achieve the level of ambition necessary to decarbonize the transportation sector. Therefore further data on learning rates and studies on innovation in battery electric vehicles are needed to enable and benefit from their decarbonization potential. In addition, research on incentives to develop appropriate technologies and policies to integrate electric vehicles into the existing electric grid infrastructure and transportation systems will inform further policy options and cost-reduction targets.

Chapter outline

* The views expressed are purely those of the authors and may not in any circumstances be regarded as stating an official position of the European Commission.

Technological Learning in the Transition to a Low-Carbon Energy System.
DOI: https://doi.org/10.1016/B978-0-12-818762-3.00009-1

9.1 Introduction

The heightened global attention toward electrification of transport underscores the potential disruption electric vehicles may have on transportation systems worldwide. Electric vehicles offer the opportunity to shift a major source of air pollution from mobile sources in densely populated urban areas to remote stationary power plants. This enables numerous benefits related to improving public health by reducing tropospheric ozone formation or smog from gasoline and diesel-powered vehicles. The shift from gasoline and diesel to electricity allows renewable and other low-carbon electricity sources to reduce the overall carbon footprint and embodied energy of transportation. Electric vehicles also offer the potential to provide distributed storage on the electric grid via coordinated charging efforts, which may help in balancing large penetrations of intermittent wind and solar generation. Though far from an environmental panacea, electric vehicles offer the potential to shift the demand for power during peak load periods for the electric utility (demand-side management). This way, they may systemically help balance and thus integrate large shares of intermittent renewable electricity on the grid.

There are now more than 5 million electric vehicles that have been deployed globally (Energy Revolution). Electric vehicles have reached 2% of new sales in the United States, 8% in The Netherlands, 7% in Ireland, 3% in Portugal, and 5% in China. In 2018, half of the new vehicles sold in Norway were electric (BNEF EVO, 2019). They are becoming affordable and available to larger segments of the population across Europe and in China.

Furthermore, cities, states, and countries are now adopting more ambitious policies to encourage the adoption of electric vehicles. These strategies are becoming more widespread and anticipate further growth and deployment. Many local and national governments around the world have increased their efforts to introduce electric vehicles into the mainstream. Several major European cities, such as London and Paris, have announced plans to ban internal combustion engine (ICE) vehicles in their city centers by 2045. Amsterdam plans to phase out internal combustion engines by 2030. More notable recent examples range from Shenzhen's full investment in electric city buses and taxis to Oslo's high electric vehicle ownership share. In California, policymakers set a target of 1 million electric vehicles on the road by 2025. Deployment measures, through policies such as tax rebates, encourage further adoption. Large-scale plans to electrify transportation remain promising—but further study is needed to understand the scale, rate, and pace of change in the cost of electric vehicles and technological learning. For instance, many expect spillover effects between steep cost declines from lithium-ion batteries in commercial and grid-scale battery applications. These spillover effects may be difficult to quantify; however, they unlock a unique opportunity to understand the rates of technological change across a variety of applications that remain critical for a low-carbon energy transition. Emerging business models are expected to drive innovation, reduce costs, promote battery waste management, improve the life cycle footprint, energy efficiency of vehicles, and promote alternative mobility plans including car sharingthat are intended to reduce private vehicle ownership.

Typically, technological learning curves have been applied to consumer electronics, renewable energy technologies, and ICE vehicles (Rubin et al. 2015). Because electric vehicles (EVs) comprise several different critical components that are all integral in achieving cost reductions, much attention has been placed on the battery pack range, cycle life, and cost. In 2018, the EV battery pack comprised about 35% of the total full EV price (BNEF, 2019). The manufacturing of electric vehicles also requires new methods of design and assembly—as the weight and density of batteries reshape manufacturing processes for vehicle structures.

Future data needs for new experience rates require integration of a variety of electric vehicle components—ranging from battery packs and chemistries to the power train, material structure of the vehicle and the battery, and design process. There are few comprehensive studies detailing the cost data for various aspects of the supply chain including raw material cost, cell manufacturing, pack components, other EV components, vehicle integration, and R&D. This type of aggregated information could inform cost development, as some current electric vehicle market leaders, such as BYD, keep most information proprietary to compete with ICE vehicles. Incorporating material or innovation in the electric vehicle manufacturing process could also inform further learning rates that go beyond total production volume or vehicle sales.

9.1.1 Description of technology

Electric vehicles broadly include any road-, rail-, sea-, or air-based vehicle that is at least partially powered by electricity. Recent advancements in battery technology have led to an expansion of the road-based electric vehicle market in the form of public transit such as buses and personal or shared vehicles. Electric road transport is generally categorized into three main types:

- Battery electric vehicles (BEVs) are fully powered by electricity. They make use of an electric propulsion system and rely on energy delivered by a battery pack. The battery is charged externally at charging stations and by recovered braking energy, called regenerative braking (Yong et al., 2015). The type of battery chemistry and design varies across different BEV models, but lithium-based batteries are currently dominating (EASE/EERA, 2017). BEVs have several advantages over the currently dominating fossil fuel–combustion engines. Besides having no tailpipe emissions and no direct reliance on fossil fuels, BEVs have higher vehicle efficiencies and better acceleration (Pollet et al., 2012; Andwari et al., 2017). Disadvantages of BEVs include high costs, low lifetime, safety concerns with flammable batteries, and relatively low driving ranges compared to ICE vehicles (due to the low energy density of the batteries that are currently available). In addition, few cities or regions have the infrastructure to support many new charging stations compared to gasoline and diesel refueling stations that are widespread globally. Based on annual sales, the weighted average battery capacity of BEV is around 39 kWh (Tsiropoulos et al., 2018).

- Hybrid electric vehicles (HEVs) combine an electricity power source with any other power source. Most commonly, an electric motor with battery storage is combined with the conventional internal combustion engine and fuel tank. The electric motor and combustion engine can be coupled in series, parallel, or series–parallel (Yong et al., 2015). Lithium-based battery packs are common for HEVs, as with BEVs, but nickel–metal hydride (Ni–MH) batteries are also used. However, the demand for new Ni–MH is fading as the largest firm supporter of Ni–MH, Toyota, has started using lithium (EASE/EERA, 2017). Plug-in hybrid electric vehicles (PHEVs) are supplemented with an external charging system for the battery and can be plugged into a power outlet, similar with BEVs. General HEV batteries are only charged by regenerative braking and the internal combustion engine (Biresselioglu et al., 2018). As PHEVs and HEVs both contain an internal combustion engine, they usually have smaller battery packs than BEVs (IRENA, 2017). Based on annual sales, the weighted average battery capacity of PHEVs is around 11 kWh (Tsiropoulos et al., 2018).
- Fuel cell electric vehicles (FCEVs), such as BEVs, rely solely on an electric propulsion system, but with the main source of energy being a fuel cell. FCEVs also contain a battery and are therefore hybrid vehicles, but the battery in FCEVs is much smaller than in BEVs and mainly used for the application of regenerative braking (OECD/IEA, 2015). Fuel cells electrochemically convert a fuel, typically hydrogen, to produce electricity and water as the only by-product, leading to zero tailpipe emissions. Fuel cells are lighter and smaller than batteries, and the vehicle can be recharged quickly, because a chemical fuel is used. However, FCEVs are currently more expensive than BEVs (also due to the production costs of H_2) and several components need to reduce in price before they are competitive (Cano et al., 2018; Staffell et al., 2019). The most common type of fuel cell in FCEVs is the polymer electrolyte membrane (Pollet et al., 2012; Pollet et al., 2019).

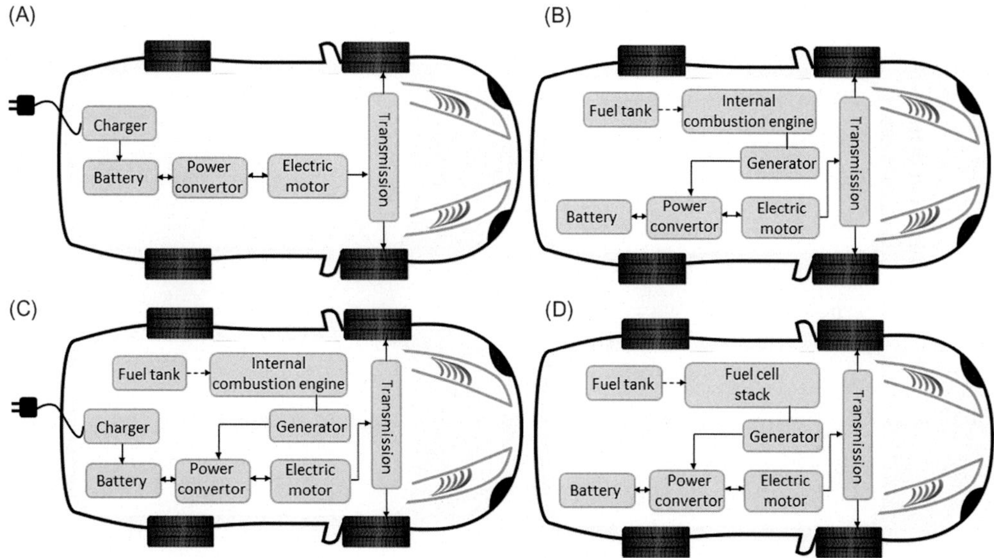

Figure 9.1

Overview of typical power train configurations for personal (A) BEV, (B) series HEV, (C) series PHEV, and (D) series FCEV. *BEV*, Battery electric vehicle; *FCEV*, fuel cell electric vehicle; *HEV*, hybrid electric vehicle; *PHEV*, plug-in hybrid electric vehicle. Source: *Based on Yong et al. (2015); Das et al. (2017)*

Fig. 9.1 shows examples of power train configurations for BEVs, HEVs, PHEVs, and FCEVs. Note that not all possible configurations (series, parallel, series–parallel) are shown.

9.1.2 Market development

Electric vehicles have been used for the past two hundred years, with several prototypes developed in the late-1800s, but not until the 2000s were highway-ready options available at widespread locations around the world. However, due to limited storage capacity in batteries and improvements in the internal combustion engines, attention for electric vehicles decreased. In the 1970s, oil crisis led to renewed interest in electric vehicles (Pollet et al., 2012). From then onwards, electric vehicles have reappeared periodically, and since the 2000s, the most recent and strongest growth is taking place (OECD/IEA, 2017c, see also Fig. 9.2 for the global historical growth of lithium-ion battery sales in main market segments). Battery electric vehicles now dominate the electric vehicle market. The emergence of companies in the United States such as Tesla has facilitated greater market penetration in addition to innovative vehicle models such as the Chevrolet Bolt and Nissan Leaf. In addition, battery electric vehicles offer better fuel efficiency compared to internal combustion engine vehicles, traveling further on less energy overall. The life cycle

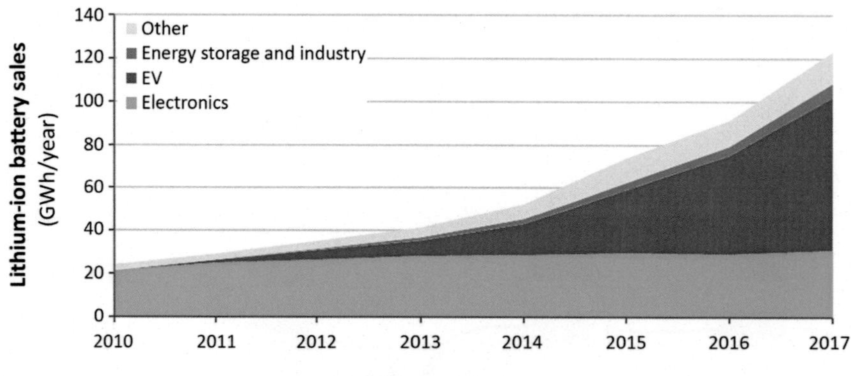

Figure 9.2
Global historical growth of lithium-ion battery sales in main market segments. Source: *Based on Tsiropoulos et al. (2018).*

greenhouse gas emissions of battery electric vehicles are less than that of internal combustion engine counterparts. These characteristics bode well for further innovations in battery electric vehicles. Alternatives to battery electric vehicles are under development as well. By 2016 Hyundai, Toyota, and Honda had FCEV passenger models available. Several other car manufacturers have announced their intention to develop FCEVs in the future (Curtin and Gangi, 2017; Cano et al., 2018), while other major car manufacturers, such as BMW, Daimler, and VW, thus far seem to solely focus on BEVs. Future FCEVs sales are seen for heavy duty passenger and freight transport in niche applications.

The global market of electric vehicles has grown rapidly in recent years. Including electric buses, there are more than 5 million electric vehicles on the road today. In the last years, growth was mainly driven by China, whose fleet increased from 11,600 vehicles in 2012 to 350,000 in 2016 and over 2 million vehicles in 2019. China has a 2020 goal of 5 million electric vehicles and is home to the world market leader for electric vehicle manufacturing BYD, followed by Renault–Nissan (France–Japan), Tesla (United States), and BMW (Germany) (REN21, 2017; BNEF, 2018a,b). The global annual sales exceeded 1 million vehicles for the first time in 2017 (excluding non-plug-in HEVs Fig. 9.3). Globally, BEVs represent two-thirds of electric vehicle sales. In specific markets, such as in Japan, PHEVs have the lion's share (two-thirds of new sales). In the European Union (EU), BEVs and PHEVs are currently sold annually in roughly equal amounts.

The growth of the electric vehicle market is mainly due to the diffusion of BEVs and PHEVs rather than fuel cell electric vehicles (OECD/IEA, 2017b). Despite rapid growth, the market share of these vehicles, in terms of overall sales of passenger vehicles, remains low in large markets, such as the United States (2%). Exceptions include Norway and Iceland. In Norway, electric vehicles have obtained a market share of 39.2% in 2017, followed by Iceland with a market share of 14.1% (EAFO, 2019). Norway's high level of

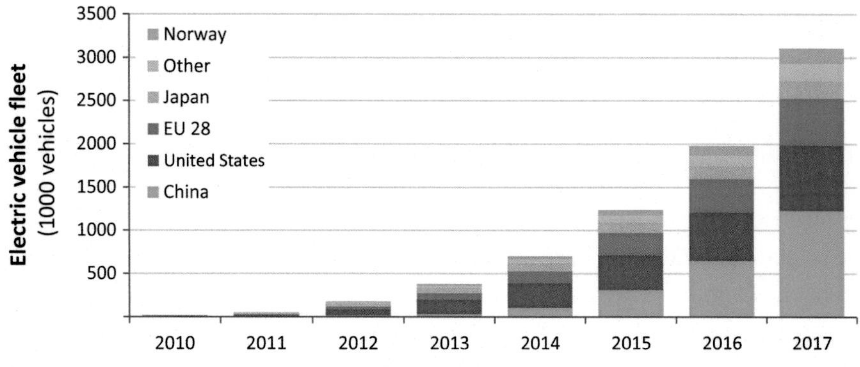

Figure 9.3
Total global electric vehicle fleet (excluding electric buses) in different regions in 2010–17. *Note*: EU 28 represents the markets of Finland, France, Germany, The Netherlands, Portugal, Sweden, and the United Kingdom. Source: *Based on Tsiropoulos et al. (2018)*

adoption has been the result of government subsidies and perks for electric vehicle owners, by waiving import duties and sales taxes and allowing for the use of bus lanes and no tolls. Within the EU, the countries with the largest market shares in 2017 were Sweden (5.3%), Belgium (2.7%), Finland (2.6%), The Netherlands (2.2%), and Austria (2.1%). In Sweden, Belgium, and Finland the majority of the vehicles are PHEVs, whereas in The Netherlands and Austria, BEVs are dominating. For the EU as a whole, the market share of electric vehicles was 1.4% in 2017 with similar shares of PHEVs and BEVs (EAFO, 2019).

The market for FCEVs remains small, and now only some studies have reported its growth (see Staffell et al., 2019 for recent database). Global sales of FCEVs represent only a small percentage of total EV sales: 0.5% in 2016 (Cano et al., 2018). According to Bloomberg New Energy Finance (BNEF), by the end of 2017, there were 6746 FCEVs, representing only 0.25% of total sales (BNEF EVO 2018). The European Alternative Fuels Observatory reported, in the first quarter of 2018, an FCEV fleet of 640 passenger cars, 85 buses, and 239 light commercial vehicles. The countries with largest FCEV deployment are the United States (7200), Japan (3500), Germany (2300), and China (BNEF, 2018b).

Deployment of BEVs, HEVs, and FCEVs are mainly dependent on advancements in battery or fuel stack technology (Cano et al., 2018). Prices for batteries have declined rapidly in recent years. Average costs for battery packs in 2015 were less than 237€/kWh for PHEVs and have dipped below 155€/kWh for BEVs (OECD/IEA, 2016b; BNEF, 2018a,b). It is generally assumed that in order to compete with internal combustion engines, battery costs should decrease further to around 88€/kWh (Nykvist and Nilsson, 2015; Kittner et al., 2017; Tsiropoulos et al., 2018). A study by Nykvist and Nilsson (2015), by combining results from different publications, show that this level is expected to be reached by 2025.

For FCEVs, total vehicle costs are estimated at 53,000€ (OECD/IEA, 2015). By 2019, many BEV costs have decreased below 27,000€.

9.2 Methodological issues and data availability

Comparing electric vehicles across makes and models to assess the experience rate suffers from a lack of standardization. There are no definitive methods or characteristics to define or measure electric vehicle cost. Many studies focus on upfront capital costs, total cost of ownership, battery pack costs, or cost per unit of distance traveled. However, electric vehicles provide transportation as a service, so many simplifying assumptions are often made. Assessing adoption and diffusion becomes more complicated, as owners of electric vehicles could purchase multiple vehicles for their own household, and whether electric vehicles serve as a direct substitute for conventional vehicles is yet to be determined. This section discusses some issues related to the functional units for electric vehicles and battery packs for electric vehicles in studies related to technological learning and cost reductions. This section also reviews challenges in data collection and availability.

The functional unit to assess the cost reductions for electric vehicles varies widely due to the nature of mobility services for electric vehicles. For instance, when focusing on cost of the service of transportation, conventional vehicles often followed $/gallon, €/L, or $/L of gasoline. More refined metrics based on kilometers traveled by the vehicle-incorporated $/km or €/km. Attempts have also been made to standardize the impact and cost related to the number of passengers in the particular vehicle. For BEVs a focus on the cost reduction for battery technologies has mostly placed an emphasis on the power in the battery pack (€/kW) or energy capacity (€/kWh) or even related to the mass of the vehicle in terms of €/kg. Geographically and temporally, these units vary as well (Tsiropoulos et al., 2018). Currency conversions also represent a methodological challenge when converting across countries and dealing with costs and prices in changing market conditions.

The cost reduction of a technology is usually the performance indicator that is used. While the cost required in producing a battery pack for an electric vehicle is an appropriate metric, it may be less relevant for other applications of lithium-ion batteries, such as stationary storage, due to the different services they provide to the system.

Previous studies have raised the issue of the challenges of using metrics such as total cost of ownership, to compare the cost effectiveness of electric vehicles with conventional counterparts (Nykvist and Nilsson, 2015; Nykvist et al., 2019). The perception of BEV and willingness to pay for range and charging time are more interlinked with the human behavioral aspects of choosing to use electric vehicles and contributing to the decline in cost. This consumer perspective is important, but also limited, to understand the overall developments across the entire electric vehicle market.

In addition, energy use, cost, and CO_2 emissions from electric vehicles are dynamic processes changing over time based on available technologies and existing electric grid infrastructure (Van Vliet et al., 2011).

Vehicle charging and grid integration costs are not often quantified in the learning rate literature. However, these transaction costs on electricity markets could make an important impact on the adoption and cost reduction potential for new vehicle technologies due to an increased need for peak capacity.

Data availability remains an issue to quantify deployment and innovation in electric vehicle technologies. However, with many electric vehicles varying across geographic regions, it becomes difficult to fully understand global manufacturing output. For instance, China has manufactured a majority of the electric vehicles to date, yet new vehicle registration data and manufacturing exports are difficult to monitor across the global supply chain with proprietary and state-owned data sources.

The basic learning curve methodology is based on the relationship between production and cost of the technology; in this case, the cost of lithium-ion batteries depends on the global manufacturing output. Mobility is the largest and growing segment that uses battery packs and most studies have focused on passenger light duty electric vehicles to estimate learning rates. However, there are several challenges around this:

- In 2018, more than 420,000 electric buses were reported to be on the road (BNEF, 2018d). While these heavy duty vehicles constituted only 10% of the global electric vehicle fleet, due to their large battery capacity (60–550 kWh) (Gao et al., 2017), the demand for lithium-ion batteries today is comparable. With future growth of passenger EVs and saturation of e-buses (even with modal shift) the demand of LDEVs will outpace that of HDEVs.
- Lithium-ion battery costs are also affected by the deployment of stationary systems and not just the number of electric vehicles produced. Data on installed capacity of stationary storage primarily reflect front-of-the-meter (e.g., industrial, commercial scale) applications. However, lithium-ion battery manufacturing for behind-the-meter storage (e.g., residential) accounted for about 20% of the total-production output.
- For EVs, battery packs are about 35% of the cost of the vehicle, cost reduction of remaining components (power train, car design) are also relevant, requiring the need for multifactor learning curves to assess whole vehicle cost.

9.2.1 Data collection and methodological issues

The main issue related to the collecting experience curve datasets for electric vehicles is that they do not refer to electric vehicles as a whole but rather describe the costs for just their battery packs or fuel cell stacks. Although the costs for the electricity storage components currently represent the majority of the price differential between EVs and conventional vehicles, it would be beneficial to gain insight into the price developments of other components specific to EVs, such as the power train and battery management system. It would also be beneficial to analyze in more detail the historical and prospective price trends of these vehicles, and to have insight in the development of costs of complete electric vehicles.

The cost of the battery pack referred in this chapter is given in €/kWh, as a function of cumulative GWh of battery packs sold. Both units do not directly relate to electric vehicles, hence, to estimate future costs of electric vehicles, assumptions need to be made on the battery pack size per electric vehicle. Assuming larger battery packs would result in faster price decline of the battery pack for the same number of electric vehicles sold. Thus the battery pack size may play an outsized role in reducing the overall cost of electric vehicles and affect other key parameters related to total range, time needed between charging periods, and peak power output.

As with the energy storage datasets, spillover effects from other applications in the electricity storage industry could likely affect BEV battery pack prices but are not taken into account due to modeling complexity. Ideally, in a more complex experience-curve modeling environment, separate experience curves should be used for lithium-ion cells, battery management and power electronics, and other components. These issues can be scaled up toward other components of EVs that are not represented in the batteries alone, including manufacturing components, materials structure and composition, and vehicle design.

9.3 Results

Several prominent and recent studies in which the experience curve method is applied are as follows:

- Nykvist and Nilsson (2015), who conducted a systematic review of historical lithium-ion EV battery pack costs to determine learning rates
- Schmidt et al. (2017), who estimated cost trajectories of lithium-ion EV battery packs, residential and utility-scale storage systems, in line with deployment that was derived from energy storage diffusion curves
- Kittner et al. (2017), who analyzed the deployment and innovation of batteries using a two-factor learning curve model
- BNEF (2019) that frequently reports estimate on learning rates of lithium-ion battery packs
- Tsiropoulos et al. (2018), who reviewed different growth trajectories of electric vehicles and stationary storage and based on literature-derived learning rates, estimate cost trajectories of lithium-ion batteries for mobility and stationary storage applications.

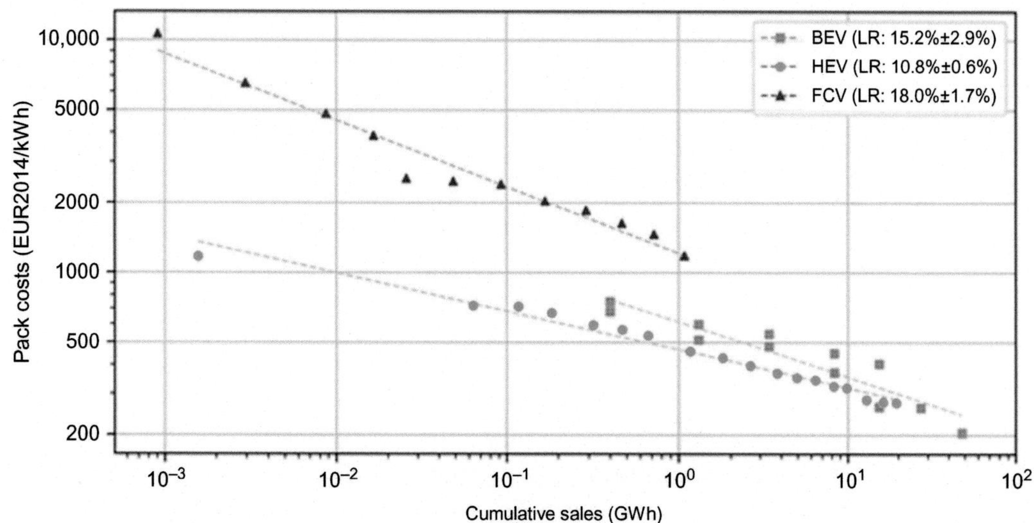

Figure 9.4
Experience curve for vehicle battery packs and fuel cell stacks. Source: *Based on Schmidt et al. (2017); Nykvist and Nilsson (2015)*

Other sources that have identified learning rates include the following:

- Weiss et al. (2012) and Weiss et al. (2019), who focused on electric mobility
- Gerssen-Gondelach and Faaij (2012), who apply learning rates from literature
- Sandalow et al. (2015), who report learning rates

The experience curves and datasets for the three vehicle types considered (BEVs, HEVs, and FCEVs) are shown in Fig. 9.4. The data represent battery pack (or fuel cell stack) costs per kWh as a function of cumulative GWh sold of each technology. Thus these curves represent only the prime mover component of the respective vehicles. Other components of electric vehicles also likely show learning effects, but since the energy storage packs account for the majority of the cost difference between electric and conventional vehicles, this chapter has examined only this component. The results indicate varying learning rates for the different vehicle types. The highest learning rate is observed for fuel cell stacks (18%), while hybrid EV batteries only have a learning rate of 10%. The data for BEV batteries is taken together from two data sources with various differences between the costs, resulting in a learning rate of 15.2% ± 2.9%.

Historical costs of battery packs for electric vehicles are provided by Nykvist and Nilsson and are summarized and complemented with more recent data in Fig. 9.5. In Europe, observed costs of EV battery packs based on BNEF have decreased from about 870€/kWh in 2010 to 170–215€/kWh in 2017. The simple mean of the reported values ranges about ± 60% compared with the volume weighted average of BNEF's price survey and declines

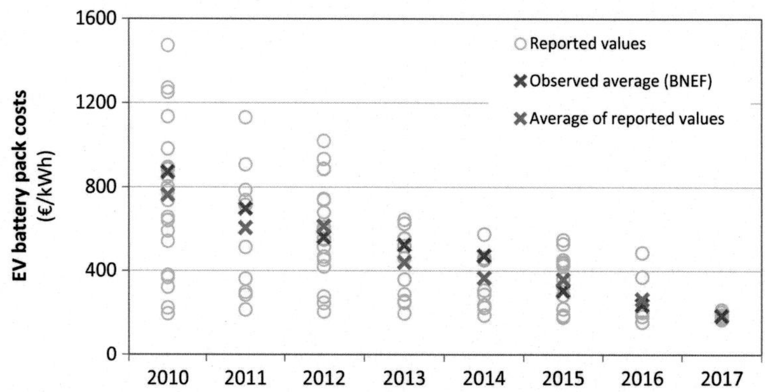

Figure 9.5
Reported lithium-ion battery pack costs for electric vehicles. *Note*: Observed average prices are based on BNEF's industry survey in 2018. Reported average is the simple average of the data included in the graph. Source: *Based on Tsiropoulos et al. (2018)*

over time (i.e., the variation is ± 15% in 2017). The lower end of the cost range of 2017 coincides with announcements of market leaders, such as Tesla, at about 170€/kWh (UCS, 2018; Tsiropoulos et al., 2018).

There are several reasons that explain the variation of historical data in either direction (higher or lower than the observed average):

- Announcements of market leaders that were typically lower than what the rest of the industry reported
- Reported metrics not being always consistent, as these may represent either production costs or market prices of battery packs (Tsiropoulos et al., 2018)
- Pricing strategies, firms which price their units either above costs (price umbrella) or below
- Different battery sizes (e.g., IEA mentions that a 70 kWh battery is expected to have a 25% lower cost per unit of energy stored than a 30 kWh battery, due to the higher cell-to-pack ratio of the former) or battery chemistry (e.g., based on IEA, the cost of a NMC-111 battery is about 5% higher than the cost of an NCA battery) (IEA, 2018)
- Cell quality, size, and format also affect the cost of a battery pack (e.g., an 18,650 cylindrical cell being about 30% cheaper than a large prismatic EV cell) (Tsiropoulos et al., 2018).

9.4 Future outlook

The growth of electric vehicles is expected to continue. When including announced policy plans, decreasing battery costs, increased charging infrastructure, and predicted trends in gas prices, it is expected that the global electric vehicle sales will reach 106 and 277 million vehicles by 2030 and 2040, respectively, according to the World Energy Outlook 2017. Even Bloomberg forecasts 500 million electric vehicles with or without meeting

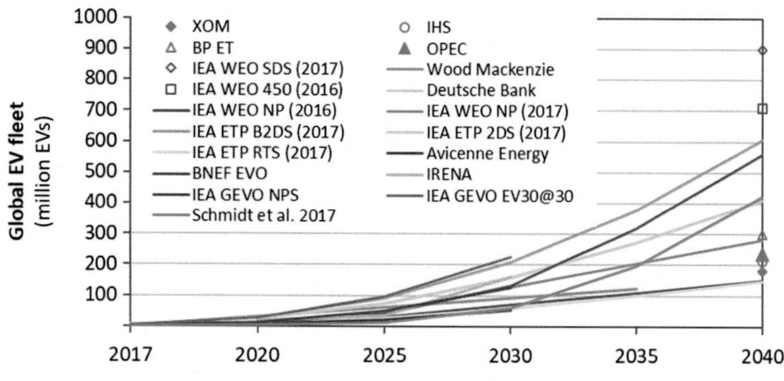

Figure 9.6
Projections of the global electric vehicle fleet over time (Tsiropoulos et al., 2018).

Table 9.1: Common assumptions in the high, moderate, and low scenarios.

	Common across scenarios				
Battery size—BEV (kWh) Battery size—PHEV (kWh)	40 18	52 19	60 20	60 20	60 20
Lifetime—EV pack Lifetime—storage	10 years 20 years				

BEV, Battery electric vehicles; *PHEV*, plug-in hybrid electric vehicles.

climate targets (BNEF, 2018a,b). Furthermore, the global lithium-ion cell manufacturing capacity is expected to increase sixfold by 2022 compared with 2017 (Tsiropoulos et al., 2018, see also Fig. 9.6). China is expected to remain the main actor in the market, accounting for 40% of global investments in electric vehicles. The global market share of electric vehicles by 2040 is expected to reach at least 60% by some reports (Bloomberg, 2018). For larger market shares and for keeping global temperature increase under two degrees, stronger policy support aimed at electrifying transportation is needed. Other modeling frameworks support this growth motivated by meeting global climate targets or through economic competitiveness alone. In a two-degree scenario, the global amount of electric vehicles increases to 243 million in 2030 and 873 million in 2040, with a 40% share of car stock in 2040 (OECD/IEA, 2017c). The IRENA two-degree road map shows a lower total electric vehicle stock projection of 160 million by 2030 (IRENA, 2019). With the current growth rate, electric vehicles are on track for the two-degree scenario (OECD/IEA, 2017a). According to IRENA, a two-degree scenario shows over 1 billion electric vehicles on the road by 2050, of which 965 million are passenger cars. In order to achieve this number, nearly all passenger vehicles sold from 2040 onward need to be electric (IRENA, 2018).These projections depend on the direction the world will take, for example, on action against climate change, or on when and how steep costs will decline (Tsiropoulos et al., 2018) (Table 9.1).

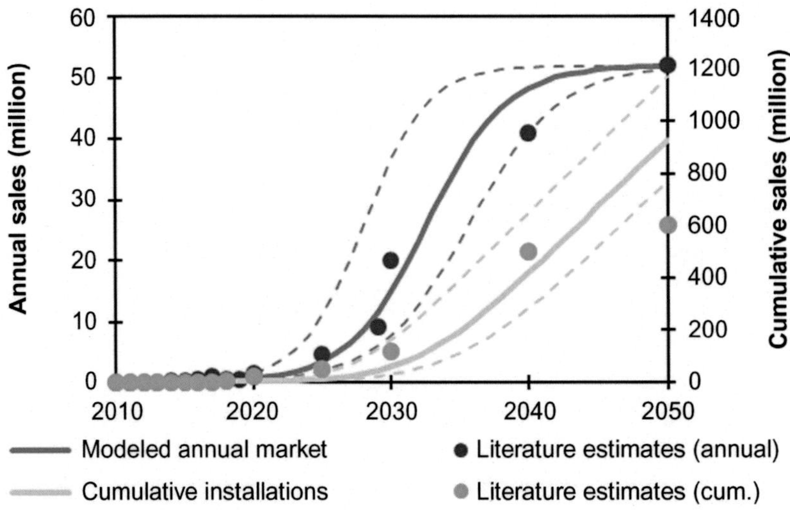

Figure 9.7
Electric vehicle deployment projection. Source: *Based on Schmidt et al. (2017).*

Table 9.2: Deployment of electric vehicles and stationary storage in the high scenario.

High **scenario**	**2020**	**2025**	**2030**	**2035**	**2040**
EVs (million EVs)	10	42	134	322	562
Stationary storage (GWh)	25	103	378	777	1 326

The global lithium-ion cell manufacturing capacity is expected to increase sixfold by 2022 compared with 2017 (Tsiropoulos et al., 2018, see also Fig. 9.6). In the longer term, future outlooks project that up to 1 billion electric vehicles may be on the road by 2040, which is about two to three orders of magnitude higher compared with today (see Fig. 9.7).

To extrapolate costs based on the learning curve method and estimate the cumulative production of battery packs, three demand forecasts for electric vehicles and stationary storage are selected. The *high* scenario is based on BNEF and sees the highest deployment of electric vehicles (BNEF, 2018b) and stationary storage (BNEF, 2018c), assuming that in the longer term, market forces drive decisions based on technology costs (Table 9.2). The *moderate* scenario is focused on CO_2-emission reduction. It sees a strong growth of electric vehicles to decarbonize the transport sector; yet lower than in the *high* scenario, possibly due to the role of biofuels in reducing road transport emissions. In the power sector the role of stationary storage is less pronounced compared with the *high* scenario. The moderate scenario is based on the 2017 IEA Energy Technology Perspectives 2DS scenario (IEA, 2017) (Table 9.3). In the *low* scenario the world is at a standstill when it comes to further action against climate change. As a result, sales of electric vehicles and stationary storage

Table 9.3: Deployment of electric vehicles and stationary storage in the moderate scenario.

***Moderate* scenario**	**2020**	**2025**	**2030**	**2035**	**2040**
EVs (million EVs) Stationary storage (GWh)	23 4	74 6	160 8	274 30	410 51

Table 9.4: Deployment of electric vehicles and stationary storage in the low scenario.

***Low* scenario**	**2020**	**2025**	**2030**	**2035**	**2040**
EVs (million EVs) Stationary storage (GWh)	9 2	27 3	58 4	101 15	146 25

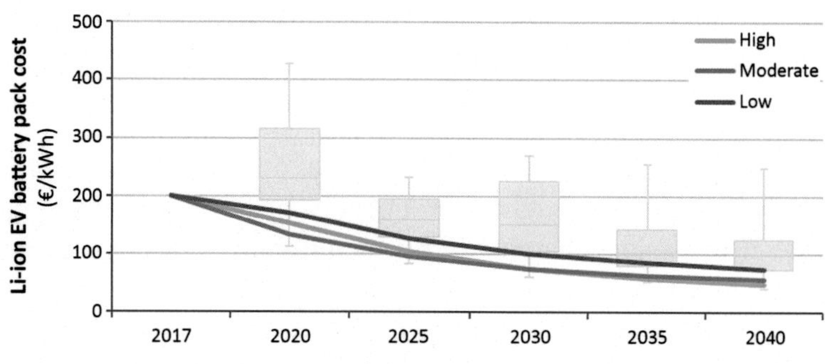

Figure 9.8

Cost trajectories of lithium-ion battery pack costs for electric vehicles based on three deployment scenarios. *Note*: Boxplots represent literature estimates. Source: *Based on Tsiropoulos et al. (2018)*

are limited. This scenario is based on the 2017 IEA Energy Technology Perspectives Reference Technology Scenario (IEA, 2017) (Table 9.4).

By 2040, almost 4 TWh would be sold annually in the *high* scenario, around 2 TWh in the *moderate* scenario, and 600 GWh would be needed in the *low* scenario, compared with about 60–70 GWh of annual sales in 2017. At these scales, next to anticipated improvements in lithium-ion cell chemistries, optimization of manufacturing processes, standardization of design and possibly vertical integration of plants, the costs of lithium-ion battery pack for EVs, and stationary storage could fall drastically, as also implied by recent literature.

The calculations show that lithium-ion EV battery pack prices could halve by 2030. By 2040 the cost could halve again, to ultimately 50€/kWh (see Fig. 9.8). These trajectories are in line with most recent estimates of other studies.

Based on inventories and the historical average of commodity prices, materials for lithium-ion battery packs may cost around 30€/kWh. In today's cost structures, materials represent between 60% and 75% of the total cost, which entails that lithium-ion battery packs may ultimately cost 40–50€/kWh. As such, the lower bound of the cost estimated in this analysis (i.e., 50€/kWh by 2040) seems feasible. Economies of scale could reduce capital cost, standardization and automation could limit operating and labor costs, and improved lithium-ion chemistries could further reduce the demand for materials, hence overall costs.

9.5 Conclusions and recommendations for science, policy, and business

Electric vehicles require a unique set of innovation and deployment tactics to compete with internal combustion engine vehicles. Conventional vehicles have system inertia, which means that behavioral shifts are necessary for consumers to substitute electric vehicles with previous gasoline-powered cars. Policies and incentives could ease the burden on customers. In addition, as part of a comprehensive climate policy toolbox, electric vehicles could significantly displace gasoline and diesel consumption that is generally based on fossil fuels.

The future of electric vehicle deployment will likely be tested in China. Cities, such as Shenzhen, have rolled out massive public investment campaigns to electrify public buses and taxis. They are also enabling charging infrastructure for personal electric vehicle use. Hotspots, such as Shenzhen, will likely contribute to the global diffusion and sales of electric vehicles in ways that are difficult to quantify in a technological learning study as the effect of knowledge spillovers, innovation diffusion, and unique geographies are poorly represented in the current modeling toolkits.

China will play a large role in future electric vehicle cost reductions, but there are windows of opportunity for the United States and the EU to also lead in terms of innovative policy deployment and manufacturing capacity. Examples include Tesla's investment in electric vehicle manufacturing within the United States and Audi and Volkswagen group's investment in transitioning their available vehicle fleet to fully electric vehicles. In Europe, the share of global cell manufacturing capacity may increase from 3% to 7%–25% before 2030, depending on whether planned announcements materialize (Tsiropoulos et al., 2018). The private sector is contributing to advanced competition for mobility as a service. Battery performance will play a large role in driving down the cost of electric vehicles, as range and degradation remain some of the key technical challenges facing electric vehicles when compared with conventional alternatives. Therefore, a majority of the learning rate studies have thus far focused on the cost-reduction potential for battery cells, packs, and modules that can be used in vehicles. Integration costs should be further studied and included into the learning rate models.

With the anticipated strong growth in uptake, our chapter suggests that by 2040, the cost could drop an additional 50% from today's level, ultimately reaching 50€/kW h. Such trajectories are feasibly based on costs of new lithium-ion cathode chemistries and other battery pack materials that are estimated at around 30€/kWh (other additional costs are estimated at around at 10–20€/kWh). Meeting policy goals such as the EU's Strategic Energy Technology Plan (SET Plan) cost target of 75€/kWh is feasible in both high and moderate growth scenarios (which entail fast ramp-up of lithium-ion manufacturing capacity, of about 300–400 GWh globally per year until 2030). This result, based on experience rates, indicates that aggressive targets may not be so difficult to meet, which can help as a decarbonization transportation strategy.

Further experience rate studies that go beyond total cost of ownership, battery pack costs, and cost per the kilometers traveled by the passenger will help better understand diffusion and adoption of electric vehicles. These include system-integration studies and costs of charging infrastructure required to enable electric vehicles on the grid. Furthermore, it requires behavioral change and utility company support to allow for fast-charging and cost-effective plans for customer rollout.

The main messages to take away from this analysis are the following:

- Widespread deployment of electric vehicles could enable a rapid decrease of lithium-ion battery costs in the near term.
- China is outpacing Europe and the United States in terms of electric vehicle deployment and innovation, but there are windows of opportunity for the United States and Europe to step in.
- European manufacturing of lithium-ion battery cells will increase its share in global production, provided that the announced plans materialize. Supplying domestic demand may prove challenging if capacity does not ramp up after 2025.
- Fuel cells are more expensive than battery packs and cannot match the rate of cost reductions, increasing uncertainty on the future of fuel-cell EVs in light-duty applications.

References

Andwari, A.M., Pesiridis, A., Rajoo, S., Martinez-Botas, R., Esfahanian, V., 2017. A review of battery electric vehicle technology and readiness levels. Renew. Sust. Energ. Rev. 78, 414–430.

BNEF, 2018a. Electric Vehicles Data Hub. Bloomberg New Energy Finance (BNEF).

BNEF, 2018b. Long-Term Electric Vehicle Outlook 2018 (EVO 2018). Bloomberg New Energy Finance (BNEF).

BNEF, 2018c. New Energy Outlook 2018 (NEO 2018). Bloomberg New Energy Finance (BNEF).

BNEF, 2018d. Cumulative Global EV Sales Hit 4 Million.

BNEF, 2019. Long-Term Electric Vehicle Outlook 2019 (EVO 2019). Bloomberg New Energy Finance (2019).

Biresselioglu, M.E., Kaplan, M.D., Yilmaz, B.K., 2018. Electric mobility in Europe: a comprehensive review of motivators and barriers in decision making processes. Transport. Res. A: Pol 109, 1–13.

Cano, Z.P., Banham, D., Ye, S., Hintennach, A., Lu, J., Fowler, M., Chen, Z., 2018. Batteries and fuel cells for emerging electric vehicle markets. Nat. Energy 3 (4), 279.

Curtin, S., Gangi, J., 2017. The Business Case for Fuel Cells: Delivering Sustainable Value. Argonne National Lab.(ANL), Argonne, IL (United States) (No. ANL-17/08).

Das, H.S., Tan, C.W., Yatim, A.H.M., 2017. Fuel cell hybrid electric vehicles: a review on power conditioning units and topologies. Renew. Sust. Energ. Rev. 76, 268–291.

EAFO, 2019. Vehicle Stats.

EASE/EERA (2017). European Energy Storage Technology Development Roadmap. https://eera-es.eu/wp-content/uploads/2016/03/EASE-EERA-Storage-Technology-Development-Roadmap-2017-HR.pdf

Gao, Z., Lin, Z., LaClair, T.J., et al., 2017. Battery capacity and recharging needs for electric buses in city transit service. Energy 122, 588–600. Available from: https://doi.org/10.1016/J.ENERGY.2017.01.101.

Gerssen-Gondelach, S.J., Faaij, A.P.C., 2012. Performance of batteries for electric vehicles on short and longer term. J. Power Sources 212, 111–129. Available from: https://doi.org/10.1016/j.jpowsour.2012.03.085.

IEA, 2018. Global EV Outlook 2018 (Towards Cross-Modal Electrification). Organisation for Economic Co-Operation and Development (OECD), International Energy Agency (IEA). Available from: <https://webstore.iea.org/download/direct/1045?filename = globalevoutlook2018.pdf>.

IEA, 2017, Energy Technology Perspectives 2017 Catalysing Energy Technology Transformations, Organisation for Economic Co-operation and Development (OECD), International Energy Agency (IEA), Paris, ISBN: 978-92-64-27597-3, pp: 1–443.

IRENA, 2017. Electric Vehicles: Technology Brief. International Renwable Energy Agency, Abu Dhabi.

IRENA, 2018. Cost of Service Tool V1-0. IRENA.

IRENA, 2019. Global Energy Transformation: A Roadmap to 2050. International Renewable Energy Agency, Abu Dhabi.

Kittner, N., Lill, F., Kammen, D.M., 2017. Energy storage deployment and innovation for the clean energy transition. Nat. Energy 2, 2017125. Available from: https://doi.org/10.1038/nenergy.2017.125.

Nykvist, B., Nilsson, M., 2015. Rapidly falling costs of battery packs for electric vehicles. Nat. Clim. Change 5 (4), 329–332. Available from: https://doi.org/10.1038/nclimate2564.

Nykvist, B., Sprei, F., Nilsson, M., 2019. Assessing the progress toward lower priced long range battery electric vehicles. Energy policy 124, 144–155.

OECD/IEA, 2015. Technology Roadmap: Hydrogen and Fuel Cells. International Energy Agency (IEA), Paris, France, Retrieved from: https://www.iea.org/publications/freepublications/publication/TechnologyRoadmapHydrogenandFuelCells.pdf.

OECD/IEA, 2016b. World Energy Outlook 2016. International Energy Agency (IEA), Paris, France.

OECD/IEA, 2017a. Energy Technology Perspectives 2017. International Energy Agency (IEA), Paris, France.

OECD/IEA, 2017b. Global EV Outlook 2017. International Energy Agency (IEA), Paris, France, Retrieved from https://www.iea.org/publications/freepublications/publication/GlobalEVOutlook2017.pdf.

OECD/IEA, 2017c. World Energy Outlook 2017. International Energy Agency, Paris, France, Online tool viewed April 4, 2018 on https://www.iea.org/weo/; Executive summary retrieved from https://www.iea.org/Textbase/npsum/weo2017SUM.pdf.

Pollet, B.G., Staffell, I., Shang, J.L., 2012. Current status of hybRid, battery and fuel cell electric vehicles: from electrochemistry to market prospects. Electrochim. Acta 84, 235–249.

Pollet, B., et al., 2019. Current status of automotive fuel cells for sustainable transport. Curr. Opin. Electrochem. 16, 90–95.

REN21 (2017). Global status report. http://www.ren21.net/gsr-2017/.

Rubin, E.S., Azevedo, I.M.L., Jaramillo, P., Yeh, S., 2015. A review of learning rates for electricity supply technologies. Energy Policy 86, 198–218. Available from: https://doi.org/10.1016/j.enpol.2015.06.011.

Sandalow, D., McCormick, C., Rowlands-Rees, T., Izadi-Najafabadi, A., Orlandi, I., 2015. Distributed Solar and Storage – ICEF Roadmap 1.0. Innovation for Cool Earth Forum (ICEF), Bloomberg New Energy Finance (BNEF).

Schmidt, O., Hawkes, A., Gambhir, A., Staffell, I., 2017. The future cost of electrical energy storage based on experience rates. Nat. Energy 6, 17110. Available from: https://doi.org/10.1038/nenergy.2017.110.

Staffell, I., Scamman, D., Abad, A.V., Balcombe, P., Dodds, P.E., Ekins, P., et al., 2019. The role of hydrogen and fuel cells in the global energy system. Energy Environ. Sci. 12 (2), 463–491.

Tsiropoulos, I., Tarvydas, D., Lebedeva, N., 2018. Li-Ion Batteries for Mobility and Stationary Storage Applications – Scenarios for Costs and Market Growth. Publications office of the European Union, Luxembourg. Available from: <https://doi.org/10.2760/87175>.

UCS, 2018. Electric Vehicle Battery: Materials, Cost, Lifespan. Union of Concerned Scientists (UCS). Available from: <https://www.ucsusa.org/clean-vehicles/electric-vehicles/electric-cars-battery-life-materials-cost> (accessed 09.08.18.).

Van Vliet, O., Brouwer, A.S., Kuramochi, T., van Den Broek, M., Faaij, A., 2011. Energy use, cost and CO2 emissions of electric cars. J. power sources 196 (4), 2298–2310.

Weiss, M., Patel, M.K., Junginger, M., Perujo, A., Bonnel, P., van Grootveld, G., 2012. On the electrification of road transport – learning rates and price forecasts for hybrid-electric and battery-electric vehicles. Energy Policy 48, 374–393. Available from: https://doi.org/10.1016/J.ENPOL.2012.05.038.

Weiss, M., Zerfass, A., Helmers, E., 2019. Fully electric and plug-in hybrid cars – an analysis of learning rates, user costs, and costs for mitigating CO_2 and air pollutant emissions. J. Clean. Prod. 212, 1478–1489. Available from: https://doi.org/10.1016/J.JCLEPRO.2018.12.019.

Yong, J.Y., Ramachandaramurthy, V.K., Tan, K.M., Mithulananthan, N., 2015. A review on the state-of-the-art technologies of electric vehicle, its impacts and prospects. Renew. Sust. Energ. Rev. 49, 365–385.

Further reading

Avicenne Energy, 2018. Battery Market for Hybrid, Plug-in & Electric Vehicles. Avicenne Energy.

Avicenne Energy, 2018. The Rechargeable Battery Market and Main Trends 2017–2025: 100 Pages Update. Avicenne Energy.

Belderbos, A., Delarue, E., Kessels, K., D'haeseleer, W., 2017. Levelized cost of storage—introducing novel metrics. Energy Econ. 67, 287–299. Available from: https://doi.org/10.1016/J.ENECO.2017.08.022.

BNEF, 2017a. 2017 Global Energy Storage Forecast. Bloomberg New Energy Finance (BNEF).

BNEF, 2017b. Storage System Costs More than Just a Battery. Bloomberg New Energy Finance (BNEF).

Comello, S., Reichelstein, S., 2019. The emergence of cost effective battery storage. Nat. Commun. 10 (1), 2038.

Fraunhofer, I.S.E., 2015. Current and Future Cost of Photovoltaics Current and Future Cost of Photovoltaics. Freiburg.

Gupta, M., 2018. U.S. Front-of-the-Meter Energy Storage System Prices 2018–2022. GMT Research. Available from: <https://www.greentechmedia.com/research/report/us-front-of-the-meter-energy-storage-system-prices-2018-2022#gs.LXJbdqE>.

Hocking, M., Kan, J., Young, P., Terry, C., Begleiter, D., 2016. Lithium 101 F.I.T.T. for Investors Welcome to the Lithium-Ion Age. Deutsche Bank Market Research. Available from: <http://www.belmontresources.com/LithiumReport.pdf>.

IRENA, 2016. The Power to Change: Solar and Wind Cost Reduction Potential to 2025. International Renewable Energy Agency (IRENA).

Junginger, M., van Sark, W., Faaij, A., 2010. In: Junginger, M., van Sark, W., Faaij, A. (Eds.), Technological Learning in the Energy Sector: Lessons for Policy, Industry and Science, 2012th ed. Edward Elgar Publishing, Cheltenham and Northampton.

Lazard, 2017. Lazard's Levelized Cost of Storage Analysis – Version 3.0. Lazard.

McKinsey and Company. A portfolio of power-trains for Europe: a fact-based analysis. In: The Role of Battery Electric Vehicles, Plug-in Hybrids and Fuel Cell Electric Vehicles. Available from: <https://www.fch.europa.eu/sites/default/files/documents/Power_trains_for_Europe.pdf>.

Sager, J., Apte, J.S., Lemoine, D.M., Kammen, D.M., 2011. Reduce growth rate of light-duty vehicle travel to meet 2050 global climate goals. Environ. Res. Lett. 6.

Staffell, I., Jansen, M., Chase, A., Cotton, E., Lewis, C., 2018. Energy Revolution: A Global Outlook. Drax, Selby.

CHAPTER 10

Power to gas (H_2): alkaline electrolysis

Subramani Krishnan[1], Matthew Fairlie[2], Philipp Andres[3], Thijs de Groot[4] and Gert Jan Kramer[1]

[1]*Copernicus Institute of Sustainable Development, Utrecht University, Utrecht, The Netherlands,* [2]*Next Hydrogen Corporation, Mississauga, ON, Canada,* [3]*Perimeter Solar Inc., ON, Canada,* [4]*Nouryon, Amsterdam, The Netherlands*

Abstract

Alkaline electrolysis as an industrial process has been around since the advent of commercial power at the beginning of the 20th century with most large-scale plants (up to 165 MW) built between the 1920s and 1980s in response to hydrogen demand for the ammonia industry. With the emergence of cheap hydrogen from steam methane reforming from the late 1980s, the production of small-scale plants (around 1 MW) dominated the electrolysis market. But in recent years, plant scale has increased (10 MW with few at 100 MW) compared to scale in the 1990s in response to increased demand for green hydrogen and moving away from hydrogen production from fossil fuels. This chapter addresses the learning rate of alkaline electrolysis systems from the period 1956–2016, issues with generating experience curves for the system as a whole as opposed to on a component basis, associated cost reduction drivers, and the future outlook of the alkaline electrolysis market until 2030.

Chapter outline

Technological Learning in the Transition to a Low-Carbon Energy System.
DOI: https://doi.org/10.1016/B978-0-12-818762-3.00010-8

10.1 Introduction

Producing hydrogen is based on the scientific principle of dissociation of water, where two molecules of water (H_2O) are separated into two molecules of hydrogen (H_2) and one molecule of oxygen (O_2). The dissociation of water is an endothermic reaction. The use of electricity for producing hydrogen is called electrolysis (Abbasi and Abbasi, 2011; Twiddel and Weir, 2015). Electrolysis is performed by applying a voltage to two electrodes that are submerged in water, to which an electrolyte is added. For electrolytes, high conductivity is important to reduce transport losses. This can be achieved by using strong acidic or strong alkaline electrolytes (Schalenbach et al., 2016a,b). Electrolysis technologies are usually categorized in low-temperature systems with alkaline cells or proton exchange membrane (PEM) cells and high-temperature systems with solid oxide cells. From these types the alkaline electrolyzer is the incumbent technology and currently the most mature and economically attractive technology (IEA, 2019). Common alkaline electrolyzers use potassium hydroxide as electrolyte. These electrolyzers do not require valuable or rare metals. Typically they use electrodes made of nickel (NREL, 2009).

An overview of hydrogen production by electrolysis at the cell level is shown in Fig. 10.1A and at the system level in Fig. 10.1B. Besides the electrolysis module, a typical system includes utilities such as demineralized water, cooling water, instrument air, hydrogen gas dryer, and purifier (NREL, 2009).

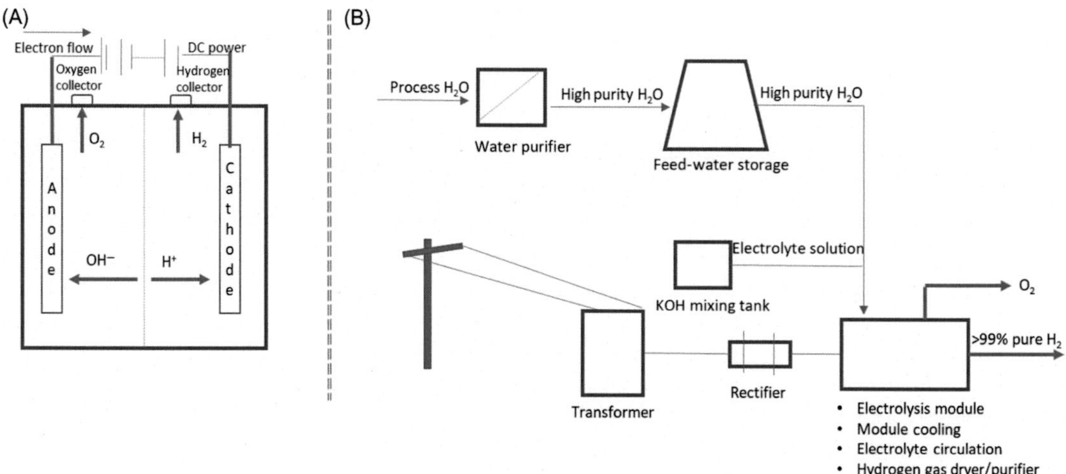

Figure 10.1
Schematic overview of alkaline electrolysis (A) and typical process for hydrogen production via electrolysis (B). Source: *Adapted from Santos et al. (2013); NREL (2009).*

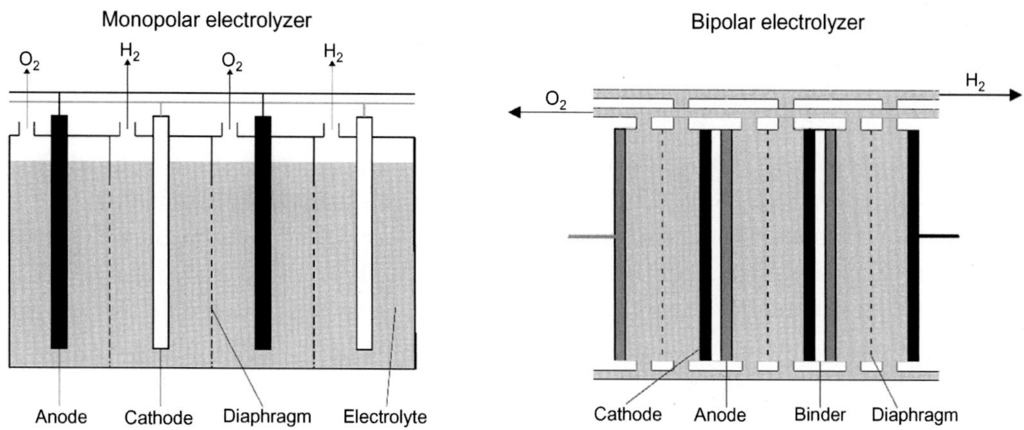

Figure 10.2
Schematic diagram of monopolar and bipolar electrolyzers. Source: *Courtesy of Farlie (2019).*

Water electrolysis was discovered in 1800 by Nicholson and Carlisle shortly after Volta invented the electric battery (Kroposki et al., 2006). Water electrolysis as an industrial process has been around since the advent of commercial power at the beginning of the 20th century. For most of the 20th century, there were two competing cell architectures: monopolar and bipolar, though bipolar is currently the dominant cell architecture.

In the monopolar electrolyzer, each cell is located in an individual tank of electrolyte and then connected to other cells by external bus bars, the term monopolar referring to each side of the cell having a single polarity being either an anode or a cathode (Fig. 10.2).

In the bipolar electrolyzer, cell stacks are assembled back to back similar to a filter press, and each side of the cell is bipolar, being a cathode on one face and an anode on the other. Although the first electrolyzers, coming out of the laboratory, were monopolar, they were closely followed by the Schmidt–Oerlikon cell, the first commercial bipolar electrolyzer, which was introduced in 1902 (Menth and Stucki, 1978). The electrolyzer industry grew substantially during the 1920s and 1930s with manufacturers Oerlikon, Norsk Hydro, and Cominco supplying plants in multimegawatt sizes. Most of these installations were near hydroelectric plants that supplied an inexpensive source of electricity (Kroposki et al., 2006).

Several alkaline electrolysis plants with capacity of up to 165 MW were built in the last century in countries with access to large hydropower resources—Canada, Egypt, India, Norway, and Zimbabwe (IEA, 2019). Commercialization of bipolar designs was largely led by Norsk Hydro that built the Rjukan plant rated at 125 MW prior to 1940. The world's largest operating electrolysis plants were built in the 1940s after World War II for ammonia production during the reconstruction of Europe (Christiansen and Andreassen, 1984) and later for ammonia plants in Africa including the hydrogen plant at Aswan, the world's largest water electrolysis plant with a nominal rating of 162 MW and a hydrogen generation capacity of 32,400 m^3/h (KIMA). Following these developments, electrolysis played a niche role in providing hydrogen in markets that were not economically serviced by merchant gas produced from fossil fuels. The introduction of liquid hydrogen, which came with the US space program in late 1950 (Sloop, 1959), further extended the reach of merchant gas supply and reduced market share for hydrogen supplied by electrolysis. This, in turn, reduced the average unit size additions of electrolyzers. With declining costs for renewable power (solar photovoltaic (PV) and wind), interest is now growing in water electrolysis for hydrogen production and in the scope for further conversion of that hydrogen into hydrogen-based fuels or feedstocks, such as synthetic hydrocarbons and ammonia, which are more compatible than hydrogen with existing infrastructure (IEA, 2019).

In countries with readily accessible renewable resources and significant dependence on natural gas imports, hydrogen produced from electrolysis connected to renewable electricity may be cheaper than producing it from natural gas. This statement is apparent in a country such as China. Currently, the share of electrolyzers being built in China is increasing (IEA, 2019). IEA's (2019) estimates that in China, for an alkaline electrolyzer with a CAPEX of 398$€_{2017}$/kW connected to hybrid onshore wind and solar PV systems running at 5000 load hours and with a electricity price of 15$€_{2017}$/MWh, the price of hydrogen ranges between 1.4 and 2.2$€_{2017}$/kgH_2. The prices for a similar system in Europe running at 5000 load hours with an electricity price of 41$€_{2017}$/MWh start from 2.3$€_{2017}$/kgH_2 and can exceed 3.5$€_{2017}$/kgH_2. The price of hydrogen from natural gas in China ranges between 1.9 and 2.5$€_{2017}$/kgH_2 (IEA, 2019). Currently, several projects are under development globally with electrolyzer sizes of 10 MW or more, with few projects aiming for 100 MW electrolyzer units (IEA, 2019).

10.1.1 Global hydrogen capacity

Currently, the global demand for pure hydrogen is around 70 Mt, which is used in oil refining and the ammonia production industry. An additional 45 Mt of hydrogen as part of a mixture of gases, such as synthesis gas, is used in methanol and steel production. Less than 0.001 Mt per year of pure hydrogen is used in fuel-cell electric vehicles, mostly derived from natural gas (IEA, 2019).

The majority of hydrogen produced today is derived from fossil fuels—mainly natural gas, and around 60% of it is produced in production facilities where hydrogen is the primary product. Other fossil fuel sources are coal, while a small fraction comes from renewable sources, namely, water electrolysis (IEA, 2019). One-third of global supply is "by-product" hydrogen, meaning that it comes from facilities and processes designed primarily to produce something else. This by-product hydrogen often needs dehydrating or other types of cleaning and can then be sent to a variety of hydrogen-using processes and facilities. Most hydrogen is currently produced near to its end use, using resources extracted in the same country (IEA, 2019).

Hydrogen can be obtained from fossil fuels and biomass, or from water, or from a mix of both.

Around 2% of the global total primary energy demand is used for the production of hydrogen today (IEA, 2019). Natural gas is currently the primary source of hydrogen production via steam methane reforming (SMR). A total of 6% of the global natural gas use is dedicated to a hydrogen production of around 52 Mt (IEA, 2019). The second dominant source is coal due to its dominant role in China. A total of 2% of the global coal use is dedicated for hydrogen production (16 Mt). Oil and electricity account for 1.4 Mt of the dedicated production of which electricity only accounts for 0.1% of the total dedicated hydrogen production which translates to around 0.07 Mt (IEA, 2019).

Besides fossil sources, hydrogen is also produced from renewable sources such as electrolysis. Less than 1% of dedicated hydrogen production globally comes from water electrolysis today, and the hydrogen produced by this means is mostly used in markets where high-purity hydrogen is necessary (e.g., electronics and polysilicon) (IEA, 2019). It is important to note that this share differs from shares of 4% reported by other authors (Bertuccioli et al., 2014; Rashid et al., 2015; IRENA, 2018). This discrepancy in share could be attributed to alkaline electrolysis being a global traded commodity, which creates an uncertainty in accounting for the total installed capacity. In addition to the hydrogen produced through water electrolysis, around 2% of total global hydrogen is created as a by-product of chloralkali electrolysis in the production of chlorine and caustic soda (IEA, 2019).

Producing all of today's global dedicated hydrogen output (70 MtH_2) from electricity would result in an electricity demand of 3600 terawatt hours (TWh), which roughly translates to 14% of the global electricity supply This is more than the total annual electricity generation of the European Union, which accounts for 13% of the global electricity supply (IEA, 2019).

Table 10.1 shows the performance of renewable hydrogen production technologies compared to conventional hydrogen production on small and large scale.

Alkaline electrolyzer is considered the most mature and economically attractive electrolyzer technology. However, it remains far more expensive than large-scale SMR. As PEM and SOEC electrolyzer technologies are less developed, they are considered to have greater potential for cost reductions and technology improvements (OECD/IEA, 2015). Furthermore, PEM are better suitable for electrolysis with intermittent electricity sources, such as wind or solar (OECD/IEA, 2015; Götz et al., 2016), due to its flexibility with

Table 10.1: Overview of hydrogen production methods.

Application	Power or capacity	Efficiency	Initial investment	Lifetime	Maturity
Steam methane reformer (large scale)	150–300 MW	70%–85% (LHV)	~805€$_{2017}$/kW	30 years	Mature
Steam methane reformer with carbon capture, utilization and storage (CCUS)	150–300 MW	65%–70% (LHV)	~1500€$_{2017}$/kW	25 years	Demonstration
Alkaline electrolyzer	Up to 150 MW	63%–70% (LHV)	440–1200€$_{2017}$/kW	60,000–90,000 h	Mature
PEM electrolyzer	0.5 MW 12 MW	50%–60% (LHV)	970–1600€$_{2017}$/kW	30,000–90,000 h	Early market
Solid oxide electrolyser cell (SOEC) electrolyzer	Lab scale	74%–81% (LHV)	~2500–5000€$_{2017}$/kW	10,000–30,000 h	R&D

CAPEX for electrolysis technology represents system costs including power electronics, gas conditioning, and balance of plant. CAPEX ranges reflect different system sizes. *LHV*, Lower heating value.
Source: *Adapted from IEA (2019); IEA Global Trends and Outlook for Hydrogen (2017).*

dynamic operation. PEM's flexibility toward dynamic operation is detailed in Section 10.4. Such electrolysis units are currently reaching demonstration scale and have not yet seen large deployments (REN21, 2017). Technology advancements are needed before being ready for widespread commercialization (IRENA, 2018).

10.2 Data availability and methodological issues

Data availability on CAPEX and capacity of alkaline electrolysis systems is highly limited, mostly attributed to the privacy suppliers place on revealing cost data. Therefore capacity and CAPEX data for the years 1956–2016 were gathered from a few sources: Braun (1978), Christiansen and Grundt (1978), Costa and Grimes (1967), Evangelista et al. (1975), NEL Hydrogen (2015a,b), Next Hydrogen Corporation (2016), Saba et al. (2017), Schmidt et al. (2017a,b), Schoots et al. (2008), and Tilak et al. (1981). Monetary data from these sources are referred to as CAPEX, but due to the lack of transparency regarding the link between CAPEX and cost or prices, it is safe to assume that the CAPEX data represents prices due to the difficulty of acquiring true manufacturing costs of alkaline electrolysis systems. Using prices as proxy for costs leads to the assumption of perfect market competition, which is not the case for alkaline electrolysis systems. There are few major manufacturers of alkaline electrolysis systems: Hydrogenics, Hydrotechnik GmbH, McPhy, NEL Hydrogen, Next Hydrogen, PERIC, Teledyne Energy Systems, and Tianjin Mainland Hydrogen Equipment. Having a few major manufacturers for a globally traded commodity indicates toward an oligopoly and not perfect market competition. Saba et al. (2017) gathered CAPEX data for alkaline electrolyzers through literature review for the past 30 years. Most of the CAPEX data collected were based on expert estimations, and a few were based on manufacturers' CAPEX. These estimations are based on electrolysis market developments and on scenarios proposed at the time of estimation. These scenarios may not always reflect market development of alkaline electrolyzers. Most of the large-scale plants were build prior to 1980 (Saba et al., 2017), while currently small-scale electrolyzers (1−10 MW) are being sold (IEA, 2019) with a great share of the manufactured alkaline electrolyzers coming from China, which probably indicates that the market is/has shifted/ shifting toward China. In combination with alkaline electrolyzers being a global traded commodity, this creates an uncertainty in accounting for the total installed capacity of alkaline electrolyzers and therefore makes it difficult to calculate the total cumulative global installed capacity of alkaline electrolysis. An issue regarding the CAPEX reported by the literature sources is the lack of transparency pertaining to the components these CAPEX data comprised. All of the gathered data analyze alkaline electrolyzer *systems* and not only the *stacks*. Some of the literature reviewed by Saba et al. provide CAPEX data that covers all plant equipment, from the rectifier to the gas holder, building of plant, office buildings, miscellaneous, and interest during construction while in other literature sources, the specifics of CAPEX are not mentioned. The same can be said for other sources such as Schmidt et al. (2017a,b) and Tilak et al (1981). Some sources do not specify if the H_2 is produced at ambient pressure or pressurized. The choice of investing in an ambience of pressurized plant can have a significant effect on the CAPEX of the electrolyzer. The CAPEX of pressurized plants (with and without installations and commissioning) is usually 20%−34% higher than for ambient systems (NEL, 2015; Smolinka et al., 2010). Therefore due to the discrepancies in the composition of acquired CAPEX data, the CAPEX data points represented in the experience curve (Fig. 10.3) are not explicitly defined, but the data points generally encompass the price of the module and the inclusion/exclusion of other system components such as the gas holder, rectifier, compressor, office buildings, and other miscellaneous capital costs. The CAPEX data is represented as €/kW_{input}.

Methodological issues can be attributed to the type of data available. CAPEX and capacity data from Saba et al. were provided in $\$_{base\ year}/kW_{input/output}$ and in N m^3/h or kg/day or MW, respectively. This data was then converted to a standard unit of $€_{2017}/kW_{input}$ for CAPEX and MW for installed capacity. Capacity data in N m^3/h was first converted into kg/h (1 N $m^3H_2 = 0.0899$ kgH_2) and then converted to MW using the higher heating value (HHV) of hydrogen ($HHV_{H_2} = 141.88$ MJ/ kg). CAPEX data in $\$_{base\ year}/kW_{input}$ was first

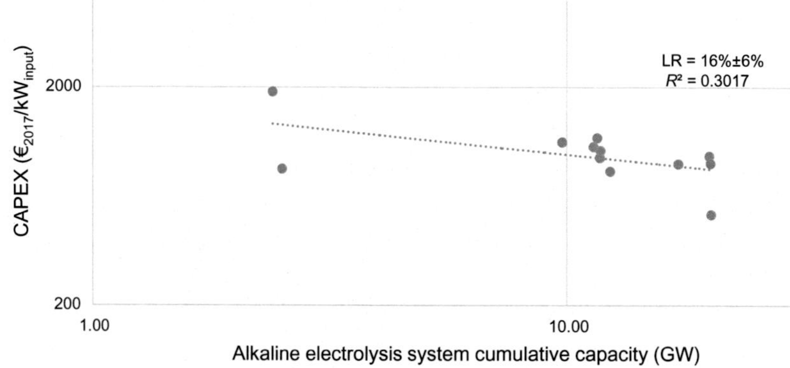

Figure 10.3
Global experience curve for the CAPEX of alkaline electrolysis systems from 1956 to 2016. Source: *Own data collection.*

corrected for inflation to $\$_{2017}$/kW$_{input}$ using the consumer price index from OECD, and a dollar to euro exchange rate from Eurostat was used to convert $\$_{2017}$/kW$_{input}$ into €$_{2017}$/kW$_{input}$. This method was also used to convert data from NEL Hydrogen and Next Hydrogen to the desired units (€$_{2017}$/kW$_{input}$ and MW). Another method would involve the conversion of the dollar value to euro in the given year and then corrected for inflation for the base year. This method was not chosen since a significant amount of data points exist prior to 1999 (introduction of the euro) and therefore circumvents the issue regarding the conversion of dollars to euros prior to the introduction of the euro. A third method would be to identify the most stable currency in the EU prior to the introduction of the euro and convert the dollar value to that currency and correct for inflation till the year 1999 and then use the euro inflation rate post 1999. This method was discarded since it is difficult to identify and provide sound reasoning for the choice of the country. All CAPEX data from Braun (1978), Christiansen and Grundt (1978), Evangelista et al. (1975), Costa and Grimes (1967), and Tilak et al. (1981) and some CAPEX data from Saba et al. (2017), Schmidt et al. (2017a,b), and Schoots et al. (2008) were provided in $/kW for years prior to the introduction of the euro (the euro was introduced in 1999). Therefore there exists an uncertainty in the dollar to euro exchange rate and euro inflation rate prior to 1999.

Cumulative installed capacity data derived from Schmidt et al. (2017a) was provided in kWh (storage). This data was reproduced as kW using a C-rate (power to energy ratio) defined by Schmidt et al. (1 kW = 10 kWh). A different C-rate (or a C-rate changing over time) would affect the cumulative installed capacity, thereby affecting the learning rate. Here, CAPEX data was provided in $\$_{2015}$/kWh and was converted to €$_{2017}$/kW$_{input}$ using the currency conversion method detailed in the previous paragraph and using the C-rate provided by Schmidt et al. (2017a).

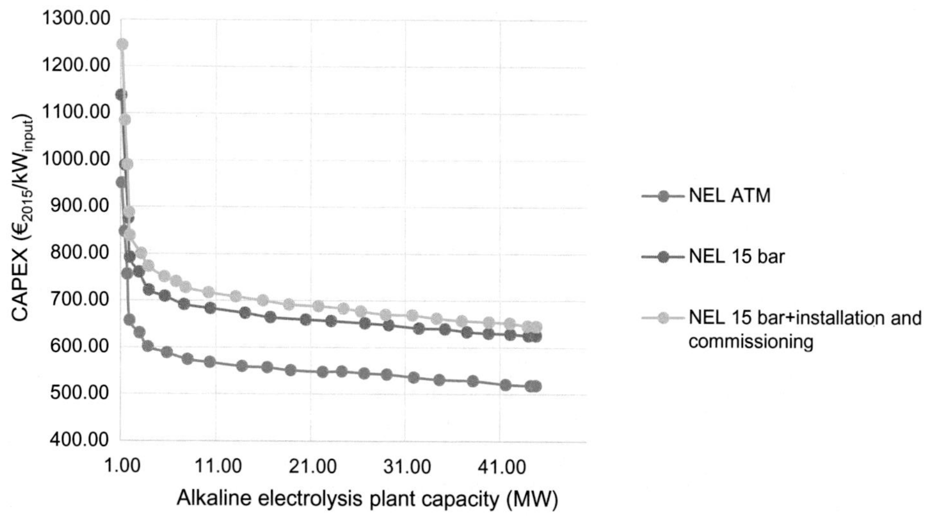

Figure 10.4
Development of CAPEX costs for large alkaline electrolyzer systems over plant capacity (1–44 MW) according to NEL. The graph shows the inclusion of pressurized systems and installations and commissioning on the CAPEX of alkaline electrolysis systems. Source: *Adapted from Saba et al. (2017).*

The CAPEX data collected from the various sources correspond to capacities that range from 1 to 100 MW. This data was not normalized to a specific capacity in order to facilitate observations on the underlying reasons for different CAPEX reported for the same capacity installed in the same year. The CAPEX for larger systems are usually lower than small systems (eg: Fig. 10.4) and therefore renders the poor distribution of the CAPEX data in Fig. 10.5.

In order to know the true cost of hydrogen, it is important to know the learning rate for the levelized cost of hydrogen ($LCOH_2$), which includes CAPEX, OPEX, and electricity/fuel cost, as it provides a better indication of the true cost of hydrogen production from alkaline electrolysis and provides a sound basis for comparison with SMR, which is the dominant technology in hydrogen production. Fig. 10.6 provides the share of CAPEX, OPEX, and electricity cost for alkaline electrolysis.

For large multimegawatt units the dominant single input in the cost of hydrogen beyond electricity is the capital cost of plant equipment (Kuckshinrichs et al., 2017). With increasing emphasis being placed on utilizing low-cost intermittent power sources, such as wind and solar, the cost of plant will play an even larger role—for two reasons: (1) electricity's share of cost is reduced and (2) with the decrease in utilization due to the intermittent nature of renewable resources, the share of CAPEX to total cost will increase.

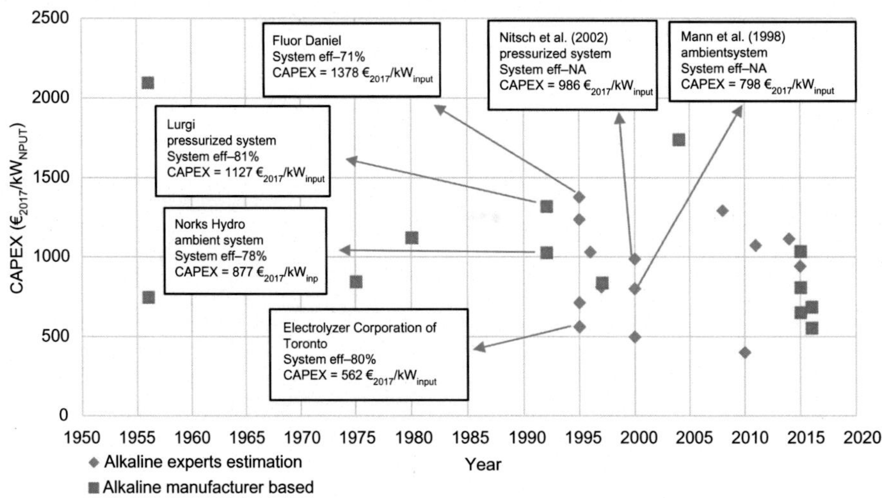

Figure 10.5

CAPEX of alkaline electrolysis systems plotted against time.

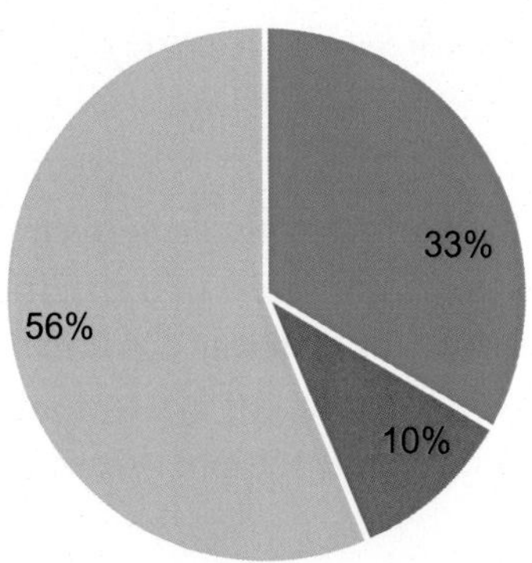

Figure 10.6

Cost breakdown of levelized cost of hydrogen ($LCOH_2$) of alkaline electrolysis. Source: *Adapted and modified from IEA (2019); Thomas (2018).*

Using the data presented in the material handling study conducted by Felgenhauer and Hamacher (2015), and assuming plant capacity of 50 kg/h, the contribution of capital cost to total cost is projected to grow from around 15% under the conditions presented (100% utilization) to 35% when the cost of power is reduced by 50% and utilization is reduced by the same amount.

10.3 Results

10.3.1 Experience curve

Fig. 10.3 shows the global experience curve for alkaline electrolysis systems. It shows the CAPEX (€$_{2017}$/kW$_{input}$) of alkaline electrolysis systems over the time frame 1956–2016. Over this period a learning rate of 16% ± 6% was established using a confidence interval of 95%. Within this dataset, the CAPEX declined from 1900€$_{2017}$/kW$_{input}$ in 1956 to a range of 500–900€$_{2017}$/kW$_{input}$ in 2016. This spread-in in the current CAPEX can be attributed to the wide range in capacity (1–100 MW). With a wide CAPEX range and cumulative capacity being in the order of 100–200 MW, the uncertainty overwhelms the expected learning in the coming decade. Another reason for the wide spread-in CAPEX is the components used in the electrolyzer system. These include unipolar or bipolar modules, pressurized or ambient systems, gas conditioning, compressor, power electronics, operating conditions, and stack configuration (Saba et al., 2017; Kuckshinrichs et al., 2017). These components can have an effect by either increasing or driving down the CAPEX of the alkaline electrolysis system. As mentioned in Section 10.2, pressurized systems usually increase the CAPEX of alkaline electrolysis systems by 20%–34%.

Fig. 10.3 shows a sharp increase in cumulative installed capacity of alkaline electrolysis systems from 2 to 9 GW. Cumulative installed capacity of 2 GW can be attributed to capacity installation between the 1920s and 1950s, which correlates with the need for hydrogen in the ammonia industry in the 1940s. The decline in installed capacity between the 1950s and 1970s can be attributed to the competition with cheap hydrogen produced from SMR due to the development of natural gas production. Between the late 1970s and the mid-1980s, with concerns about security of supply of fossil fuels, significant development of bipolar electrolyzer technology was undertaken and hydrogen production from electrolysis increased. Table 10.2 shows a list of the major large-scale alkaline electrolyzer plants built between the 1920s and 1980s.

After 1990, with cheap natural gas prices, the hydrogen cost from SMR was much cheaper than hydrogen from alkaline electrolysis (IEA, 2019). Therefore since 1990, it can be

Table 10.2: Major large-scale plants between 1920 and 1980.

Plant location	Country	Supplier	Capacity (MW)	Year of construction
Rjunkan	Norway	Norsk Hydro (now NEL)	165	1927
Glomfjord	Norway	Norsk Hydro (now NEL)	135	1953
Trail	Canada	Cominco	90	1939
Nagal	India	Denora	125	1958
Aaswan	Egypt	BBC	162	1977

Source: *Adapted and modified from Smolinka et al. (2010); Sunde (2012).*

assumed that this cost difference led to the scaling down of alkaline electrolysis with most producers selling between 1 and 10 MW electrolyzer systems in the order of a few dozen per year and a few large-scale electrolyzer systems sold in the order of one to two per year (NEL Hydrogen, 2015a; Saba et al., 2017; IEA, 2019). The smaller plants play a niche role in providing limited amounts of hydrogen when SMR is not possible.

Therefore based on the wide spread of CAPEX, cumulative capacity being in the order of 100–200 MW and the different components that drive up or down the CAPEX of alkaline electrolysis systems, the poor R^2 of 0.3017 can be explained.

Fig. 10.5 represents the same data as Fig. 10.3 and shows the CAPEX of alkaline electrolysis systems plotted against time. It shows CAPEX derived from expert estimations and manufacturers' prices. It shows that the CAPEX of alkaline electrolysis systems, estimated by experts, falls in the range of manufacturers' prices for the years 1996 and 2015, which aids in adding validity to the expert estimates. From this graph, it can be observed that CAPEX of alkaline electrolysis systems shows a decline from a range of 2100–750$€_{2017}$/kW_{input} (average of 1900$€_{2017}$/kW_{input}) in 1956 to a range of 900–500$€_{2017}$/kW_{input} (average of 700$€_{2017}$/kW_{input}) in 2016, which correlates with the CAPEX decline seen in the experience curve (Fig. 10.3). The shape of the data plot in Fig. 10.5 is slightly wedge-shaped, a consequence of the wide spread of CAPEX and corresponding capacity data acquired from expert estimates and manufacturers' prices. The CAPEX in 2016 was 63%–73% lower than the CAPEX in 1956. The cost reduction in the past decade was more moderate. This could be explained with the maturity of alkaline electrolysis technology (Saba et al., 2017) since alkaline electrolysis systems have been available in the MW scale since the 1950s. Drivers for these cost reductions are detailed in Section 10.3.2.

Though the graph shows a partial decline, the wide spread of CAPEX and capacity data points within the same year adds to the uncertainty of predicting the trend of CAPEX for alkaline electrolysis systems and therefore bolsters the uncertainty of generating a reliable learning rate for the alkaline electrolysis system.

Fig. 10.5 also shows the CAPEX of three pairs of alkaline electrolysis systems of the same size and built in the same year. Further on, we illustrate the underlying reasons for the variation in cost: In 1992 the graph shows two 100 MW bipolar plants built by Norsk Hydro (efficiency—78%, ambient pressure) and Lurgi (efficiency—81%, 31 bar) with a CAPEX of 877$€_{2017}$/kW_{input} and 1127$€_{2017}$/kW_{input}, respectively. The 28% increase in the Lurgi plant CAPEX can be attributed to the difference in pressure (ambient vs 31 bar) and efficiency (78% vs 81%). This 28% CAPEX increase falls within the range (20%–34%) reported by Smolinka et al. (2010) and NEL (2015).

For the year 1995, Thomas and Kuhn (1995) reported a CAPEX of 1378$€_{2017}$/kW_{input} estimated by Fluor Daniel, Inc. in 1991 (based on forecast) and a CAPEX of 562$€_{2017}$/kW_{input}

estimated by the Electrolyser Corporation of Toronto (estimation made in 1995), both for 100 MW systems. The CAPEX estimated by the Electrolyser Corporation of Toronto excludes components, such as the compressor, storage tank facilities, and engineering, while these components are included in the estimation made by Fluor Daniel, Inc., who report that compressor, storage tank, and engineering account for 36% of the CAPEX. Therefore the high CAPEX reported by Fluor Daniel, Inc. can be attributed to the inclusion of the previously mentioned components and the year the estimation was made.

For the year 2000, Mann et al. (1998) and Nitsch et al. (2002) reported the CAPEX of two 2 MW systems at 798€$_{2017}$/kW$_{input}$ (based on forecast) and 986€$_{2017}$/kW$_{input}$ (estimation based on historical data). The CAPEX estimated by Nitsch et al. (2002) is 23% higher, which can likely be attributed to the different pressure conditions [ambient (Mann et al., 1998) vs 30 bar (Nitsch et al., 2002)], and the year of estimation [1998 (Mann et al., 1998) vs 2002 (Nitsch et al., 2002)].

Therefore the wide range in CAPEX for the same capacity can be attributed to the dynamics of the system and inclusion/exclusion of different components. These include choices made on the type of cells (unipolar or bipolar), whether the system is pressurized or ambient, and the inclusion of components, such as compressor, gas rectifier, storage tank, and other engineering and balance of plant (BOP) costs in the CAPEX breakdown. Therefore it can be concluded that the CAPEX has a significant dependency on the dynamics and components that comprise the system and therefore generating experience curves for the entire system is not the soundest method to apply as it leads to significant uncertainty.

Fig. 10.4 retrieved for Saba et al. (2017), which was derived from NEL (2015), shows the CAPEX of large-scale alkaline electrolyzer systems from 1 to 41 MW capacity. Information on the methodology used by NEL to derive this curve was not available. It can be assumed that probably a fixed cost for a 250 kW system was used as the base line. The plot was then extrapolated to 41 MW plant capacity assuming scaling effects. Therefore the data may not represent the true cost of alkaline electrolysis systems at their respective capacity. NEL currently only produces ambient systems, a decision which could be based on its low costs as seen in Fig. 10.4. The three trend lines show CAPEX of the system at ambient, pressurized, and pressurized with installation and commissioning. Pressurized alkaline electrolysis systems show around a 20% increase in CAPEX when compared to ambient systems and a further 28% increase when installations and commissioning are included in the CAPEX. Therefore it can be assumed that for an ambient system, installation and commissioning will drive up the CAPEX by 8%. This finding affirms the earlier statement made that CAPEX of alkaline electrolysis systems significantly depends on the dynamics and components that surround the alkaline electrolyzer stacks. The graph also shows a steep decline of more than 60% when plant size is increased from 250 kW$_{input}$ to 2.5 MW$_{input}$ and then approaches a value of 516€$_{2017}$/kW$_{input}$ (Saba et al., 2017). These sharp reductions in CAPEX in the small capacity range can be attributed to the peripheral costs, which are independent of the capacity (Smolinka et al., 2010).

From the analysis, it can be concluded that the CAPEX of alkaline electrolysis system for a given capacity depends on the type of cells (unipolar or bipolar); pressurized or ambient; inclusion/exclusion of gas conditioning, power electronics, compressor, and other installation costs and operating conditions, which in turn determines the choice of materials used and the cost components included in CAPEX calculation. These added components along with the uncertainty surrounding the composition of the cost breakdown of alkaline electrolysis system data; it can be deduced that generating experience curves for the entire alkaline electrolysis system can lead to much uncertainty.

From this analysis, it can be recommended to study the learning rate of the electrolyzer stacks and other components, such as the rectifier and gas chamber, separately and analyze if a decline in CAPEX is a result of scaling up the technology or mass fabrication and trying to achieve optimal configuration of the electrolyzer stacks via multistacking (e.g., one 40 MW stack or twenty 2 MW stacks) (IEA, 2019). So far, however, lack of publicly available data has prevented a more detailed analysis of the individual electrolyzer components.

10.3.2 Drivers

Early developments that led to cost reduction in alkaline electrolyzers can be linked to developments in cell architecture.

Monopolar cells used steel or cement tanks and heavy metal electrodes connected by copper bus bar and were favored for their reliability and ease of maintenance, it being easy to replace a cell, but were generally of lower efficiency compared to bipolar designs. Monopolar electrolysis plants need lower voltage and higher current configurations, which result in significantly higher rectification costs. Advanced monopolar designs were commercialized in 1980 (Crawford and Benzimra, 1986), but bipolar stack design ultimately prevailed as bipolar stack costs came down, and emphasis was placed on packaging plants in shipping containers (Fairlie and Scott, 2000). Today, almost all electrolyzers have a bipolar stack design.

The commercialization of bipolar designs in the 1940s (Christiansen and Andreassen, 1984) and significant developments of the design in the late 1970s and early 1980s, with concerns about security of supply of fossil fuels, were undertaken as part of a large initiative to position hydrogen as an alternative transportation fuel (Vandenborre et al., 1982). Employing more efficient electrode materials and separators, significant gains in efficiency were made with current density increasing from 0.15–0.2 to 0.2–0.4 A/cm^2

(Schmidt et al., 2017a,b). These developments led to smaller, more compact systems and the introduction of packaged or containerized plants, which were cheaper to site and erect. Membrane materials used to keep hydrogen and oxygen separate in the cell evolved from the use of asbestos to Ryton PPS to current separators based on inorganic chemistry (Vermeiren et al., 1998).

The technical development of alkaline electrolysis over the past 120 years has largely focused on increasing current density and a general improvement in cell efficiency, largely made possible with improved electrode materials and cell membranes. These improvements have led to more compact plants requiring fewer cell parts and lower stack costs (LeRoy and Stuart, 1978; Schmidt et al., 2017a,b). BOP costs have also decreased and this can be attributed to developments in cells working at near atmospheric pressure that are controlled using water seals and gasholders (LeRoy and Stuart, 1978). These have evolved with improved materials and sealing technology to pressurized cell designs, lowering downstream gas conditioning and compression costs (Schmidt et al., 2017a,b).

Ursua et al. (2012) reports that most price reductions can be attributed to technological improvements at a modular level since all electrolyzers consist of an electrolyzer stack, comprising up to 100 cells, and the BOP. Stacks can be mounted in parallel using the same BOP infrastructure, which is why electrolyzers are highly modular systems. Some of the improvements mentioned are as follows:

- Minimizing the space between the electrodes in order to reduce the ohmic losses thus making it possible to work with higher current densities. Currently, distances among the electrodes less than 1 mm are typical, which is referred to as zero-gap configuration.
- Development of new advanced materials to be used as diaphragms replacing the previous ones made of asbestos. In this regard the use of porous separators, made of polysulfide possibly combined with ZrO_2 (zircon) has become widespread.
- Development of high-temperature alkaline water electrolyzers: working temperatures up to 150°C increase the electrolyte conductivity and promote the kinetics of the electrochemical reactions on the electrode surface, although this has not yet entered the market due to issues related to corrosion.
- Development of advanced electrocatalytic materials to reduce the electrode over voltages.

10.4 Future developments

10.4.1 Current and future markets

Section 10.2 highlights the effect capacity factor has on the levelized cost of hydrogen. The future competitiveness of low-carbon hydrogen produced from natural gas with CCUS or from renewable electricity (from solar PV or onshore wind) thus mainly depends on gas and electricity prices. At low gas prices, renewable electricity must reach a cost range less than 8.7€$_{2017}$/MWh for electrolysis to become cost-competitive with natural gas with CCUS. Higher gas prices would make higher cost renewable electricity cost-competitive: at a gas price of 33€$_{2017}$/MWh, renewable electricity would be competitive at up to around 26–39€$_{2017}$/MWh (IEA, 2019).

Increasing shares of variable renewables in grid electricity leads to intermittency and surplus electricity, which may be available at low cost and can be used to produce hydrogen, although a low availability of surplus electricity is unlikely to have a major impact in driving down costs. Running the electrolyzer at high full-load hours and paying for the additional electricity is maybe cheaper than running with low full-load hours (IEA, 2019). The relationship between electricity costs and operating hours becomes apparent when looking at electrolyzers that use grid electricity for hydrogen production. Very low-cost electricity is generally available only for a limited number of hours within a year, which implies low capacity factor and high hydrogen costs, which reflect CAPEX costs (IEA, 2019). With increasing load hours, electricity costs increase, but the high capacity factor leads to a decline in the cost of producing a unit of hydrogen up to an optimum level at around 3000–6000 equivalent full-load hours (IEA, 2019). Beyond that, higher electricity prices during peak hours lead to an increase in hydrogen unit production costs.

Dedicated electricity generation from renewables offers an alternative to the use of grid electricity for hydrogen production. With declining costs for solar PV and wind generation, building electrolyzers at locations with excellent renewable resource conditions could become a low-cost supply option for hydrogen, even after taking into account the transmission and distribution costs of transporting hydrogen from (often remote) renewable locations to the end users (IEA, 2019)

Historically, alkaline electrolyzers were connected to hydropower dams (Kroposki et al., 2006) and could run on constant mode. In more recent years alkaline electrolyzers have been connected to the grid with a share of renewables or to dedicated renewable electricity supply (IEA, 2019). This could lead to issues of intermittency and therefore requires the electrolysis system to be flexible by accommodating dynamic operation. Current alkaline electrolysis systems do not possess the flexibility to run at dynamic operation. They typically operate at 20%–100% rated power (Lehner et al., 2014). Operating in the lower half of the range results in significantly reduced gas quality and increasingly reduced system efficiency (Lehner et al., 2014). Alkaline electrolysis systems tend to have long startup times and have difficulty to adjust to rapid changes in power inputs (Lehner et al., 2014). PEM electrolyzers, on the other hand, offer flexible operation since they can operate from 0% to 160% rated power (IEA, 2019) and are able to respond to power fluctuations within 100 ms (Lehner et al., 2014). But the downside to this is that PEM electrolyzers require expensive electrode catalysts such as platinum or iridium and membrane materials. They also have shorter lifetimes than alkaline electrolysis and are more expensive (Schmidt et al., 2017b; IEA, 2019). With an increasing share of renewables, and the assumption that PEM electrolyzers will undergo cost reductions (Schmidt et al., 2017b; IRENA, 2018; IEA, 2019), the potential of PEM electrolyzers in the future might be promising and therefore could be interesting to generate and assess learning curves for PEM.

Asahi Kasei Corporation produces hydrogen via chloralkali process, which is a form of electrolysis (brine) and is powered by electricity. Currently, up to 0.5 MtH_2 worldwide is vented to the air and 22 MtH_2 is used for low-value applications, such as heat and power generation, without purification (IEA, 2019). This potential, in addition to hydrogen from alkaline electrolysis, could be tapped and used as feedstock in the iron and steel and chemical industry. But chloralkali is not as attractive as electrolytic hydrogen since it is more expensive and is only available as a by-product, therefore not being able to meet global hydrogen demand.

Due to China's dependency on imported natural gas and abundant availability of renewable resources, hydrogen production from electrolysis is gathering attraction. In some places in China, renewable energy has been deployed at a rapid rate that electricity networks have had difficulty adapting in real time (IEA, 2019). This has provided an opportunity for producers of hydrogen and hydrogen-rich chemicals to tap into renewable resources. China is the world's largest user of nitrogen fertilizers, consuming 46 Mt/year (IEA, 2019). This provides further opportunity for China to tap into the electrolysis market for hydrogen production, which is a feedstock for the ammonia industry. IEA conducted a detailed economic assessment on hydrogen production from electrolysis, based on hourly solar and wind data over a year in five locations across different provinces—Xinjiang, Qinghai, Tibet, Hebei, and Fujian (IEA, 2019). The assessment suggests that hydrogen can be produced at a cost of $1.7-2€_{2017}/kgH_2$, which is much cheaper than costs in Europe ($>4€_{2017}/kgH_2$) (IEA, 2019). In some provinces the lowest production costs are reached by using only solar (Qinghai) or wind (Hebei and Fujian), while in Xinjiang and Tibet, performance is best with a combination of the two (IEA, 2019). Though curtailment in China is declining, there still exists a large curtailment potential. In 2017 over 100 TWh of solar, wind, and hydropower output were curtailed, which is roughly equivalent to electricity consumption in The Netherlands (IEA, 2019). Installing electrolyzer plants in regions with excess renewable electricity potential could aid in the growth of cheap hydrogen produced from electrolysis. China's major source of primary energy is coal. In Northeast China, inflexibility of coal plants caused by heat loads was a factor in the curtailment of 40 TWh of wind power in 2017 (IEA, 2019). Until other sources of energy are implemented to meet this heat demand, electrolytic hydrogen at full-load hours and cheap electricity prices could potentially be used to avoid curtailment of either coal or renewable energy. These opportunities will probably aid in China becoming the dominant market for electrolysis system production.

Accounting for the assumption that in the future, low-cost dedicated renewable electricity will be available at a level to ensure the electrolyzer can operate at relatively high full-load hours is therefore essential for the production of low-cost hydrogen. This also highlights that in the future, CAPEX will have a major share in the cost/kg of hydrogen produced from electrolysis.

10.4.2 Future drivers in CAPEX reduction

The two factors that will drive future cost reduction of electrolyzers are likely: (1) scale of manufacturing and (2) incremental future innovation. Manufacturing cost reductions can be achieved by increasing the scale of production by moving to assembly line manufacturing, investing in tooling, such as injection molding of parts, material substitution, and ultimately automation.

Incremental innovation will be led by increasing cell current density, which has the effect of reducing the number of parts needed, and changes in architecture, to facilitate scale-up and reduction of BOP costs. Advanced alkaline designs, introduced by manufacturers, embody a number of these advances, including scaling up production per unit with single stacks up to 10 MW (Asahi Kasei Corporation), multistacking, which is shown to be cheaper than single large-scale stacks (IEA, 2019), internal gas–liquid separators/gas coolers that avoid need for external vessels, and electrolyte circulation that relies on gas-lift circulation rather than external pumps (Next Hydrogen Corporation, 2016) and pressurization (up to 35 bar) that reduces downstream gas drying and compression costs (Green Hydrogen, 2019). The need to increase operating flexibility to take advantage of lower electricity costs and to provide ancillary services to the grid, such as grid regulation services, will also have an impact on cell design and stack architecture. Development of RuO/IrO-based catalysts with improved kinetics can achieve current densities up to 1 amp/cm^2 increasing the operating range and ultimately reducing stack costs (Schmidt et al., 2017a,b), although it is important to note that Ir is one of the rarest elements in the Earth's crust, with annual production averaging around 3 million tonnes (USGS Platinum Group Statistics, 2019). Therefore large-scale use of Ir could drive up the cost in the future. Other developments, such as moving toward zero-gap membrane electrode assemblies using thinner separators and microporous electrodes, are projected to achieve better efficiency than those of PEM electrolyzers with acidic Nafion membranes (Schalenbach et al., 2016a,b).

10.4.3 Future CAPEX of alkaline electrolysis systems

The current electrolysis annual market size, estimated to be 100 MW (Hebling et al., 2018), is expected to grow at a 6% compound annual growth rate (CAGR) during the period 2017–23 (Water Electrolysis Market, 2018) with the European Union and a major consortium of car companies and energy companies expecting electrolysis to expand into a multibillion dollar industry by 2030 (Hydrogen Road Map Europe, 2019). At 6% yearly growth rate between 2017 and 2023, the cumulative capacity additions will reach around only 1 GW by 2023. Extrapolating the experience curve from Fig. 10.3, using this growth rate, provides a cumulative capacity of 20.01 GW with an average CAPEX of $692€_{2017}/kW_{input}$. A recent survey of manufacturers and industry experts showed that industry could ramp up capacity to 2 GW/year by 2020 (Hebling et al., 2018). But with hydrogen from SMR being the most inexpensive ($1.3€_{2017}/kgH_2$) (IEA, 2019) production route due to cheap gas prices, this assumed ramp-up capacity seems too optimistic.

According to the expert elicitation method used by Schmidt et al. (2017b), increasing R&D support alone is seen to have a small effect on reducing capital cost, but R&D support along with production scale-up can lead to a 27% drive-down of CAPEX by 2030. This drop is mostly attributed to production scale-up through improved manufacturing methods and automation as well as increased operational experience leading to optimized system design (new electrocoating methods, increased current density, and more stable electrodes for high-temperature operating conditions) (Schmidt et al., 2017b). A 27% drop in CAPEX by 2030 from current levels translates to an average CAPEX of around $500€_{2017}/kW_{input}$, which is lower than the average CAPEX ($600€_{2017}/kW_{input}$) quoted by Saba et al. (2017), which was based on expert estimates. Extrapolating the experience curve in Fig. 10.3 requires a cumulative installed capacity of 80 GW by 2030 to achieve $500€_{2017}/kW_{input}$. This roughly translates to two doublings, or a capacity addition of 60 GW between 2023 and 2030 to achieve this cost reduction. The projected cost reported by the IEA (2019) lies within a range of $750-350€_{2017}/kW_{input}$. These findings highlight that increasing R&D support alone is likely going to have a small effect on reducing capital cost. Arguably, the benefits of these advances are hidden in the metric of cost/kW, which does not account for increases in efficiency, and reduction in BOP costs, which will impact the total cost of hydrogen. It also highlights that scaling up, mass fabrication, and multistacking will play major roles to drive down the CAPEX of alkaline electrolysis systems

10.5 Summary and conclusion

The experience curve for alkaline electrolysis system between 1956 and 2016 shows a learning rate of $16\% \pm 6\%$ with CAPEX decreasing from 2100 to $750€_{2017}/kW_{input}$ in 1956 to a range between 900 and $500€_{2017}/kW_{input}$ in 2016. These values fall in the range ($440-1200€_{2017}/kW_{input}$) reported by the IEA's (2019). The observed reductions in CAPEX can be attributed to developments in cell architecture, increasing current density, and improving cell efficiency. The experience curve for alkaline electrolysis systems shows a poor fit ($R^2 = 0.307$), which can be attributed to discrepancies in the CAPEX composition of the gathered data and the wide spread in capacity (1–100 MW). Systems with the same capacity and manufactured in the same year have different CAPEX, which can be explained by

1. The inclusion/exclusion of CAPEX of other system components, such as rectifier, dryer, compressors, and other engineering costs, which could account for around a 36% increase in CAPEX.
2. The pressure at which the system operates with pressurized systems being 20%–34% more expensive than ambient systems.
3. Installation and commissioning, which could drive up the CAPEX by 8%.

Engineering and installation could be used as interchangeable terms and therefore overlap might exist in the % increase in CAPEX presented in (1) and (3).

It can therefore be recommended to avoid extrapolating the learning curve of an alkaline electrolysis system to estimate the future costs of stacks and the future costs of the system. It is recommended to generate experience curves for each component that makes up the system to know the true CAPEX and learning rate of alkaline electrolysis systems. With increasing share of renewable electricity and intermittency, PEM, due to its flexibility with dynamic operation, might play an important role in hydrogen production in the future. Similarly to alkaline electrolysis system, it is recommended to generate experience curves for PEM systems on a component level (stack, gas dryer, compressor, etc.). The two factors that will drive future cost reduction of electrolyzers will be scale of manufacturing by increasing the scale of production by moving to assembly line manufacturing and investing in tooling, such as injection molding of parts, material substitution, and ultimately automation and incremental future innovation. With the current electrolysis market estimated to be 100 MW/year and is expected to grow at 6% CAGR during the period 2017–23, the CAPEX of alkaline electrolysis systems will drop to around $692€_{2017}/kW_{input}$ at a cumulative capacity of 20.01 GW. Though Schmidt et al. (2017b), through expert elicitation process, estimated a 27% drop in CAPEX by 2030 due to R&D and production scale-up. This translates to a cumulative capacity addition of 80 GW with CAPEX of the system around $500€_{2017}/kW_{input}$. In order to achieve this target, around 60 GW or two doublings of cumulative capacity is required, which highlights the major role production scale-up will play in driving down the CAPEX of alkaline electrolysis systems.

The levelized cost of hydrogen does mainly determines the total cost of hydrogen, but with decreasing cost of renewable electricity (wind and solar PV) and electrolyzer plants operating between 3000 and 5000 load hours to match renewable electricity supply; the share of CAPEX in the cost of hydrogen will likely be the dominant factor in driving down costs. The future cost of hydrogen will be dependent on: (1) the cost reduction potential in the CAPEX and all the components it encompasses, (2) gas and renewable electricity prices, and (3) availability of renewable electricity for hydrogen production from electrolysis, which in turn affects the capacity factor and the levelized cost of hydrogen.

Acknowledgments

This chapter was written as part of the REFLEX project. The authors gratefully acknowledge the financial support of the European Union's Horizon 2020 research and innovation program under grant agreement number 691685 (REFLEX—analysis of the European energy system under the aspects of flexibility and technological progress). This chapter was also written as part of the ISPT 1 GW Electrolysis project. The GW Electrolysis design project is an initiative by the Institute for Sustainable Process Technology, in which Nouryon, Dow, Shell, OCI Nitrogen, Yara, Frames, Orsted, Imperial College London, Utrecht University and ECN part of TNO are partners.

References

Abbasi, T., Abbasi, S.A., 2011. Renewable hydrogen: prospects and challenges. Renew. Sustain. Energy Rev. 15, 3034–3040.

Bertuccioli, L., Chan, A., Hart, D., Lehner, F., Madden, B., Standen, E., 2014. Study on development of water electrolysis in the EU. In: Final Report. E4tech and Element Energy. Prepared for the Fuel Cells and Hydrogen Joint Undertaking (FCHJU).

Braun, M.J., 1978. Brown boveri electrolyzers today and in the near future. In: Proceedings Symposium Industrial Water Electrolysis.

Christiansen, K., Andreassen, K., 1984. Proceedings of 5th World Hydrogen Conference, Toronto, pp. 715, 727.

Christiansen, K., Grundt, T., 1978. Proceedings Symposium Industrial Water Electrolysis.

Costa, R.L., Grimes, P.G., 1967. Electrolysis as a source of hydrogen and oxygen. Chem. Eng. Prog. 63 (4), 56–58.

Crawford, G.A., Benzimra, S., 1986. Int. J. Hydrogen Energy 11 (11), 691–701. Available from: https://doi.org/10.1016/0360-3199(86)90137-0.

Evangelista, J., Phillips, B., Gordon, L., 1975. Electrolytic hydrogen production: an analysis and review. In: NASA Memorandum. Report Number: NASA-TM-X-71856.

Fairlie, M., Scott, P., 2000. Proceedings of 2000 Hydrogen Program Review, NREL/CP-570-28890.

Farlie, M., 2019. Personal Communication with Matthew Farlie, Next Hydrogen Corp. Ontario, Canada.

Felgenhauer, M., Hamacher, T., 2015. Int. J. Hydrogen Energy 40, 2084–2090. Available from: https://doi.org/10.1016/j.ijhydene.2014.12.043.

Götz, M., Lefebvre, J., Mörs, F., Koch, A.M., Graf, F., Bajohr, S., et al., 2016. Renewable power-to-gas: a technological and economic review. Renew. Energy 85, 1371–1390.

Green Hydrogen, 2019. Kolding, Denmark. Available from: http://greenhydrogen.dk/solutions/power-to-gas/.

Hebling, C., Kiemel, S., Lehnert, F., Smolinka, T., Wiebe, N., 2018. Industrialization of Water Electrolysis in Germany. Mission Innovation, Berlin.

Hydrogen Road Map Europe, 2019. Hydrogen and Fuel Cells Joint Undertaking, Publication Office of the European Union, Luxembourg, 2019.

IEA, 2017. Global Trends and Outlook for Hydrogen. IEA Hydrogen Technology Collaboration Program (TCP). http://ieahydrogen.org/pdfs/Global-Outlook-and-Trends-for-Hydrogen_Dec2017_WEB.aspx.

IEA, 2019. The future of hydrogen: seizing today's opportunities. Paris. Available from: https://webstore.iea.org/download/direct/2803?fileName = The_Future_of_Hydrogen.pdf.

IRENA, 2018. Hydrogen From Renewable Power: Technology Outlook for the Energy Transition. ISBN: 9789292600778.

KIMA. The Egyptian Chemical Industries. Aswan, El Sad El Ally Road, Egypt. Available from: http://41.222.168.85/CompanySections.aspx.

Kroposki, B., Levene, J., Harrison, K., Sen, P.K., Novachek, F., 2006. Electrolysis: information and opportunities for electric power utilities. In: NREL/TP-581-40605.

Kuckshinrichs, W., Ketelaer, T., Koj, J.C., 2017. Front. Energy Res. Available from: https://doi.org/10.3389/fenrg.2017.00001.

LeRoy, R.L., Stuart, A.K., 1978. Proceedings of the Second World Hydrogen Conference. Zurich, 21–24 August, pp. 359–377.

Lehner, M., Tichler, R., Steinmuller, H., Kopper, M., 2014. Power-to-Gas: Technology and Business Models. Springer Briefs in Energy.

Mann, M.K., Spath, P.L., Amos, W.A., 1998. Technoeconomic analysis of different options for the production of hydrogen from sunlight, wind, and biomass. In: Proceedings of the 1998 DOE U.S. DOE Hydrogen Program Review. Report number: NREL/CP-570e25315. National Renewable Energy Laboratory.

Menth, A., Stucki, S., 1978. Proceedings of the Second World Hydrogen Conference. Zurich, 21–24 August, pp. 55–63.

NEL Hydrogen, 2015a. Large Scale Hydrogen Production. Trondheim, Norway.

NEL Hydrogen, 2015b. NEL Hydrogen Electrolyser Reference List.

NREL, 2009. Current (2009) State-of-the-Art Hydrogen Production Cost Estimate Using Water Electrolysis. Retrieved from: <https://www.hydrogen.energy.gov/pdfs/46676.pdf>.

Next Hydrogen Corporation, 2016. Toronto, Canada. Personal Communication with Matthew Farlie.

Nitsch, J., Fischedick, M., Eine, 2002. vollstandig regenerative Energieversorgung mit Wasserstoff—illusion oder realistische Perspektive. Wasserstofftag Essen, 12e14. Essen.

OECD/IEA, 2015. Technology Roadmap: Hydrogen and Fuel Cells. International Energy Agency(IEA), Paris. Retrieved from: <https://www.iea.org/publications/freepublications/publication/TechnologyRoadmap HydrogenandFuelCells.pdf>.

REN21, 2017. Renewables 2017 Global Status Report. REN21 Secretariat, Paris.

Rashid, M., Mesfer, M.K., Naseem, H., Danish, M., 2015. Hydrogen production by water electrolysis: a review of alkaline water electrolysis, PEM water electrolysis and high temperature water electrolysis. Int. J. Eng. Adv. Technol. (IJEAT) 4 (3), ISSN: 2249-8958.

Ryton PPS. Chevron Phillips Chemical Company LLC. https://www.solvay.com/en/brands/ryton-pps#Applications.

Saba, S.M., Muller, M., Robinius, M., Stolten, D., 2017. Investment costs of electrolysis—a comparison of costs studies from the past 30 years. Int. J. Hydrogen Energy 43 (2018), I209–I223.

Santos, D.M.F., Sequeira, C.A.C., Figueiredo, J.L., 2013. Hydrogen production by alkaline water electrolysis. Quimica Nova 36 (8), 1176–1193.

Schalenbach, M., Tjarks, G., Lueke, W., Mueller, M., Stolten, D., 2016a. J. Electrochem. Soc. 163 (11), F3197–F3208.

Schalenbach, M., Tjarks, G., Carmo, M., Lueke, W., Mueller, M., Stolten, D., 2016b. Acidic or alkaline? Towards a new perspective on the efficiency of water electrolysis. J. Electrochem. Soc. 163 (11), F3197–F3208.

Schmidt, O., Hawkes, A., Gambhir, A., Staffell, I., 2017a. The future cost of electrical energy storage based on experience rates. Nat. Energy 2 (8), 17110.

Schmidt, O., Gambhir, A., Staffell, I., Hawkes, A., Nelson, J., Few, S., 2017b. Future cost and performance of water electrolysis: an expert elicitation study. Int. J. Hydrogen Energy 42, 30470–30492. Available from: https://doi.org/10.1016/j.ijhydene.2017.10.

Schoots, K., Ferioli, F., Kramer, G.J., van der Zwaan, B.C.C., 2008. Learning curves for hydrogen production technology: an assessment of observed cost reductions. Int. J. Hydrogen Energy . Available from: https://doi.org/10.1016/j.ijhydene.2008.03.011.

Sloop, J.M., 1959. Liquid Hydrogen as a Propulsion Fuel. 1945–1959, NASA. https//history.nasa.gov/SP-4404/contents.htm.

Smolinka, T., Gunther, M., Garche, J., 2010. NOW-Studie: "Stand und Entwicklungspotenzial der Wasserelektrolyse zur Herstellung von Wasserstoff aus regenerativen Energien".

Sunde, S., 2012. Water electrolysis technology—concepts and performance. In: Sushgen Spring School, "Fuel Cells and Hydrogen Technology". NTNU.

Thomas, C.E., Kuhn Jr., I.F., 1995. Electrolytic hydrogen production infrastructure options evaluation. In: Final Subcontract Report to National Renewable Energy Laboratory in US for Contract No. DE-AC36e83CH10093.

Thomas, D., 2018. Cost Reduction Potential for Electrolysis Technology. Hydrogenics Europe N.V, Berlin. Available from: <http://europeanpowertogas.com/wp-content/uploads/2018/06/20180619_Hydrogenics_EU-P2G-Platform_for-distribution.pdf>.

Tilak, B.V., Lu, P.W.T., Coleman, J.E., Shrinivasan, S., 1981. Electrolytic production of hydrogen. In: Comprehensive Treaties of Electrochemistry.

Twiddel, J., Weir, T., 2015. Renewable Energy Resources, third ed. Routledge, New York.

Ursua, A., Gandia, L.M., Sanchis, P., 2012. Hydrogen production from water electrolysis: current status and future trends. Proc. IEEE 100 (2), 410–426.

USGS Platinum Group Statistics, 2019. https://www.usgs.gov/centers/nmic/platinum-group-metals-statistics-and-information.

Vandenborre, H., Leyson, R., Nackaerts, H., Van Asbroeck, P., 1982. Proceedings of Fourth World Hydrogen Conference. California, June, pp. 107–117.

Vermeiren, W., Adriansens, J.P., Moreels, R., 1998. Int. J. Hydrogen Energy 23, 321–324.

Water Electrolysis Market, 2018. Available from: <www.marketresearch.com>.

Further reading

IEA, 2018. Hydrogen From Renewable Power Technology Outlook for the Transition. International Renewable Energy Agency, Abu Dhabi. ISBN 978-92-9260-077-8. Langas, H. Large Scale Hydrogen Production. International Workshop on Renewable Energy.

CHAPTER 11

Heating and cooling in the built environment

Martin Jakob[1], Ulrich Reiter[1], Subramani Krishnan[2], Atse Louwen[2,3] and Martin Junginger[2]

[1]TEP Energy GmbH, Zurich, Switzerland, [2]Copernicus Institute of Sustainable Development, Utrecht University, Utrecht, The Netherlands, [3]Institute for Renewable Energy, Eurac Research, Bolzano, Italy

Abstract

In this chapter the market development and the techno-economic learning are investigated for two competing technologies to assess their relative dynamics in terms of techno-economic performance: heat pump (HP) as an exemplary technology to tap renewable energy sources and (condensing) gas boilers as the reference system.

Long-term historical data series are particularly important when using the experience curve method. Motivated by the data availability of such data The Netherlands and Switzerland are chosen to perform two case studies for which further original and secondary data was collected. Specific methodological issues related to experience curves of end use technologies are discussed.

Although the empirical basis is by far not complete, it can be affirmed that learning rates (LR) from previously studies reconfirmed by recently gathered data. The learning rate of the main cost components of HP is 12%–22% and that of the system as a whole about 20% while their utility in terms of less noise emission, system integration, and energy-efficiency improved. This is equal or higher as compared to the LR of condensing gas boiler (13%). This holds for both countries assessed, although it should be kept in mind that the time series is quite short in the Dutch case. Moreover, a LR for the coefficient of performance was found which is 5% in the case of ground source HP and 9% for air-to-water HP. Thus HP offer a high potential to improve their cost-effectiveness relative to reference systems in terms of cost of delivered heating energy which depend on both specific investment costs and on the energy efficiency.

Chapter Outline

Technological Learning in the Transition to a Low-Carbon Energy System.
DOI: https://doi.org/10.1016/B978-0-12-818762-3.00011-X

11.1 Introduction

Most of the technologies in the chapters focus on renewable electricity production technologies, storage of energy in the form of chemical energy (H_2, batteries), and on mobility. One other major end use sector is heat, which has been historically much harder to decarbonize than the electricity sector. Especially, the built environment with demand for low-temperature heat poses significant challenges to decarbonize.

Final energy demand for heating and cooling purposes in buildings is determined basically by the following factors:

- Thermal losses: energy-efficiency of the envelope and air exchange
- Thermal passive gains: solar gains (mainly through windows) and internal heat loads (persons, appliances)
- End use system: heating and cooling systems

Thus in technical terms, final energy demand may be managed by passive measures reducing net useful energy demand (insulation, glazing) or active systems to tap renewable energy demand (solar thermal systems, heating systems). Within the context of REFLEX, heat pump (HP) is one of the key technologies in terms of sector coupling to tap renewable energy sources (RES) (power to heat) and flexibility. Accordingly, the focus of this chapter is on HPs. For comparative reasons, condensing boilers (that might be fueled by biogas or power to gas in the future) are considered as well, also because these boilers are in many countries the current-dominant heating technologies and can be considered a reference (business as usual) technology in some countries. This chapter first describes the development and deployment of HPs and natural gas boilers in the EU and then zooms in on case studies in The Netherlands and Switzerland. From these two countries, techno-economic development and learning curves are presented.

11.2 Technology description of heat pumps

A HP is a device that transfers heat from a lower temperature level (source, typically ambient heat from air, water, or ground) to a higher temperature level (sink, typically the buildings' energy and/or hot water system) by adding mechanical energy to the system. The mechanism is described as an inverse Carnot cycle or refrigeration cycle, as shown in Fig. 11.1A (Çengel and Boles, 2015). Work ($W_{\text{net,in}}$) can be added in the form of electric or thermal energy (REN21, 2017). Fig. 11.1B shows the inverse Carnot cycle for cooling (left) and heating purposes (right). HPs are able to deliver more thermal energy than mechanical energy required to operate them. In standard conditions, they can deliver three to five times more energy than consumed (REN21, 2017).

Two types of HPs are typically distinguished: air source HPs (ASHPs) and ground source HPs (GSHPs). For ASHPs the HP unit is normally fitted to the side of the building, and this type is commonly used in densely populated urban areas. There are two main types of ASHPs: air-to-air systems that directly heat the air of the rooms and can also function as air-conditioning units and air-to-water systems that are connected to the (water-based) central heating system of a house and can provide both space heating/cooling and hot water.

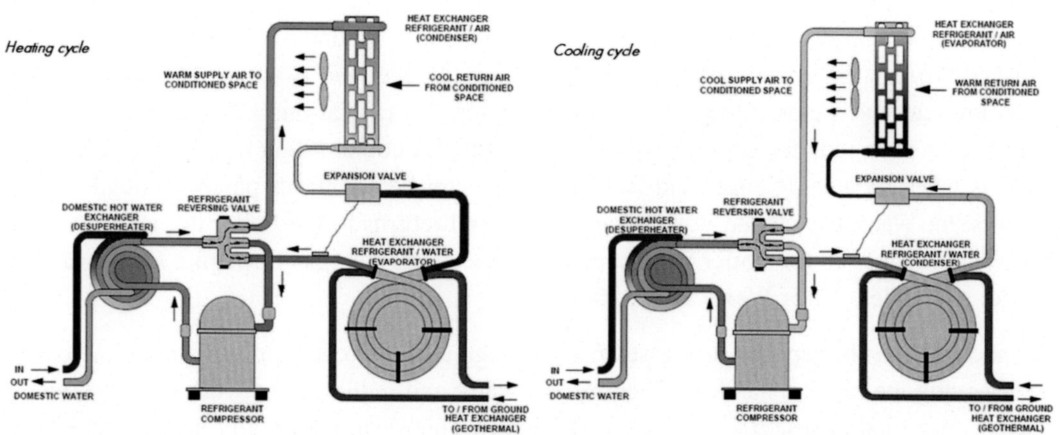

Figure 11.1
Schematic overviews of (A) the inverse Carnot cycle and (B) applications of the inverse Carnot cycle for cooling and heating. *Source: Adapted from Çengel and Boles (2015).*

For GSHPs a heat exchanger unit is installed in underground. This enables the system to reach a higher quality source of heat but also leads to higher costs (Staffell et al., 2012). Because of the stability of the underground temperature, GHSPs are very efficient all year round. ASHPs, however, have much lower efficiency when the outside air temperature is lower. In many cases, ASHPs commonly require an additional boiler (electrical or condensing gas) as backup for the winter months.

Other types of HPs include water source HPs (extracting heat from water such as groundwater layers, lakes, rivers, or the seas), systems that utilize waste heat from industrial processes, sewage water or buildings, and hybrid systems that combine different heat sources (REN21, 2017).

HPs are mainly used for space heating, cooling, and providing hot tap water in the residential buildings or commercial sectors (REN21, 2017). Also in the industrial sector, HP applications are found, to increase the temperature of industrial waste heat so it can be reused in the process. Industrial HPs are more advanced than HPs in the building sector, because they need to deliver higher temperature heat at 100°C–250°C, with a difference of up to 100°C between source and sink (Kleefkens and Spoelstra, 2014).

11.3 Technology description of condensing boilers

Conventional boilers use the sensible heat that is generated by burning fuels such as gas, oil, and wood. It can be used for space heating or heating domestic hot tap water or a combination of both (space heating + domestic hot tap water). A natural gas boiler burns the gas in the combustion chamber of the boiler to generate hot jets that move through a heat exchanger usually made of copper. The heat exchanger helps transfer heat from the gas to the water contained in the heat exchanger and heats it to around 60°C. The heated water is pushed through the system using an electric pump (Çengel and Boles, 2015). The heated water flows around a closed loop inside each radiator, entering at one side and leaving at the other. As each radiator is giving off heat, the water is cooler when it leaves a radiator than it is when it enters (Çengel and Boles, 2015). After heated water passes through all the radiators, the water has cooled down significantly and returns to the boiler to pick up more heat. Waste heat from the boiler is dispersed into the air as flue gas through a smokestack (Çengel and Boles, 2015; Weiss et al., 2008).

In condensing gas boilers the flue gases pass through a heat exchanger that warms the cold water leaving the radiators and thus heating the water and reducing the work done by the boiler (Çengel and Boles, 2015; Weiss et al., 2008). This system basically taps the latent heat of vaporization by all or part of the condensing water vapor in the exhaust gases (Weiss et al., 2009). As such condensing boilers achieve higher efficiencies (can be more than 90%) as compared to conventional boilers that do not use the latent heat.

Combi gas boilers are used to provide heat for space heating and hot tap water. They typically have two independent heat exchangers (Çengel and Boles, 2015). One exchanger is connected to the radiators for space heating and the other for hot water supply (Çengel and Boles, 2015). Combi boilers provide instant hot water, by constantly being on standby when there is no water demand. When the request for hot water is trigged by someone turning on the hot tap, this signals the boiler to start heating water inside of the system. The heat exchanger transfers the majority of the heat from the burnt gas inside the boiler to the cold water and then delivers it to the taps as required (Çengel and Boles, 2015). Inside combi boilers, the control valves operate in different directions to either allow water to flow through the central heating system or divert it to the appropriate hot water tap as required.

11.4 Market development of heating technologies in the EU

Currently, the major heating technologies in the EU are natural gas condensing/noncondensing boilers, HPs, electric boilers, and oil jet burners (see Fig. 11.2A). Yet, the heating technology market is structured quite differently across the various European countries, see Fig. 11.2B that represents the shares of the heating technology unit sales by type per country in the EU for the year 2014. For obvious reasons the heating technology market also differs by size. Below, we describe the current market developments of gas boilers and HPs.

11.4.1 Current market developments of gas boilers

Since the beginning of this century, gas boilers have been the dominating heating technologies in Europe. This becomes apparent from Fig. 11.2A that represents the heating technology unit sales by type in the EU between 2004 and 2016 (excluding district heating). Noncondensing gas boilers were the dominant heating technology in 2004 with the share of natural gas condensing boilers being 18%. Since 2004 this share has increased to 80% in 2016, and since 2007 natural gas condensing gas boilers has overtaken noncondensing gas boilers as the dominant heating technology due to (1) its high efficiency (90%) (Kemna et al., 2019), (2) technological developments in noncondensing boiler manufacturing such as introduction of new generations of modulating burners that require more expensive closed boiler systems (historically they were open boiler systems) (Weiss et al., 2009), and (3) production and additional cost for installation and maintenance of condensing gas boilers continued to decrease compared to noncondensing boilers (Weiss et al., 2009; Kemna et al., 2019).

The most pronounced shift between condensing and noncondensing boilers occurred between 2015 and 2016 which can be attributed to the implementation of the Ecodesign and energy label regulation coupled with the end users' choice to purchase gas condensing

(A)

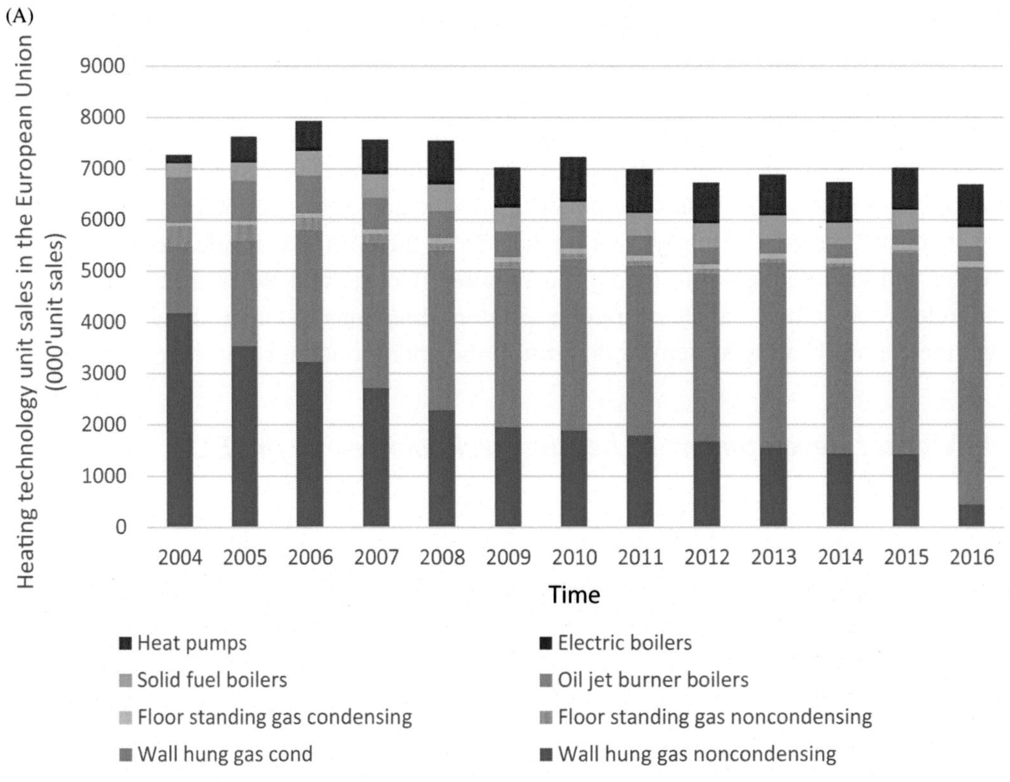

(B)

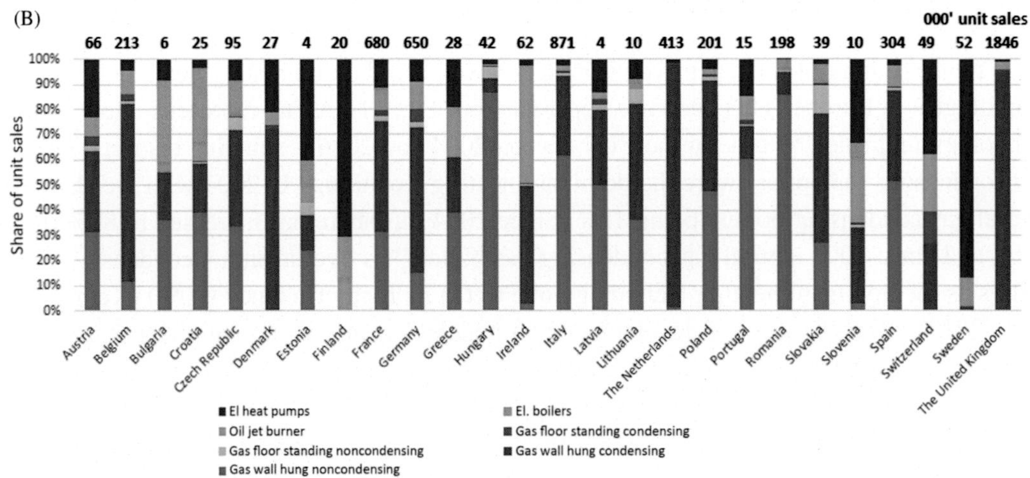

Figure 11.2

(A) Heating technology unit sales in the EU by type from 2004 to 2016 retrieved from Kemna et al. (2019), based on data from Eurostat. The figure does not include district heating. Information on the fuel type for the solid fuel boilers (e.g., wood, other biomass, and coal) was not available. (B) Share of heating technology unit sales in the EU by type and per member state in 2014. *Source: Adapted from EHPA (2016); FWS (2014); Kemna et al., (2019).*

boilers over less efficient noncondensing boilers (Kemna et al., 2019). This regulation could also be a possible reason for the increase in HP sales in 2016 as compared to 2015. Despite the implementation of the Ecodesign regulation, there still exists a share of 7% of noncondensing boilers in 2016. This could be explained by (1) the preexisting stock of noncondensing gas boilers at wholesalers and retailers that could still be sold after the implementation of the Ecodesign regulation (Kemna et al., 2019) and (2) noncondensing boilers with rated heat output ≤ 10 kW and noncondensing combination boilers with rated heat output ≤ 30 kW that can still be sold for connection to a flue shared between multiple dwellings in existing buildings (Kemna et al., 2019).

Regarding the trend in condensing gas boiler sales, an increase in share of wall hung condensing gas boilers can be seen in Fig. 11.2A when compared to standing floor gas condensing boilers. This is further highlighted for the year 2014 in Fig. 11.2B. It shows that in most EU countries, wall hung condensing gas boilers have the highest share among the types of boilers sold.

In 2014 the United Kingdom had the highest boiler sales in the EU (1.77 million units) (Kemna et al., 2019) with wall hung condensing boilers accounting for 96% of the United Kingdom's heating technology mix. These numbers could illustrate the short product lifetime of the United Kingdom boilers which leads to higher replacement rates and increased sales (Kemna et al., 2019). Italy has the second largest sales of boilers (830,000 units) (Kemna et al., 2019) in the EU in 2014 with noncondensing boilers making up 63% of the Italian heating technology mix. Though, this share has dropped and replaced with condensing gas boilers post 2014 with the Italian government's incentive by offering tax relief of 55% or 65% for upgrading efficiency of existing buildings (Italian Energy Efficiency Action Plan, 2017). Although this scheme has been in place since 2007, its effect has been negligible due to the financial crises in 2009 (Italian Energy Efficiency Action Plan, 2017). But with the Italian government making it impossible to place noncondensing gas boilers in the market from October 2015, the share of condensing gas boilers in existing buildings probably increased (Italian Energy Efficiency Action Plan, 2017). France and Germany have the third and fourth largest sales of boiler units (540,000 and 520,000, respectively) (Kemna et al., 2019) in the EU, with boiler units accounting for 80% the respective countries' heating technology mix. In both countries, wall hung condensing boilers dominates the market due to ease of installation and lower installation costs than floor standing boilers (Kemna et al., 2019). Therefore from Fig. 11.2A and B, it can concluded that natural gas condensing gas boilers are the dominant heating technology in the EU.

The share of other boiler technologies (mentioned in Fig. 11.2A and B) is relatively constant and small compared to the condensing and noncondensing gas boilers with the exception of HPs whose share increased from 2% in 2004 to 11% in 2010, followed by an increase to 12% in 2016. In the EU the total stock of HPs is estimated to have reached almost 8.5 million units in 2016, with annual sales figures reaching around 800,000 units per year. The EU market is dominated by ASHPs, representing more than 80% of the market with the largest growth is seen in air-to-water HPs since they provide heating for both space heating and sanitary hot tap water (EHPA, 2018). The increased share in HP sales (2% in 2004 to 12% in 2016) could probably be attributed to (1) countries moving away from fossil-based heating technologies to meet future climate goals, (2) government subsidies to promote the use of HPs. [e.g., Investeringssubsidie Duurzame Energie (ISDE) scheme in The Netherlands, energy efficiency action plan (EEAP) in Italy], and (3) a decrease of sales prices and an improvement of efficiencies which will be illustrated by two country case studies later.

11.4.2 Current market developments of heat pumps

The scale of the current HP market can be assessed based on the technology associations such as european heat pump association (EHPA). Annual reports give a broad overview of the current sales statistics and the market development. However, there is a lack of data and dataset inconsistency, especially in terms of technology differentiation and cost development. The first is caused by classification issues in different countries, as the main function of HPs differs per climate type: heating in cold climates or cooling in moderate climates. The latter is caused by insufficient data gathering, small- to mid-sized manufacturer structure with different market environments and subsidy schemes. Therefore availability of consistent global data for HP sales and cost development is limited (REN21, 2017).

In the EU the total stock of HPs is estimated to have reached almost 9.5 million units in 2017, with annual sales figures reaching 1 million units per year.

The EU market is dominated by ASHPs, representing more than 80% of the market. The largest growth is seen in the sanitary hot water type that combines a HP with a hot water storage tank (air-to-water system). Another trend observed by the European Heat Pump Association is that larger HPs for industrial applications are gaining popularity [+ 22% in 2017 compared to 2016 (EHPA, 2018)]. In 2017 the strongest relative growth in HP sales was seen in Austria and Lithuania, followed by Denmark and The Netherlands (EHPA, 2018). For 2017 numbers on HPs in the EU show a growth of about 13% for ASHP and GSHP. An estimated 92 GW_{th} was installed by the end of 2017. The market is led by Sweden, accounting for a total capacity of 5.6 GW_{th} (REN21, 2017). Even though the market is expanding, HPs delivered less than 1% of total final heating and cooling demand in the EU in 2015. For the residential sector, this share was a little higher but remained lower than 2% (Fleiter et al., 2017).

In relative term, HPs have a high market relevance in quite different. The highest share of HPs in their respective country's heating technology sales mix is observed in Sweden with a market share of 87% in 2004, followed by Finland (71%), Estonia (40%), Switzerland (37%), Slovenia (33%), and Austria (23%). Particularly Sweden and Switzerland have a long-term experience in introducing HPs into the heating market, see Kiss et al. (2012).

Numbers for HP markets outside of Europe are uncertain. The global amount of HPs installed in 2015 is estimated at 20 million units in buildings and 0.2 million units in the industrial sector (IRENA, 2018a). Heat consumption from HPs is estimated to have grown by 7% in 2017 compared to 2010, with the largest growth of 50% in China (OECD/IEA, 2017a). For GSHPs only the United States and China had a capacity of 16.8 and 11.8 GW_{th} at the end of 2014, respectively. Other important markets in Asia are Japan and the Republic of Korea (REN21, 2017). For 2017 international renewable energy agency (IRENA) reported a record increase in installed HPs globally (IRENA, 2018a).

11.5 Current market developments of gas boilers and heat pumps—a case study for The Netherlands

As outlined earlier, the natural gas condensing combi boiler is one of the most important current residential heating technologies in the EU and worldwide and accounts for 90% of the market share for heating technologies in the EU (Kemna, et al., 2019).

One front-running country is The Netherlands, which have—due to the abundance domestic natural gas—pushed the development and market deployment of these boilers since the late 1970s. The yearly sales trend of natural gas noncondensing and condensing combi gas boilers in The Netherlands for the period 1981–2017 are shown in Fig. 11.3. Condensing gas combi boiler sales in The Netherlands were around 13,000 units, while noncondensing gas boilers being the dominant technology had sales of around 167,000 units in 1981. By 1996 condensing combi boiler sales increased to 93,000 units and by 2008 reached to around 388,000, after which the sales trend seems to stabilize with the emergence of district heating and alternative heating sources such as HPs (Natuur & Milieu, 2018a; Kemna et al., 2019). Until the late 1970s noncondensing gas boilers with relatively low efficiencies of around 80% were the standard technology for space heating (Weiss et al., 2009). The increasing energy prices after the first oil crises in 1973 triggered the development of alternate sources of heating—the condensing gas combi boiler that uses the latent heat of evaporation contained in the water vapor of the flue gas. The relative low amount of sales in the 1980s can be attributed to lack of training and experience of installers with the new technology, additional requirements on household infrastructure, less reliable and more expensive than noncondensing gas boilers, and introduction of more efficient (88%) noncondensing boilers to the market (Weiss et al., 2009). From 1990 to 2004 the

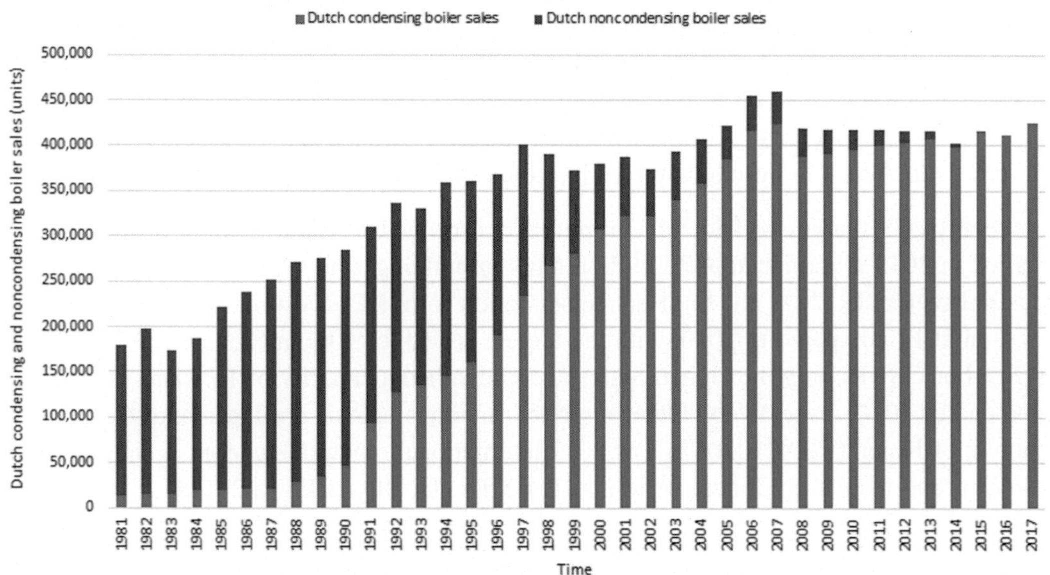

Figure 11.3

Development of natural gas noncondensing and condensing combi boiler sales in The Netherlands from 1981 to 2017. *Source: Data from Weiss et al., 2009 (original source: Weber et al., 2000; BRG Consult, 2009); Natuur & Milieu, 2018a; Kemna et al., 2019.*

condensing gas combi boiler sales were the dominant boiling technology with yearly sales increasing by around 40% compared to the previous year. This rapid market distribution can be attributed to technological developments in conventional noncondensing gas boilers that led to switch from inexpensive open to more expensive closed boiler systems, thereby making condensing gas combi boilers increasingly attractive (Weiss et al., 2009; Kemna et al., 2019).

Currently, around 90% of the 7.7 million Dutch homes have a connection to the natural gas network, while the aim is to be gas free by 2050 (Natuur & Milieu, 2018a,b). However, since 2018 the Dutch government has completely reversed its view on the deployment on natural gas. This ambition stems from the Paris Climate Agreement and the desire to reduce the risk of earthquakes in Groningen by natural gas (Natuur & Milieu, 2018a,b). Recently, the Dutch government has an ambition to "close the gas valve" for 30,000–50,000 houses every year from 2021. Therefore a transition is required in the built environment, to move away from gas sources for heating and hot tap water to electrical and more sustainable source for heating.

In order to move away from gas as a source of heating, the Dutch government introduced the ISDE scheme in 2016. This scheme provides subsidies for the purchase of solar boilers, HPs, biomass boilers, and pellet stoves for both private and business users (RVO, 2019). The available subsidy for business users and private individuals is around 100€ million in 2019. The ISDE is a multiyear plan that opened on January 1, 2016 and runs until December 31, 2020 (RVO, 2019). With the introduction of the ISDE scheme, HPs have become financially attractive and thus providing a solid boost in the market (Dutch New Energy Research, 2018). The HP sales in The Netherlands have increased from 36,000 in 2010 to 84,000 in 2017 (see Fig. 11.4). The growth was particularly high after the introduction of the ISDE scheme is 2016: + 44% in 2016 as compared to 2015. In 2017 the growth rate decreased to 15% whereas sales of natural gas condensing combi boilers have been growing at a low rate of 3% per year. However, despite the introduction of the ISDE scheme, the sales of HPs are significantly lower than the sales of natural gas condensing combi boilers (84,800 HP units versus 425,000 natural gas condensing boiler units in 2017). Note that sanitary hot water is not included in the Dutch HP unit sales due to its negligible share (< 1%).

The first HPs in The Netherlands were installed during the early 1980s for space heating, but the growth was affected by low sales due to factors such as low oil prices, lack of policy support, and technical problems (Weiss et al., 2008). In the early to mid-1990s, HPs began to receive attention as a result of increasing energy prices and discussions of anthropogenic climate change (Weiss et al., 2008). The introduction of new energy standards in 1995 aided in increasing the demand for HPs (Weiss et al., 2008) (67 units sold) (CBS, 2019). By 2000 the yearly number of units sold increased to 930 units, though sales figures and market growth prior to 2010 were low (16,700 units in 2009 and 4000 units in 2005) (CBS, 2019). The low sales figures could be attributed to (1) the availability

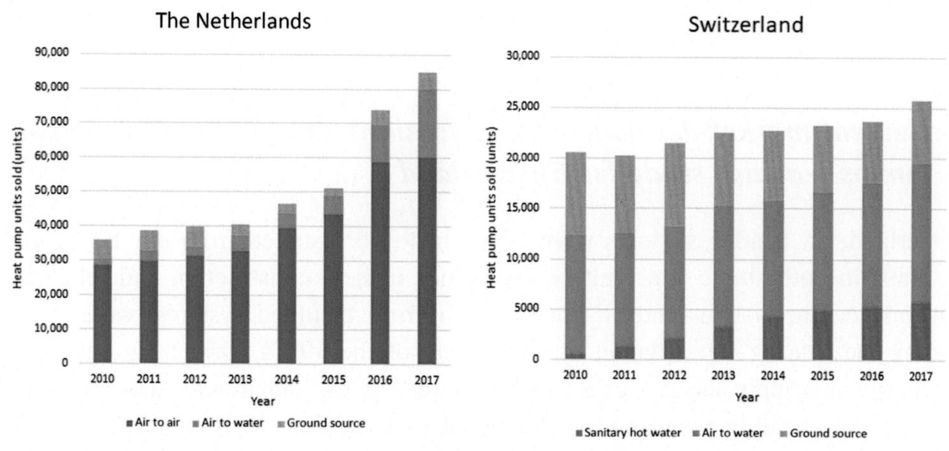

Figure 11.4

Heat pump sales by type in The Netherlands and Switzerland between the period 2010 and 2017. *Source: Data from FWS (2014); EHPA (2016); Natuur & Milieu (2018a); CBS (2019); Kemna et al. (2019).*

of highly efficient cheap natural gas boilers and cheap gas supply for heating, (2) lack of experience of installation companies in installing HPs, and (3) earlier HP systems that were suitable of newly built houses but generally not for the replacement market of the existing building stock (Weiss et al., 2008).

The CBS (Centraal Bureau voor de Statistiek) excluded air-to-air HP unit sales before 2010. According to CBS and RVOs Renewable Energy Monitoring Protocol (Protocol Monitoring Hernieuwbare Energie) (2015), air-to-air HPs sold prior to 2010 are assumed to not comply with the energy performance standards of the European Commission (2013), and therefore data on air-to-air sales was not available. From 2010 air-to-air HPs have the highest share of units sold in the observed years (81% in 2013 and 71% in 2017), while GSHPs have the lowest share (8% in 2013 and 6% in 2017) since GSHPs usually have difficulty in installing in densely populated areas (such as The Netherlands) due to a specific requirement for spacing between installations, and they are the most expensive type of HPs (Kieft et al., 2017, Dutch New Energy, 2018). Air-to-air source HP only requires a source for air and is the cheapest among the three types of HPs with an average system cost of 6800€ (Dutch New Energy, 2018). The advantage of using air-to-water HPs over air-to-air HPs is that they can provide heating for tap water (Forsén et al., 2005). The increase in share of air-to-water HPs (11% in 2013 and 23% in 2017) in The Netherlands can probably be attributed to the multifunctionality of the system to provide heat for space heating and hot tap water. The ISDE subsidy scheme introduced in 2016 does not provide subsidies for cooling technologies (RVO, 2019). Since air-to-air HPs are mostly used for air conditioning in The Netherlands, the ISDE scheme does not cover this technology and therefore could explain the 8% drop in share of air-to-air HPs and 8% rise in share of air-to-water HP units sold between 2016 and 2017.

11.6 Current market developments of fossil heating systems and heat pumps—a case study for Switzerland

In Switzerland, oil heating systems were common to be installed from the 1950s und up to the 1980s in both single and multifamily homes in new construction and retrofits. As a result, oil heating systems and oil burners (to retrofit installed systems) were dominating the heating sales market up to the end of the 1990s. Driven by concerns about security of supply the gas grid infrastructure was expanded and switched to natural gas to serve also heating purposes. As compared to other European countries, particularly The Netherlands, gas took off relatively late, also due to topographical reasons. Next to security of supply local air pollution was a further concern that impacted energy policy in Switzerland from the mid-1980s. In the case of oil and gas heating systems, technical developments toward low-emission burners and boilers were fostered by respective technical requirements. The low-NO_x and condensing technologies were first introduced for gas heating systems from the mid-1990s, followed by similar developments in the case of oil systems and later on also for wood technologies. This time lag is also explained by the technical challenge that was the lowest in the case of gas.

Referring to security of supply, diversification, clean air, and availability of nonfossil and moderately priced electricity HPs was promoted in Switzerland from the early/mid-1990s in a concerted action: the electricity industry offered special tariffs, some Cantons subsidized HP for a certain period, the Cantons' building codes and the energy-efficiency label *Minergie* were incentivizing HPs, and trust in the technology was created by building up a national HP test center (see Kiss et al., 2012 for more details). These favorable framework conditions drove a continuous increase of the HP market share, starting with the segment of new single family houses. In this market segment, HP reached a share of more than 50% from the early/mid-2000s and currently the predominating system (source: Swiss Federal Statistical Office). Further segments such as HP in building retrofits were slowly developing from the late 1990s/early 2000s and more recently also larger buildings such as multi family houses (MFHs) are equipped with HPs. As a result of these developments the technology sales gradually shifted from fossil heating systems toward HPs and since about 2008 HPs is the top ranked sales technology (see Fig. 11.5).

In the 1990s and still in 2000 the market shares of air-to-water and GSHP were quite similar, with somewhat higher shares for air-to-water HP (50%–55%, measured in terms

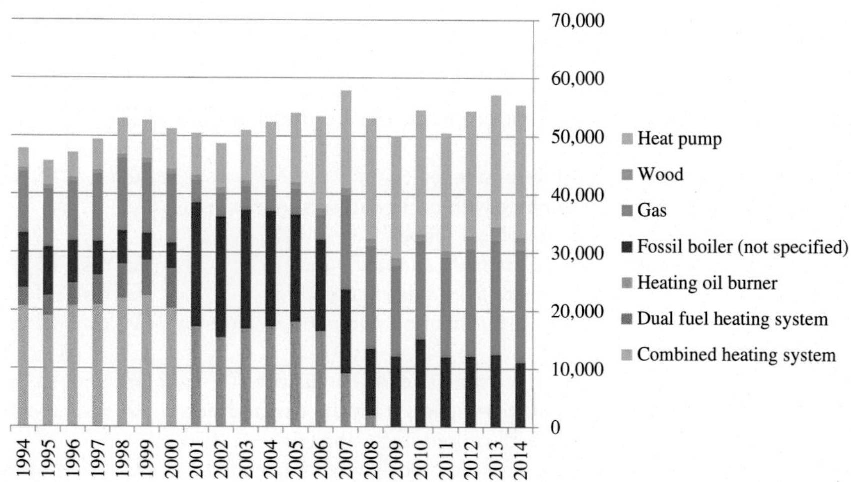

Figure 11.5
Heating technology sales by type in Switzerland between 1994 and 2014. *Source: Data from FWS (2006, 2014); TEP Energy.*

of units sold). The market share of water-to-water HP that tap groundwater was quite low, also due to legal restrictions. Water heat sources including rivers and lakes have rather been tapped with larger HP. Since the mid-2000s the relative market share of air-to-water HP (including sanitary hot water HP) gradually increased at the cost of GSHP [see Fig. 11.4 that shows the Swiss HP sales by type between 2010 and 2017; air-to-air HPs are excluded in the Swiss HP in this figure due to its negligible share (0.3% in 2010, 0.1% in 2011, 0.3% in 2013 and 0% for the remaining years)]. This shift toward air as a heat source is explained by the lower costs of these systems and the improved energy performance both in terms of coefficient of performance (COP) and seasonal energy-efficiency rating.

11.7 Methodological issues and data availability

An overview of the general data collection issues applicable to HPs is given in Table 11.1 and is discussed as follows:

- Data is not for cost but for price: This is a common issue in learning curve assessments. In the case of competitive markets price, data can be used as a proxy indicator for costs. We assume that this is a valid approach for heating technologies, albeit that in an early phase of the market introduction an extra premium is charged to tap the specific willingness-to-pay of early (demand side) adopters. This issue may be verified by

Table 11.1: General data collection issues for natural gas boilers and heat pumps.

Issue	Resolution	Applicability	
		Boilers	Heat pumps
Data is not for cost but for price	Use price data as indicator for costs	☑ Yes	☑ Yes
Data not available for desired cost unit	Data might be converted or transformed	☐	☐
Data is valid for limited geographical scope	Convert currency	☑	☑ Yes
	Combine with other datasets from various geographical scopes	☑	
Cumulative production figures not available	Calculate from annual figures	☑ Yes	☑ Yes
	Calculate from sales figures	☑ Yes	☑ Yes
Data is in incorrect currency year or currency	Correct for inflation and convert currency	☑ Yes	☑ Yes
Early cumulative production figures are not clear or available	Roughly estimate early cumulative production figures from the demand side and perform sensitivity analysis	☑ ☑ Yes	☑ Yes (NL (Netherlands))
Supply/demand affecting costs significantly	Data discarded for final recommended learning rate	☑	Yes (NL (Netherlands))
Lack of empirical (commercial scale) data	Check alternative data sources Run specific surveys		

specific expert interviews, and long-term data series might reveal such effects (in terms of particularly steep learning curves) after the termination of such phases, as outlined in Chapter 2.

- Various geographical scopes: Technologies that are composed by different production and cost elements might be produced in and traded between different countries or regions which could make it nontrivial to choose the appropriate functional unit. In such cases a geographical scope should be chosen that is predominantly determining the prices (if prices are used as a proxy). In the case of demand side and end use technologies that need to be planned and installed in buildings, the appropriate choice might be the final clients' country rather than the country of production. Indeed, learning takes place very much also at the local scale, as pointed out by Neij et al. (2017) for the case of photo voltaic (PV) installations. Thus if possible the costs are decomposed and reference units covering different geographical scopes are chosen. To aggregate the different elements a weighted index might be considered. For the case of HP the costs are decomposed (see later), but the same geographical scope is used for all elements.
- Cumulative production figures not available: This drawback might be overcome by calculating cumulative figures from annual (sales) figures. When doing so, it is important to estimate historical (sales) figures. An underestimation of the (historical) cumulative figures results in an overestimation of the learning rate (LR).
- Inflation and currency: If long-term data from count with different currencies is to be reported and compared, there is a choice on how inflation and currency conversion is performed. As end use technologies such as heating systems are usually purchased by national actors (building owners, installers) on a national market, we first deflate the data at a national scale and then to convert the real price time series using the conversion rate of the base year (e.g., the year of the most recently available data). This was done to compare price data and derived experience curves between Switzerland, The Netherlands, and other European countries: converting nominal Swiss Francs to real CHF with a common base year (e.g., 2017), using the consumer price index (CPI). Swiss Francs are then converted to Euro, using an exchange rate between CHF and EUR of that base year. With this approach, we take into account that the purchasers of HP are seeing prices in their national currency, and we avoid any deformation of the time series that would be the case if current exchange rates were used (due to the volatility of currency markets).
- Early cumulative production figures/data from the early phase are not clear or available: especially in case in which the market introduction of a technology is already a long time in the past and if the market is quite developed, early cumulative production figures should be estimated, even only roughly. Such estimates can be done from the demand side perspective if no production or (apparent) supply side data can be tapped. In case of doubt, sensitivity analysis should be performed (which is done for the Swiss

case). If the time series covers a late period only and if it is short (less than one or two doublings), one should refrain from estimating LRs and rather perform a descriptive analysis (done for the case of Dutch HP).

- Supply/demand affecting costs significantly: If such effects cannot be isolated, for example, through interviews or through long-term data series, LRs should not be estimated and used for policy recommendation.
- Lack of empirical (commercial scale) data: Alternative data sources such as statistical offices or associations that publish price indices or cost indicators. Alternatively or in addition, dedicated expert interviews or specified surveys could be performed. This approach was chosen for the case of Swiss HP.

Additional methodological issues that are specific for building related end use technologies in general and for HP or condensing gas boilers in particular are discussed in the following sections.

11.7.1 Methodological issues for heat pumps

Ideally learning and experience curves assessments cover the considered systems as a whole as energy performance, and costs very much depend on system configuration and building integration. Due to data availability and data gathering costs, such assessments can possibly be decomposed in different elements. In the case of HP systems, cost components that occur for building owners can be summarized in three groups which might of comparable cost relevance (which also implies that the HP device as such might only amount to half or even less of the total system cost):

- HP device
- Tapping the energy source [e.g., air—water, water—water, or borehole heat exchangers (BHEs)]
- Ancillary systems (e.g., controls, technical storage, heating, and hot water mixer), installation (to install and connect HP device, source tapping device, and heating distribution system), and other up-front costs (grid connection, structural adjustments, planning, etc.)
- Energy costs to operate the HP system

Such cost categories are differentiated analogously for gas heating systems which in addition to the boiler need components such as a grid connection and exhaust air installation. Referring to these cost components, learning and experience curve effects in the case of HP heating systems (thus including tapping the RES) can be attributed to the following effects:

1. Decreasing costs (price) per unit (at a given technical performance)
2. Improved energy efficiency and technical peak performance at constant (real) prices
3. General technical improvements or additional features not directly linked to energy-efficiency

In the ideal case, these effects are considered at ones in an integrated approach. Effects 1 and 2 can be integrated by adopting a life cycle-cost approach with an appropriate functional unit such as costs of delivered heat per unit of heat floor area (similarly to the cost-of-saved energy concept presented in Jakob and Madlener, 2004).

Nonenergy aspects could be integrated by adopting a multivariate learning curve approach [as proposed for instance by Badiru (1991, 1992)] to attribute costs to different utility components (similarly to hedonic price functions that elicit the economic value of such attributes).

If experience curve assessments are done for different cost elements changing boundary condition and interaction between these elements and between the different learning attributes need to be taken into account. Concretely this relates to as follows:

- A higher degree of integration: HP systems in the 1990s or the early 2000s are characterized by quite some components that needed to be integrated in the planning and installation processes, for example, MSC (metering, sensors, and controls), connectors, and heat source exchangers. Much of these devices and functionalities are now integrated the HP device as such, which means that the utility of the device has been increased (which might compensate part of the cost reduction from learning, if only the device is considered).
- Noise mitigation: particularly ASHPs have been facing noise issues, either internally of the building (compact HP) or against neighboring buildings (split configurations). As such issues often hindered the HP option against other heating systems, the HP industry improved the design of their products to reduce noise emissions.
- Generally, the quality of the HP systems including the planning and installation process (to which costs and efficiency of HP systems are particularly vulnerable) has been improved over the past years, particularly related to integrated controls including remote and self-auditing.

11.7.1.1 Data availability and data collection

Data availability long-term techno-economic data series of end use systems such as HP or gas heating systems are generally scarce. For these reasons, we focused on two cases for which historical data was available and could be extended within the framework of the REFLEX project. To do so, original data was collected from industry stakeholders and experts in Switzerland and in The Netherlands.

11.7.1.2 Data collection for heat pumps

Prices and technical data were collected by approaching key market players both in Switzerland and in The Netherlands. This included both the HP suppliers (representatives from large European manufacturers) as well as borehole companies and planers. In addition, data from the cost element indicator database of the Swiss Federal Office of Statistics was collected and used (BFS, 2019). Data collection included of a selection of implemented projects, generic cost indicators, and price catalogs from which also technical data was considered (energy-efficiency, rated power, noise emissions. etc.). In addition, data from the Swiss HP center is used to characterize the development of the energy-efficiency of the HPs. Stiebel Eltron provided price data for The Netherlands for different HP capacities from the year 2011 to 2016. The capacities of these residential HPs range from 3 to 30 kW. This price data needed to be harmonized with the 7.5 kW HP capacity selected for the Swiss HP learning curve by Weiss et al. (2008). For each year (2011–16), price data was plotted against their respective capacity and a power curve was fitted to the graph. The equation of the power curve was then used to calculate the price per kW for a 7.6 kW HP. The equation of the power curve provides an indication of the economies of scale for HP capacity for the years 2011–16. The economy of scale is quite pronounced: with a doubling of unit capacity, the price of a HP unit on average drops by 27%.

11.7.2 Methodological issues and data collection for gas boilers

Price and sales data of Natural Gas Combi boilers from the year 1981 till 2007 were gathered from Weiss et al. (2009). Weiss et al. acquired the price data from Consumentenbond (1981–2006) and Warmteservice (2007) and acquired sales data for The Netherlands from Aproot and Meijnen (1993). They uniformly used market prices and Dutch market sales data as proxies for actual production costs and actual natural gas combi boiler production. Using prices as proxies for production cost assumes a competitive market where prices closely follow production costs (BCG, 1972) which is also the case for Natural Gas Combi boilers (Weiss et al, 2009). The result of this research is a continuation/updating of the experience curve generated by Weiss et al. (2009).

To maintain consistency, prices and Dutch sales were used as proxies for production costs and actual boiler production respectively also for the research presented hereafter. For the updated experience curve, natural gas combi boiler price data was gathered from Consumentenbond (2015–18), and Dutch sales data was acquired from Natuur & Milieu (2018a). Boiler price and sales data between 2007 and 2013 were not readily available. Therefore sales data prior to 2013 was interpolated using the average yearly change in sales between 2013 and 2018. This was done in order to calculate the cumulative boiler sales from 2007 till 2013. The price of natural gas combi boilers excludes VAT. Weiss et al. reported the price of natural gas combi boilers in $€_{2006}$/kW. This was converted to $€_{2017}$/kW using the EU CPI from Eurostat (2019). Based on the price data, for each year, the average prices were calculated and used to generate the experience curve. In their analysis, Weiss et al. (2009) only included natural gas combi boilers with a capacity less than or equal to 30 kW_{th} because these boilers are typically used for central heating and hot water production in residential buildings (CBS, 2019; Weiss et al., 2009). Therefore to maintain consistency with Weiss et al. (2009) analysis, only boilers within the same range ($\leq 30\ kW_{th}$) were considered.

11.8 Techno-economic progress and experience curves

Techno-economic progress and experience curves for the case of HP are illustrated along the following dimensions:

- Decreasing costs (price) per unit (at a given technical performance)
- Improved energy efficiency
- General technical improvements or additional features not directly linked to energy-efficiency

11.8.1 Decreasing costs (prices) per unit for the case of heat pumps

The investment costs for HPs with BHEs continuously decreased in Switzerland since 1890 (Fig. 11.6), both for the case of single family houses (data available up to 2008) and for multifamily houses (data available from 2012). For comparability reasons, data was transferred into specific costs per m^2 using a standard design approach (about 40 W/m^2). The cost level of HP for the case of The Netherlands is considerably lower (40–50€/m^2) as

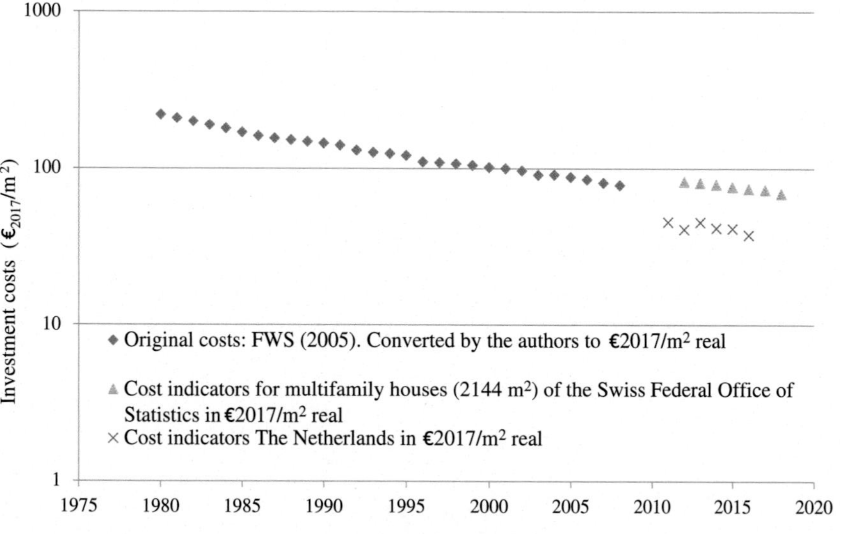

Figure 11.6
A specific investment for complete heat pumps systems with borehole heat exchangers for multifamily buildings in Switzerland (*gray triangle*) and for air-to-water residential heat pumps in The Netherlands (*red cross*).

compared to the Swiss one (in 2017 about 65€/m^2). There are several reasons for this: which is explained by the different types of HP and scope that in the case of the NL only includes the HP module and inverter and not a BHE. The borehole covers about 50%–70% of the cost. Second, the Swiss data includes installation and piping (another 20%). Last but not the least, price levels of all consumer goods, labor, etc. are more expensive in Switzerland than in the EU, including homogenous products such as household appliances and heating systems.

Fig. 11.7 shows two experience curves for a 7.6 kW residential HP sold in Switzerland and The Netherlands. The country's specific selling prices were plotted against cumulative HP sales of each of the respective countries and costs were converted as per m^2 of heated floor area using a standard conversion factor of 40 W/m^2. The experience curve for the Swiss HPs is derived from the dataset of FWS (2006), and Kiss et al. (2012) show a LR of 18% for a time series of 28 years (1980–2008). Weiss et al. (2008) estimated a LR of 26% for the period 1980–2004, but probably carried out inflation correction twice. The LR for the case of Swiss MFH HP of the same type (see BFS, 2019 for details) for the period of 2012–18 is considerably higher (27%) which is mainly explained by the learning effects for BHEs (see Fig. 11.8), which supports this result by independent data. Combining the

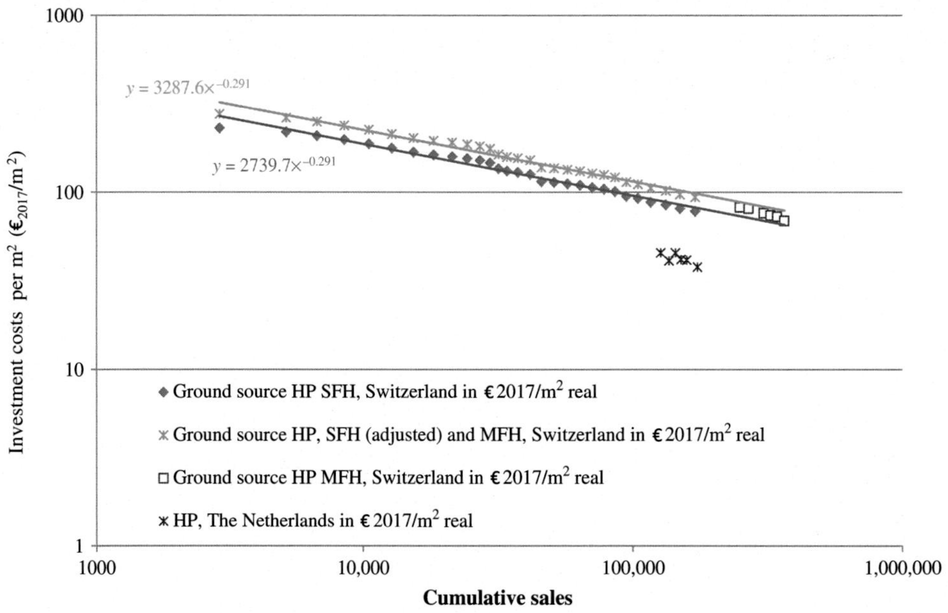

Figure 11.7

Experience curves for heat pumps. Blue data points stem from FWS (2006) and Kiss et al. (2012), green data points are level-adjusted of the latter, red data points are for MFH (source SFOS), violet data points were gathered for this report, and were scaled to costs for heat pump with a thermal capacity of 7.6 kW and converted into €/m^2 using 40 W/m^2.

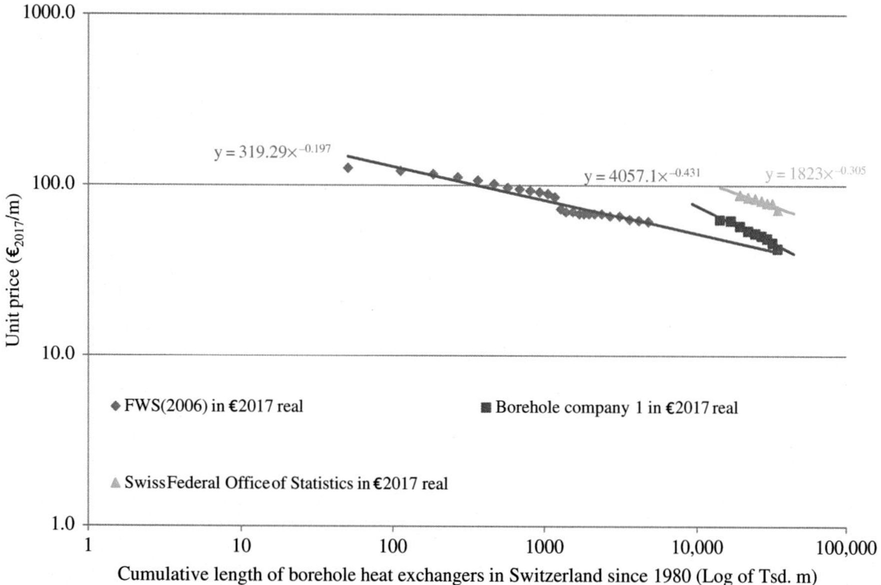

Figure 11.8
Experience curves for borehole heat exchangers for heat pump with a thermal capacity of 7.6 kW (case of new single-family houses in Switzerland). Blue diamonds were taken from FWS (2006), orange squares were gathered by TEP Energy for this report.

two series (level adjusted) yields a long-term LR of 19%. A sensitivity analysis was performed in terms of cumulative sales of the period before 1980 (assuming 3000 units) which almost did not affect the LR result. The LR of the Dutch data for the period 2011–17 is estimated to 19%. Although this result is similar to the one of the Swiss case, it should be kept in mind that the data series are quite short and covers less than one doubling of the sales, and the R^2 is very poor.

The observed cost reductions can be attributed to technological learning in the manufacturing and installation of both HPs and related system components (Weiss et al., 2008). In the past decades, heat exchangers became both smaller and cheaper. The main components of a HP (e.g., the vapor compression cycle and heat exchangers) are used in the cooling industry, and therefore the major driver for cost reductions in HPs can be attributed to technological learning in HP assembling and system integration. Note that compressors are imported with increasing shares from Asia, and production in Europe is decreasing. Cost reductions were also achieved by economies of scale, including manufacturing costs, purchasing costs, sales costs, and possibly other cost items (Weiss et al., 2008). Further learning potential could be expected from other optimizations in the design and production processes when HPs are produced in larger numbers above 100,000 units sold per manufacturer.

A second experience curve for Dutch HP prices is also shown in Fig. 11.7, based on own data collection. The curve for the Dutch HPs shows a LR of 18% for a time series of 5 years (2011–16). We used price lists for different HP capacities from the year 2011 to 2016. This price data was harmonized with the 7.6 kW HP capacity selected for the Swiss HP learning curve by Weiss et al. For all years together (2011–16), price data was plotted against their respective capacity on a linear axis. For each year the average price was taken and plotted above. From this dataset the data for the year 2012 was omitted since it was determined to be an outlier.

11.8.2 Decreasing costs (prices) per unit for the case of heat pump borehole heat exchangers and heat pump modules

HPs have a long-term tradition in Switzerland which was part of the first phase of the development of the technology. To foster energy-efficiency, to diversify the energy supply mix, and sales of domestic electricity production HP were strongly promoted from the mid-1990s (see Kiss et al., 2012 for an overview). HPs with vertical BHE to tap geothermal energy had relatively high market shares, especially in the steep market development phase, mainly due to energy-efficiency reasons. Later on, although air–water HPs have been working more efficiently in recent years (Fig. 11.9), some saturation of the installed amount

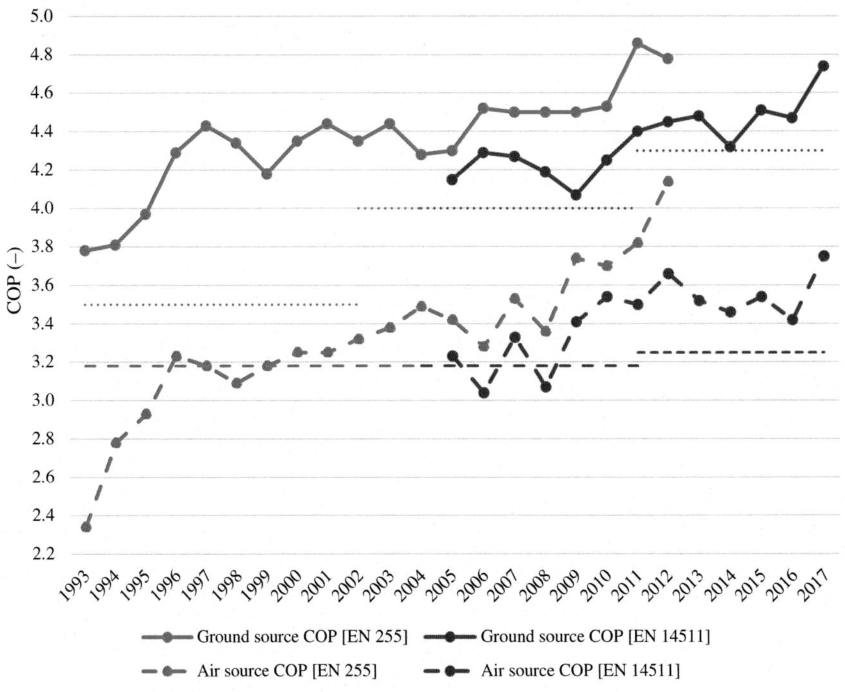

Figure 11.9
Development of COP of HP with borehole heat exchangers and for air source HP in Switzerland (Eschmann, 2017). *COP*, Coefficients of performances; *HP*, heat pump.

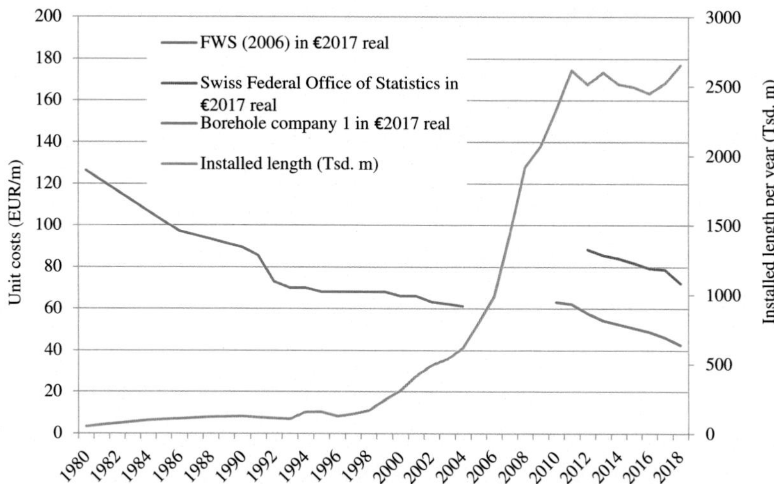

Figure 11.10

Unit costs and installed length per year (in thousand m) for borehole heat exchangers (based on a case of HP with a thermal capacity of 7.6 kW for new single family houses in Switzerland). Unit costs up to 2004 were real prices (base year 2004) and were taken from FWS (2006), those from 2010 were gathered by TEP Energy for this report (*green*) and by the Federal Office of Statistics (*red*). All price data in CHF is first deflated to real CHF (base year 2017) and then converted to EUR (exchange rate 2017: 1€ = 1.11 CHF). *Source: Data from (FWS, 2000, 2006, 2018).*

is observed (Fig. 11.10). Alongside with gaining experience and higher demand for BHE, specific costs dropped continuously over the past years (Fig. 11.10). Note that unit cost data is missing from the years 2005 to 2009, and that unit costs stem from three different sources which might explain the different level.

Cost respectively price reductions are related to the cumulative installed borehole length, as shown in Fig. 11.8 that also displays the estimated learning curve in a log–log representation. LRs are about 12% for the first data source (FWS, 2006), about 26% (Borehole company 1) and 19% (Swiss Federal Office of Statistics). The reasons for the price decrease since 2010 are attributed to the following reasons:

- Economy of scale (larger drilling machines with greater power).
- Higher utilization rate of borehole equipment entailing lower unit costs.
- Equipment (drilling machines) is mostly paid off and thus prices are related to variable costs, which hampers drilling companies to reinvest.
- Increased competition though overcapacity caused by new market players who focus on low prices (as the overall demand has been quite constant during the last few years, see Fig. 11.8).
- Increased competition against other types of HP, particularly air-to-water HP, due to technical improvements (higher COP) and low(er) prices.

The LRs may not be extrapolated for longer time scales as some of these factors are specific to the recent and current market situation.

Similarly, LRs were estimated separately for HP modules (the HP appliance as sold by the manufacturers and suppliers) and for installation and piping. Independent data from FWS (2006) covering a long-term period and from the Swiss Federal Office of Statistics covering some recent years BFS (2019) was used. The LR of HP appliances was estimated to 22% for piping and to 14% for installation.

11.8.2.1 Improved energy efficiency in the case of heat pumps

In Switzerland (new), HP models are tested by an independent test center. Since 2005 the first year the COP was measured according to EN 14511, the COP increased on average by 0.8% per year (Eschmann, 2017). Before 2005 also the COP increased considerably (see Fig. 11.9). A considerable increase of the COP was achieved with the introduction of the HP quality certificate in 2011. In order to receive this certificate, a COP of 4.4 needed to be achieved from 2011.

The learning and experience curve concept may also be applied to the technical development. To apply the usual experience curve concept, the (log of the) cumulative sales (e.g., in terms of sold units) are related to the (log of the) inverse of the COP. A LR of 5% is estimated for the transformed variable 1/COP. This means that the COP increased with 5% with each doubling of the sales (Fig. 11.11).

11.8.3 Decreasing costs (price) per unit in the case of gas boilers

The experience curve for natural gas condensing combi boilers from the period 1981 to 2018 shows a curve with an R^2 of 0.8645 and a LR of 13.1% (see Fig. 11.12). This is 1% less than the LR reported by Weiss et al. (14%) from the period 1981 to 2007 but still lies

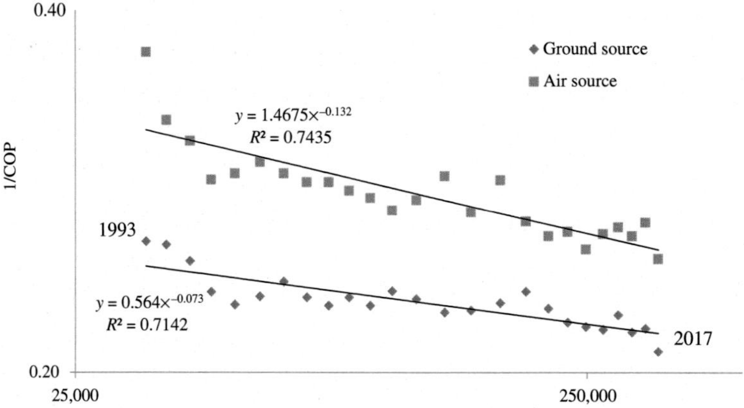

Figure 11.11
COP of ground source and air source HP as a function of cumulative sales of HP in Switzerland. *COP*, Coefficients of performances; *HP*, heat pump.

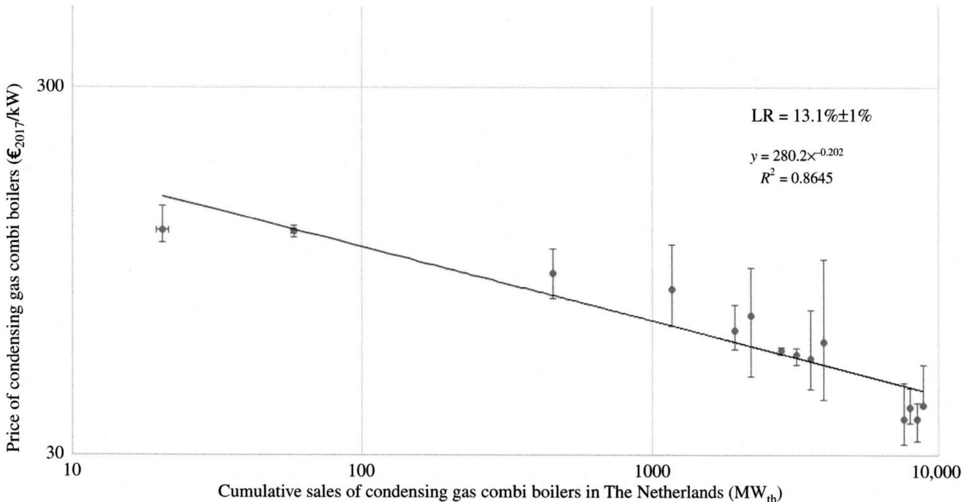

Figure 11.12
Experience curve for natural gas condensing combi boilers in The Netherlands between 1981 and 2017.

within the reported range of $\pm$ 1%. Prices of natural gas condensing combi boilers reduced from 122€$_{2017}$/kW in the 1980s to around 40€$_{2017}$/kW in 2018. Between the years 1981 and 2007, Weiss et al. (2009) identified (1) economies of scale (2) increased specialization and automation of production process, and (3) outsourcing of production to low-wage regions as the main drivers for price reduction. They reported that the major changes related to individual components refer specially to heat exchangers and control electronics. The share of cost of heat exchangers has drastically reduced from 30% in the past 25 years. Currently, control electronics are the most important cost component (Weiss et al 2009). They conducted a cost–benefit analysis on natural gas condensing combi boilers for the period 1981–2007. The analysis showed that these boilers were not profitable for consumers several years after market introduction. It required more than 10 years to break through the market. Their analysis further showed the increased cost effectiveness in saving nonrenewable energy resources and CO_2 emissions after 1999. The high natural gas prices between 1999 and 2006 aided in the improved cost effectiveness of condensing combi boilers. The price reduction from 55€$_{2017}$/kW in 2007 to 40€$_{2017}$/kW in 2018 can be attributed to similar drivers (Natuur & Milieu, 2018a,b; Navigant, 2018) as mentioned by Weiss et al. (2009).

11.9 Future market trends of heat pumps in The Netherlands and Switzerland

Currently, investment costs for GSHPs are still significantly higher than natural gas boilers in countries, such as The Netherlands, Italy, the United Kingdom, Germany, which could be attributed to the maturity of the natural gas boiler market in these countries and high installation costs of GSHPs (Kemna et al., 2019). With low power prices, GSHPs could become a cost-effective option at higher levels of space heating because of the relatively low operation costs. In terms of costs for delivered thermal energy, ASHPs are already competitive with condensing gas boilers in some European countries (EHPA, 2018; Iten et al., 2017). Its further diffusion is hampered by the fact that they are recommended for very well-insulated buildings only and require backup in colder climates, among other issues. For low-temperature industry heat, it is expected that costs of HPs will be similar to costs of natural gas boilers around 2040. This expectation is driven by increasing operational costs of gas boiler costs due to increasing CO_2 taxes and fuel costs, and maintaining similar costs for HPs due to higher electricity costs opposing technological learning (OECD/IEA, 2016b).

The future market size of HPs remains uncertain, but HPs play a critical role as source of renewable heat in scenarios of both the international energy agency (IEA) and IRENA for keeping global temperature rise under two degrees. The IEA scenario shows an increasing trend of HP units installed toward approximately 25% of heating (space and water) equipment in buildings by 2060 (OECD/IEA, 2017a). The IRENA scenario expects the stock of HPs to increase to 253 million units in buildings, accounting for a share of 27% of the heat demand, and 80 million units in the industrial sector by 2050 (IRENA, 2018a). With the current growth rate, HP technology is not on track of meeting the two degrees scenario (OECD/IEA, 2017a).

In The Netherlands the current Dutch subsidy support scheme ISDE has aided the (near) doubling of sales in The Netherlands between 2013 and 2017. This trend is expected to continue at least until 2020. Also, it is likely that the HP is the dominant technology in the new building with a market share of at least 60%. Under these circumstances, it is expected that the growth of HPs in new housing will grow within a few years to 30,000 pumps per year (Dutch New Energy, 2018). But with current sales of over 20,000 HPs annually versus nearly 400,000 natural gas condensing combi boilers (mainly for replacement in existing homes), the current market penetration is still only around 5%. Installing HPs in existing buildings is more limited than in the new building because of infrastructural issues and the need to first take adequate insulation measures installation (Dutch New Energy, 2018; Bettgenhauser et al., 2013). A possible transition solution could involve hybrid concepts with an HP as a basic facility in conjunction with a gas or biomass (pellet) boiler as peak supply and providing hot water (Dutch New Energy, 2018; Natuur & Milieu, 2018a,b; Bettgenhauser et al., 2013). These techniques require less or no major changes to the building shell and can dramatically save on gas. The Dutch government's ambition to revamp the entire housing stock as sustainable by 2050 means that in the short term 200,000 homes per year will have to be sustainable (near zero energy) (Dutch New Energy, 2018; Bettgenhauser et al., 2013).

We refrain from using this expected growth rate to extrapolate the Dutch HP experience curve due to the uncertainty surrounding the experience curve data. The experience curve covers a short time series (2011–16) with a poor R^2 which renders the LR to be highly uncertain. However, even though price data for the Dutch HPs relate to air-to-water HPs and the price data encompasses the price of the HP and inverter and the Swiss data relates to GSHPs, we can use the Swiss LR of the HP device (w/o borehole heat exchanger) to extrapolate the Dutch HP experience curve till the projected cumulative sales in 2030. Using the Swiss LR of 19% and an expected yearly installations of Dutch HPs of 30,000 units till 2030, the learning curve for the Dutch case is extrapolated over the Dutch cumulative HP sales. It is estimated that the price of air-to-water HPs in The Netherlands drops to around $28€_{2017}/m^2$ by 2030. It is important to note that using the Swiss LR as proxy to extrapolate the learning curve for the Dutch case leads to high uncertainty due to (1) different types of HP s, (2) different time series, and (3) heterogeneity of cost components included in the price/m^2, so these findings should be considered indicative at best.

In Switzerland, it is currently explored at several levels how carbon emissions from buildings could be cut significantly or even to zero by 2050 (or even shorter). The Federal Office of Environment for instance considered a subsidiary ban of fossil heating systems if other policy instruments would not be sufficient to achieve the targets of the Swiss government (− 50% by 2027/28). Common to all these initiatives is the prominent role that is attributed to HP. Up to 2030 the cumulative sales of small HP for single family house (SFH) could be doubled (from about 370,000 to about 730,000). If medium and large HPs were included cumulative sales (weighted by its thermal power to account for their size), the cumulative sales up to 2030 could be tripled. Thus prices for HP could drop by 20% to more than 30% while their energy-efficiency could increase by 7%–12% or even more if targeted research and development is launched (note that the physical and technical limits are not reached by far), especially in terms of system integration.

11.10 Summary and conclusions

Altering the European building stock and switching from fossil-based heating systems to technologies that rely on RES is the key if ambitious climate change mitigation goals should be achieved. A competitive techno-economic performance would decisively support a significant diffusion of such technologies. Yet, as in other cases, RES technologies are newer and initially more costly as compared to fossil systems that are long term deployed to the market place. In the previous sections, we have investigated the market development and the techno-economic learning for two competing technologies to assess their relative dynamics in terms of techno-economic performance: HP as an exemplary technology to tap RES and (condensing) gas boilers as the reference system to be substituted (or to be fueled by renewable gas).

Table 11.2: Summary of estimated learning rates (LRs) for natural gas boilers and heat pumps.

	Gas condensing boilers		Heat pumps	
	NL (%)	CH	NL (%)	CH (%)
Specific investment costs				
Appliance (boiler or HP module)	13	NA	18	22
Heat source tapping	–	–		12–20
Balance of system	NA	NA		14
Total system	NA	NA		20
Technical performance				
Air-to-water HP				9
Ground source HP				5
Total system (specific cost of generated energy)				20–25

HP, Heat pump; *NL*, The Netherlands; *CH*, Switzerland.

Although the empirical basis is by far not complete, it can be affirmed that LRs from previously gathered historical data are confirmed by more recent data gathered in the context of the REFLEX project. As compared to the condensing gas boiler generally, the LR is equal or higher for all specific cost components of HP (see Table 11.2). This holds for both countries assessed, although it should be kept in mind that the time series is quite short in the Dutch case.

Overall, the relatively high LRs for HPs compared to natural gas imply that HPs have the potential to become increasingly competitive in the coming years. Indeed, given the historical production and sales of the two systems, the cost reduction of HP will be more dynamic as the cumulative production (or sales) will double faster, particularly in a climate mitigation scenario. Moreover, for HP, there is still a considerable technical potential to improve peak performance and energy-efficiency (as opposed to gas condensing boilers where the technical potential is basically tapped). Therefore competitiveness particularly will be improved from a life cycle-cost perspective, and from this perspective HP are competitive already for different use cases in many countries, also depending on the framework conditions. Indeed, as also pointed out in Chapter 14, diffusion will also depend on other factors, such as electricity and natural gas prices, effective policy instruments (e.g., CO_2-tax, differentiated energy taxes, ban of fossil energies), the rate of replacement of heating systems in existing buildings, and other soft factors (such as consumer preferences: installers' and planners' acceptance and skills).

To conclude, decreasing costs could contribute to the deployment of HP and thus to climate change mitigation and reversely stringent climate change mitigation policy instruments could foster the diffusion of HP which entails a cost decrease that supports climate change mitigation (self-enforcing feedback loop).

References

Aproot, R., Meijnen, A.J., 1993. Ontwikkeling in de c.v.-ketel techniek. Het milieu en de consument gaan erop vooruit. Verwarming en Ventilatie 1, 19–28.

BFS, 2019. Cost element indicator database of the Swiss Federal Office of Statistics. Available from: <https://www.bfs.admin.ch/bfs/de/home/statistiken/preise/baupreise/kostenkennwerte-berechnungselemente-ekg.html> (last accessed 21.06.19 for current data, personal communication for historical data).

Badiru, A.B., 1991. Manufacturing cost estimation: a multivariate learning curve approach. J. Manuf. Syst. 10 (6), 431–441. Available from: https://doi.org/10.1016/0278-6125(91)90001-I.

Badiru, A.B., 1992. Multivariate learning curve model for manufacturing economic analysis. In: Parsaei, H. R., Mital, A. (Eds.), Economics of Advanced Manufacturing Systems. Springer, Boston, MA, pp. 141–162.

Bettgenhauser, K., Offermann, M., Boermans, T., Bosquet, M., Grozinger, J., Manteuffel, B.V., Surmeli, N., 2013. Heat Pump Implementation Scenarios Until 2030. Ecofys, Koln, Germany.

BCG – Boston Consulting Group, 1972. Perspectives on experience. Boston ConsultingGroup Inc.

CBS, 2019. Statline. CBS—Central Bureau voor de Statistiek (Statistics), The Netherlands. Available from: <www.cbs.nl> (accessed 08.07.19.).

Çengel, Y.A., Boles, M.A., 2015. Thermodynamics, an Engineering Approach, eighth ed. McGraw-Hill Education, New York.

Consumentenbond, 1981–2006. Condumentengids. Maandblad van de Nederlandse Consumentenbond. Consumentenbond, The Hague, The Netherlands.

Dutch New Energy, 2018. National WarmtePomp Trendrapport 2018.

EHPA, 2016. European Heat Pump Market and Statistics Report 2015. European Heat Pump Association.

EHPA, 2018. European Heat Pump Market and Statistics Report 2017.

Eschmann M., 2017. Qualitätsüberwachung von Kleinwärmepumpen und statistische Auswertung der Prüfresultate 2017. NTP i.A. Bundesamt für Energie, Bern.

Eurostat, 2019. Statistics on EU consumer price index (CPI). Available from: <www.ec.europa.eu> (accessed 10.04.19).

FWS, 2000. Statistiken (Statistics) 2000. In: FWS, Fördergemeinschaft Wärmepumpen Schweiz (Ed.), Swiss Association for the Promotion of Heat Pumps. Available from: <www.fws.ch> (last accessed 01.10.01.).

FWS, 2006. FWS Wärmepumpenstatistik (Heat Pump Statistics) 2006. In: Fördergemeinschaft Wärmepumpen Schweiz (Ed.), Swiss Association for the Promotion of Heat Pumps.

FWS, 2014. Statistik 2014. In: FWS, Fördergemeinschaft Wärmepumpen Schweiz (Statistics 2014. In: FWS, Swiss Association for the Promotion of Heat Pumps). Available from: <https://www.fws.ch/unsere-dienstleistungen/statistiken/> and <https://www.fws.ch/wp-content/uploads/2018/04/fws-statistiken-2014.pdf>.

FWS, 2018. Statistik 2018. In: FWS, Fördergemeinschaft Wärmepumpen Schweiz (Statistics 2018. In: FWS, Swiss Association for the Promotion of Heat Pumps), 2010–18 eds. Available from: <https://www.fws.ch/wp-content/uploads/2019/06/fws-statistiken-2018_V02.pdf> (last assessed 07.07.19) and <https://www.fws.ch/unsere-dienstleistungen/statistiken/>.

Fleiter, T., Elsland, R., Rehfeldt, M., Steinbach, J., Reiter, U., Catenazzi, G., Stabat, P., 2017. Profile of heating and cooling demand in 2015. In: Deliverable of the Project Heat Roadmap Europe. Retrieved from: <http://heatroadmap.eu/resources/>.

Forsén, M., Boeswarth, R., Dubuisson, X., Sandström, B., 2005. Heat Pumps: Technology and Environmental Impact. Swedish Heat Pump Association, SVEPR.

IRENA, 2018a. Global Energy Transformation: A Roadmap to 2050. International Renewable Energy Agency (IRENA), Abu Dhabi. Available from: <https://www.irena.org/-/media/Files/IRENA/Agency/Publication/2018/Apr/IRENA_Report_GET_2018.pdf>.

Italian Energy Efficiency Action Plan, 2017. European Commission. https://ec.europa.eu/energy/sites/ener/files/documents/it_neeap_2017_en.pdf.

Iten, R., Jakob, M., Wunderlich, A., Sigrist, D., Catenazzi, G., Reiter, U., 2017. Auswirkungen eines subsidiären Verbots fossiler Heizungen – Grundlagenbericht für die Klimapolitik nach 2020. Infras, TEP Energy i.A. Bundesamt für Umwelt (BAFU), Zürich/Bern.

Jakob, M., Madlener, R., 2004. Riding down the experience curve for energy-efficient building envelopes: the Swiss case for 1970–2020. Int. J. Energy Technol. Policy 2 (1–2), 153–178.

Kemna, R., van Elburg, M., Corso, A., 2019. Review study task 2: market analysis draft final report (energy label) space and combination heaters. In: Ecodesign and Energy Labeling. Available from: <www.ecoboiler-review.eu>.

Kieft, A., Harmsen, R., Hekkert, M.P., 2017. Chapter 3: Perceptions of problems and solutions in innovation system: institutional logics in the Dutch case of renovating houses energy efficiently. Stimulating Technological Innovation: Problem Identification and Intervention Formulation With the Technological Innovation Systems Framework (Ph.D. thesis). Utrecht University, Alco Kieft. Available from: <https://home.deds.nl/~alcokieft/Kieft%20-%20Dissertation.pdf>.

Kiss, B., Neij, L., Jakob, M., 2012. Chapter 24: Heat pumps: a comparative assessment of innovation and diffusion policies in Sweden and Switzerland. Historical case studies of energy technology innovation. In: Grubler, A., Aguayo, F., Gallagher, K.S., Hekkert, M., Jiang, K., Mytelka, L., et al., The Global Energy Assessment. Cambridge University Press, Cambridge.

Kleefkens, O., Spoelstra, S., 2014. R&D on industrial heat pumps. The Netherlands: Energy Research Center (ECN), Petten.

Natuur & Milieu, 2018a. Gasmonitor: Marktcijfers warmtetechnieken. p. 23. Available from: <https://www.natuurenmilieu.nl/wp-content/uploads/2018/08/NM-Gasmonitor-2018-rapport.pdf>.

Natuur & Milieu, 2018b. Verkenning Energienetten. Vervolg op Gasmonitor 2018. p. 25. Available from: <https://www.natuurenmilieu.nl/wp-content/uploads/2018/12/NM-rapport-Verkenning-Energienetten-2018.pdf>.

Navigant, 2018. Water heating, boiler, and furnace cost study (RES 19). In: The Electric and Gas Program Administrators of Massachusetts Part of the Residential Evaluation Program Area. Ref No: 183406. p. 44. Available from: <http://ma-eeac.org/wordpress/wp-content/uploads/RES19_Task5_FinalReport_v3.0_clean.pdf>.

Neij, L., Heiskanen, E., Strupeita, L., 2017. The deployment of new energy technologies and the need for local learning. Energy Policy 101, 274–283. Available from: https://doi.org/10.1016/j.enpol.2016.11.029.

OECD/IEA, 2016b. World Energy Outlook 2016. International Energy Agency (IEA), Paris.

OECD/IEA, 2017a. Energy Technology Perspectives 2017. International Energy Agency (IEA), Paris.

REN21, 2017. Renewables 2017 Global Status Report. REN21 Secretariat, Paris.

RVO, 2019. Investeringsubsidie duurzame energie ISDE. Available from: <https://www.rvo.nl/subsidies-regelingen/investeringssubsidie-duurzame-energie-isde> (accessed 15.04.19.).

Staffell, I., Brett, D., Brandon, N., Hawkes, A., 2012. A review of domestic heat pumps. Energ. Environ. Sci. 5 (11), 9291.

Warmteservice, 2007. Verwarming. Available from: <www.warmteservice.nl> (accessed 23.04.19.).

Weiss, M., Junginger, H.M., Patel, M.K., 2008. Learning Energy Efficiency. Report prepared at Utrecht University for the Ministry of Economics, Utrecht, The Netherlands. Available through the UU repository from: <https://dspace.library.uu.nl/handle/1874/32937>.

Weiss, M., Dittmar, L., Junginger, M., Patel, M.K., Blok, K., 2009. Market diffusion, technological learning, and cost-benefit dynamics of condensing gas boilers in the Netherlands. Energy Policy 37 (8), 2962–2976. Available from: https://doi.org/10.1016/j.enpol.2009.03.038.

Further reading

FWS, 2009. FWS heat pump statistics [Online]. In: FWS, Fördergemeinschaft Wärmepumpen Schweiz (Swiss Association for the Promotion of Heat Pumps). Available from: <http://www.fws.ch/zahlen_04.html> (accessed 11.06.09.).

Iten, R., Wunderlich, A., Sigrist, D., Jakob, M., Catenazzi, G., Reiter, U., 2017. Auswirkungen eines subsidiären Verbots fossiler Heizungen – Grundlagenbericht für die Klimapolitik nach 2020. Zürich/Bern. p. 104. Available from: <https://www.aramis.admin.ch/Texte/?ProjectID = 42619>.

RVO (Rijksdienst voor Ondernemend), 2019. ISDE (Investeringssubsidie Duurzame Energie). Available from: <https://www.rvo.nl/subsidies-regelingen/investeringssubsidie-duurzame-energie-isde/apparaten-isde/warmtepompen>.

Stiebel Eltron, 2017. Dutch Air-to-Water Price List (2011–2016).

CHAPTER 12

Concentrating solar power

Wilfried van Sark and Blanca Corona

Copernicus Institute of Sustainable Development, Utrecht University, Utrecht, The Netherlands

Abstract

Learning effects in concentrating solar power (CSP) technology have been limited in the past decade due to a low deployment of CSP plants. The technology development suffered mainly from the consequences of a financial and regulatory stagnation in locations with reasonable direct normal irradiance availability as well as from the fast decrease in cost of the competing solar photovoltaic technology. Nevertheless, a learning rate of about 10% can be assumed. The benefit of integrated thermal storage in CSP should be valued in a future fully renewable energy system.

Chapter outline

12.1 Introduction

Concentrating solar power (CSP) is based on the principle of concentration of photons that are directly coming from the sun, without being scattered by the Earth's atmosphere. The primary energy resource of the technology is direct normal irradiance (DNI), which is typically available in subtropic regions and/or high altitudes, see Fig. 12.1.

The concentration of direct sunlight takes place via reflective surfaces that are able to track the Sun in either two or three dimensions. The redirected photons subsequently heat up a

Technological Learning in the Transition to a Low-Carbon Energy System.
DOI: https://doi.org/10.1016/B978-0-12-818762-3.00012-1

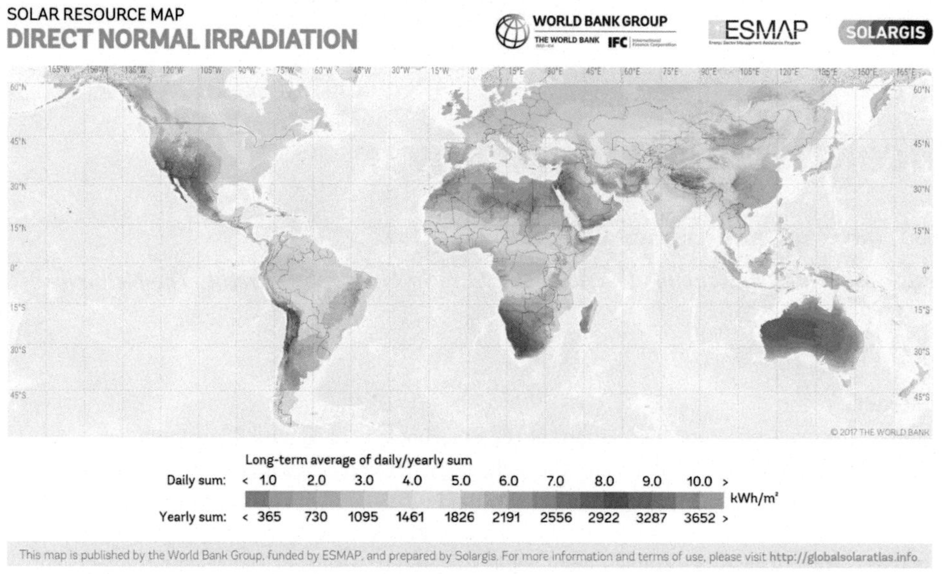

Figure 12.1

Direct normal irradiation world map. Solar resource data obtained from the Global Solar Atlas, owned by the World Bank Group and provided by Solargis. Source: *From WBG (2016).*

fluid that is used to drive a heat engine for the generation of electricity. Therefore CSP is also referred to as solar thermal electricity. It can easily be coupled to thermal energy storage (TES), so that energy collected during periods of high solar irradiance can be used to improve dispatchability (by compensating cloudy or overnight periods).

The CSP operation principle is similar to that of a magnifying glass (De Laquil et al., 1993), with which many youngsters have experimented. Light is concentrated on a heat absorber that contains a heat transfer fluid (HTF) that is heated to temperatures between 600°C and 1200°C, depending on the technology. As a consequence of these high temperatures, thermodynamic energy conversion efficiencies are high: Carnot efficiencies are theoretically about 66% and 80% for these two temperatures, respectively. The HTF typically runs in a closed circuit through the solar receiver tubes and transfers the heat to the power block. Molten salts and thermal organic or synthetic oils meet the necessary conditions as HTF (low melting point and very high boiling point). Alternatively, the use of water/steam as heat transfer medium has also been investigated and is recently under development (de Sá et al., 2018). The power block converts the heat into electricity, generally through a thermodynamic Rankine cycle, using organic or water-based liquid as working fluid, while in parabolic dish technology, Brayton and/or Stirling cycles are used. As a heat sink in the cycle, wet cooling or dry cooling towers are used. Although water is

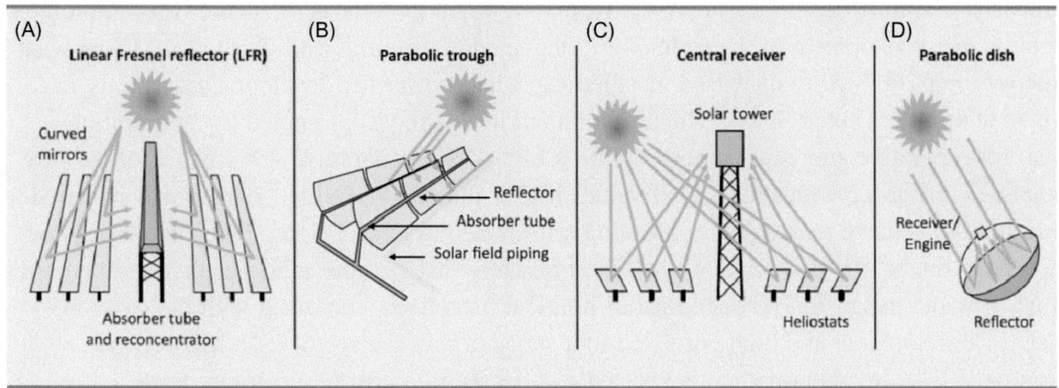

Figure 12.2
CSP technology options. (A) linear Fresnel reflector, (B) parabolic trough, (C) central receiver, (D) parabolic dish. Source: *Reprinted from Fernández et al. (2019), with permission from Elsevier.*

usually not abundant in places with high solar irradiance, wet cooling towers are sometimes chosen due to their lower costs and higher efficiencies than air cooling towers (Carter and Campbell, 2009).

Commercial CSP technology can be generally subdivided in the following four main types, see Fig. 12.2: linear Fresnel reflector, parabolic trough collector, central receiver, and parabolic dish (Fernández et al., 2019). In the first two types, light is concentrated on a linear receiver, and these two types are therefore denoted as line-focus systems, with a maximum 2D concentration ratio of 210 × (Twidell and Weir, 2015). The receiver typically is a steel tube inside an evacuated glass container for isolation. Typical generated temperatures are ~400°C with thermal oil as working fluid (Pitz-Paal et al., 2004), while conversion efficiencies range from 8% to 18%. In the other types, light is concentrated to a point receiver and has a maximum 3D concentration ratio of 46,000 × (Twidell and Weir, 2015), with typical temperatures in the receiver of ~800°C (parabolic dish) and 600°C–1200°C (central receiver), with molten salts or thermal oils as working media (Pitz-Paal et al., 2004). Efficiencies range from 20% to 40%, which come at the expense of needing more accurate Sun tracking. Recent reviews of CSP are provided by Gil et al. (2010), Teske et al. (2016), Gauche et al. (2017), Tasbirul Islam et al. (2018), and Fernández et al. (2019).

Typical installation sizes are between 30 and 200 MW. The first installations were commissioned in the 1980s in southwestern United States, for example, the solar electricity generating systems (SEGS) plants in California (SEGS I to SEGS IX), amounting to a total capacity of 344 MW (Fernández et al., 2019). The first development of the technology was primarily driven by the 1970s oil crisis. However, when the effects of such crisis were left behind, the development of the technology stagnated (Lovegrove and Csiro, 2012). Renewed interest after 2007 led to increased installed capacity and further developments, mainly in Spain and in the United States. The development in Spain was prompted by an ambitious plan for renewable energies, combined with a high DNI availability. Spain is one of the few countries in Europe with annual DNI values higher than 2000 kWh/m^2, which is considered a threshold to achieve reasonable economic performance (Viebahn et al., 2008). However, the European financial crisis forced Spain to cut down in 2013 on the subsidies that had favored the CSP rapid expansion. The absence of political incentives, combined with the still low state of development and high costs, led to a stagnation of the sector (San Miguel and Corona, 2018). In addition, the unexpected fast decrease in cost of electricity from photovoltaic (PV) solar energy decreased the economic competitiveness of the technology (Gauche et al., 2017).

Today, total installed power is estimated at 5.5 GW, with 2.5 GW under construction and 1.5 GW in development (SolarPACES, 2019). Spain (2.3 GW) and the United States (1.7 GW) together make up for 72% of the total global installed capacity, while new construction is mainly taking place in China (1.2 GW) and the MENA region (1.1 GW). Chile shows most new developments (1.1 GW). A detailed geographical breakdown is provided by SolarPACES (2019).

Present costs of new CSP plants range from 3500 to 6000$/kW for systems without TES 6000–9000$/kW for systems with 6-hour thermal storage (Fernández et al., 2019). This leads to power purchase agreements, which typically are used to secure investment, at 0.12–0.40$/kWh, which is considerably higher than for PV. It is noteworthy that some 15 years ago, cost of electricity was estimated at 0.20–0.30$/kWh depending on location (Pitz-Paal et al., 2004; Sargent and Lundy LLC Consulting Group, 2003).

12.2 Methodological issues and data availability

In many studies, it is recognized that estimation of cost development of CSP is very difficult due to the limited amount of realized plants till date, hence an inherent lack of data, as well as the fact that cost information of those plants is hard to acquire, next to having four different technologies of choice as well as difference in storage capacities (Samadi, 2018). Also, CSP plants consist of different components with different amounts of modularity, that is, the power block is usually one large unit, while the concentration devices are modular. The major contribution to the overall cost is due to the concentration devices, while learning by doing may be possible in those devices rather than the more or less standard power block devices (Platzer and Dinter, 2016). One can envisage that learning rates for the modular concentrating devices will be larger than for the power block. One may thus observe that a learning rate actually may become smaller as the technology progresses, that is, when the relative cost of the concentration devices will be decreasing compared to the cost of the power block. Similar arguments will hold for the TES facilities.

12.3 Results

12.3.1 Time trend analyses

Given the difficulty in acquiring data, the International Renewable Energy Agency and the International Energy Agency have nevertheless set up databases in which data on costs and characteristics of CSP plants is collected (IRENA, 2018, SolarPACES, 2019). Their analysis over time does not show a clear change in cost over the past decade; however, it does show a rather fluctuating downward trend. Interestingly, not including TES, parabolic trough systems have somewhat declined in cost, or rather the cost range has become narrower (IRENA, 2018). TES increases the cost of CSP plants not only because of additional capital investment for TES components, but also because the solar field capacity has to be increased to provide heat for both the power block and the charging of the storage tanks. However, a TES system provides dispatchability and higher revenues due to increased electricity generation, which makes it cost-efficient. For instance, including a 7.5-hour TES into a 50 MW CSP plant with parabolic troughs increases the operation capacity from 1785 to 2800 equivalent full-load hours (Corona et al., 2014). Adding heat storage to CSP plants has become current practice (IRENA, 2019a).

Feldman et al. (2016) reported that plotting cost data for all four CSP technologies versus time shows no reduction of cost. Another approach is to calculate the levelized cost of electricity (LCOE). Dowling et al. (2017) have reviewed economic aspects of CSP technologies and showed a large range in LCOE of 20–400US$/kWh, irrespective of the specific technology. Zhao et al. performed a similar technology study for China, arriving at LCOE of 0.19 euros/kWh (1.45 Yuan/kWh). San Miguel and Corona (2018) show a range in LCOE of 0.167–0.182 euros/kWh, depending on additional electricity generation by added gas input (up to 30% additional electricity generation). At these values, San Miguel and Corona (2018) state that “this type of CSP is highly unprofitable,” leading to a complete stall on new installations.

IRENA (2018) reported that the LCOE values were relatively stable between 2009 and 2012, see Fig. 12.3. After that, deployment increased, especially in Spain, however not only leading to decrease of LCOE but also an increase in the range of LCOE values, as also reported by Dowling et al. (2017). Note that San Miguel and Corona (2018) stated that the growth of CSP in Spain was realized by replicating existing technology only, without any learning effects occurring. Such replication was due to a regulatory limit of suboptimal 50 MW of power output (del Río et al., 2018). Lilliestam et al. (2017) also note that lower LCOE was likely caused by installations at locations with higher DNI, rather than being due to learning effects, which is also corroborated by IRENA (2019a). For the period

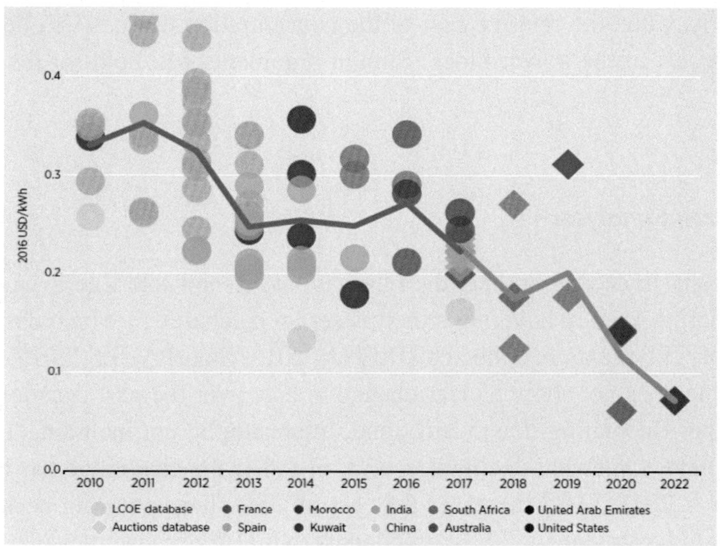

Figure 12.3
Development of levelized cost of electricity since 2010 and expected values up to 2022. Note that each bubble represents a renewable energy project. The center of the bubble is the winning bid price in that year. Source: *Reproduced with permission from IRENA (2018).*

2013–16, LCOE values are stable at 0.22 US$/kWh, on average, while for more recent projects, a decrease in LCOE is found or expected [Fig. 12.2, IRENA (2018)]. Note that such projections of LCOE have been made in the past as well (Pitz-Paal et al., 2004), but did not happen. Nevertheless, increased competitiveness can be expected (IRENA, 2018).

12.3.2 Learning curve analyses

Although solar thermal power plants are in existence since the 1980s, due to the limited amount of installations, only a few studies have been performed using experience curves primarily to analyze future cost reduction trends. A study by Enermodal (1999) has used data of installed capital costs for the SEGS plants in California (SEGS I to SEGS IX). From their experience curve a learning rate of 12% was inferred. Future technology development at that time was estimated to show a learning rate range of 8%–15% for parabolic trough and central receiver technology. Neij (2008) derives a learning rate of 20% from planned projects in Spain. The New Energy Externalities Developments for Sustainability project analyzed future global projections of CSP considering learning curves and different technology scenarios up to 2050 (Viebahn et al., 2008). They initially defined a learning rate of 12% for storage and collector field, and 5% for power block, leading to a future cost of pure electricity generation of 0.043–0.055 euros/kWh (for 2050, and reasonable DNI locations).

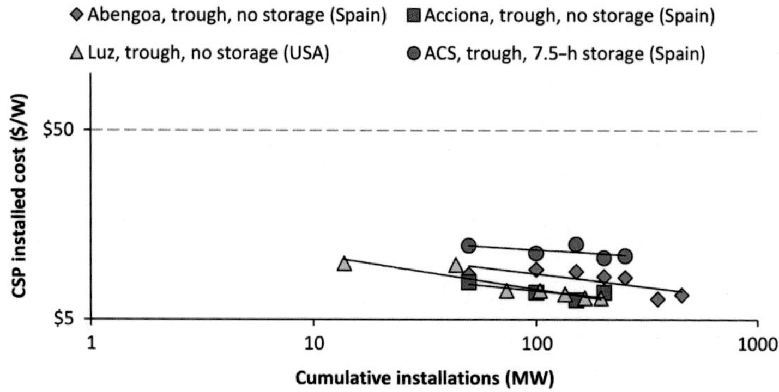

Figure 12.4
CSP experience curve for different projects, developer, location, and storage size. Sources: *Reproduced with permission from Feldman et al. (2016).*

Of a more recent date are studies from Hernández-Moro and Martínez-Duart (2013) and Pietzcker et al. (2014), who have used data from the US plants as well as data from recent plants installed in the United States and Spain up to 2013. A learning rate of 10% and 11% is found, where the latter value is for systems from Spain. In particular, Pietzcker et al. (2014) use a learning rate of 7%–16%, based on developments between 2002 and 2013, as input for future scenario analyses. Another study (Platzer and Dinter, 2016) finds a learning rate of 16% for parabolic trough plants built in Spain between 2006 and 2011. Interestingly, learning rates of 10% and 20% are used in assessment of current and future potential of CSP and PV for electricity generation, where 10% is denoted as slow learning, and 20% as fast learning (Köberle et al., 2015).

Feldman et al. (2016) state that as CSP deployment was much slower than anticipated, cost development does not seem to show learning effects, contrary to the more cost-effective utility scale PV. In fact, in a graph in which costs of all four CSP technologies over time are combined, no cost reduction is observed. del Río et al. (2018) suggested that an increase in material costs could have contributed to the low cost reduction. More detailed analysis differentiating per technology and country does reveal learning effects, as shown in Fig. 12.4: derived learning rates are between 5% and 12%, averaging at 8.5%. Recently, Lilliestam et al. (2017) have presented learning rates ranging from 5.2% up to even 25.2%; however, they most recently have presented corrected learning rates of 2.7%–6.8% (Lilliestam et al., 2019).

12.3.3 Main drivers of the price decline

According to IRENA (2018), clear possibilities exist for future cost reductions and in fact are needed to ensure that CSP will contribute significantly to a mid-century renewable energy generation portfolio. It is expected that for various parts of a CSP system, technological improvements are possible. These pertain to elements of the solar field, such as collectors and mirrors, but also to potential cost reductions in installation and engineering. As the thermal power generation units used in CSP are similar to those used in conventional power plants, learning is limited for these. With the expected growth in installations, indirect costs, such as investments and cost of capital, are expected to decrease due to increased trust of financing institutions. Pitz-Paal (2017) further argues that additional technology advances can be expected in the storage system and sheer economies of scale, while optimized and standardized system design will lead to lower cost as well. In addition, San Miguel and Corona (2018), based on their study in Spain, of which one of the conclusions was that costs were too high, argue that reduction of fixed operation costs (through higher capacities and more automated plants) would be also required to achieve financial viability under absence of public incentives.

12.4 Future outlook

Although predicted LCOE will be lower than 0.10$/kWh, the rather low learning rates of about 10% (Breyer et al., 2017), which can be inferred from the review of learning rates above, will lead to an expected 2% contribution to globally generated electricity in the REmap scenario of IRENA (2019b, 2019c), see Fig. 12.5. This is based on an expected LCOE of 0.06$/kWh in 2050 (IRENA, 2019c), with LCOE of 0.216$/kWh in 2018. These high LCOE values may lead to delayed deployment of CSP in the coming decades. In contrast, Lunz et al. (2016) argue that CSP including integrated thermal storage is of great

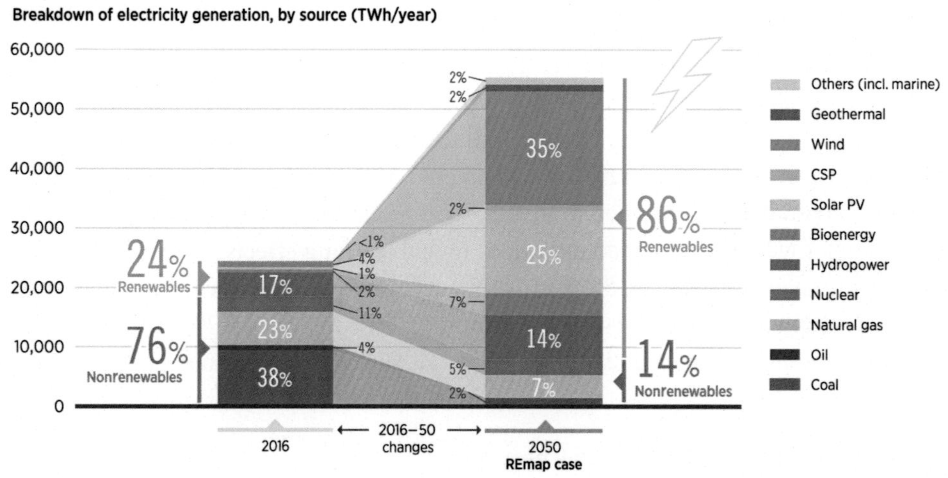

Figure 12.5
An example of a future scenario on electricity generation illustrating the importance of solar and wind energy in a 2050 scenario. Source: *Reproduced with permission from IRENA (2019c).*

value in a system with variable renewables such as wind and PV. They report on a renewable scenario for Germany in 2050, with economic benefits for a CSP/PV/wind system limiting PV and wind share to 70% at maximum. However, Breyer et al. (2017) state that LCOE of PV combined with battery systems may be lower than CSP with storage from 2030 onward, providing strong competition.

12.5 Conclusions and recommendations for science, policy, and business

In the past decade the expected learning rate of about 20% for CSP was not realized, and deployment of CSP installations suffered from a combination of drawbacks: high competitiveness from PV installations, high investment costs, low resource availability in developed countries, and a financial and regulatory stagnation in Spain. It now seems that a learning rate of about 10% for CSP is seen as more realistic, while the integrated storage option has an additional value and in fact is required in a fully renewable electricity generated system with large amount of variable wind and PV installations. Although technological improvements can be expected, especially on the component level, perhaps CSP suffers from a rather low-tech level compared to PV and therefore lower learning rates. It is recommended that CSP should have a role in conjunction with wind and PV in a fully renewable system, and policies should be in force that supports that.

Acknowledgments

This chapter was written as part of the REFLEX project. The authors gratefully acknowledge the financial support of the European Union's Horizon 2020 research and innovation program under grant agreement number 691685 (REFLEX—Analysis of the European energy system under the aspects of flexibility and technological progress).

References

Breyer, C., Afanasyeva, S., Brakemeier, D., Engelhard, M., Giuliano, S., Puppe, M., Schenk, J., Hirsch, T., Moser, M., 2017. Assessment of mid-term growth assumptions and learning rates for comparative studies of CSP and hybrid PV-battery power plants. AIP Conf. Proc. 1850, 160001.

Carter, N.T., Campbell, R.J., 2009. Water Issues of Concentrating Solar Power (CSP) Electricity in the US Southwest. Congressional Research Service, Library of Congress. Available from: <https://www.circleofblue.org/wp-content/uploads/2010/08/Solar-Water-Use-Issues-in-Southwest.pdf> (last accessed 21.06.19.).

Corona, B., San Miguel, G., Cerrajero, E., 2014. Life cycle assessment of concentrated solar power (CSP) and the influence of hybridising with natural gas. Int. J. Life Cycle Assess. 19, 1264–1275.

De Laquil III, P., Kearney, D., Geyer, M., Diver, R., 1993. Solar-thermal electric technology. In: Johansson, T. B., Kelly, H., Reddy, A.K.N., Williams, R.H. (Eds.), Renewable Energy—Sources for Fuels and Electricity. Earthscan, Island Press, London, Washington, DC, pp. 213–296.

de Sá, A.B., Pigozzo Filho, V.C., Tadrist, L., Passos, J.C., 2018. Direct steam generation in linear solar concentration: experimental and modeling investigation—a review. Renew. Sustain. Energy Rev. 90, 910–936.

del Río, P., Peñasco, C., Mir-Artigues, P., 2018. An overview of drivers and barriers to concentrated solar power in the European Union. Renew. Sustain. Energy Rev. 81 (2018), 1019–1029.

Dowling, A.W., Zheng, T., Zavala, V.M., 2017. Economic assessment of concentrated solar power technologies: a review. Renew. Sustain. Energy Rev. 72, 1019–1032.

Enermodal, 1999. Cost Reduction Study for Solar Thermal Power Plants. Enermodal Engineering Limited, Marbek Resource Consultants Ltd., Kitchener and Ottawa, ON, Canada.

Feldman, D., Margolis, R., Denholm, P., Stekli, J., August 2016. Exploring the potential competitiveness of utility-scale photovoltaics plus batteries with concentrating solar power, 2015–2030. In: Technical Report, NREL/TP-6A20-66592. Available from: <https://www.nrel.gov/docs/fy16osti/66592.pdf> (last accessed 23.04.19.).

Fernández, A.G., Gomez-Vidal, J., Oró, E., Kruizenga, A., Solé, A., Cabeza, L.F., 2019. Mainstreaming commercial CSP systems: a technology review. Renew. Energy 140, 152–176.

Gauché, P., Rudman, J., Mabaso, M., Landman, W.A., von Backström, T.W., Brent, A.C., 2017. System value and progress of CSP. Sol. Energy 152, 106–139.

Gil, A., Medrano, M., Martorell, I., Lázaro, A., Dolado, P., Zalba, B., et al., 2010. State of the art on high temperature thermal energy storage for power generation. Part 1—Concepts, materials and modellization. Renew. Sustain. Energy Rev. 14, 31–55.

Hernández-Moro, J., Martínez-Duart, J.M., 2013. Analytical model for solar PV and CSP electricity costs: present LCOE values and their future evolution. Renew. Sustain. Energy Rev. 20, 119–132.

IRENA, 2018. Renewable Power Generation Costs in 2017. International Renewable Energy Agency, Abu Dhabi.

IRENA, 2019a. Renewable Power Generation Costs in 2018. International Renewable Energy Agency, Abu Dhabi.

IRENA, 2019b. Global Energy Transformation: A Roadmap to 2050, 2019 Editon. International Renewable Energy Agency, Abu Dhabi.

IRENA, 2019c. Global Energy Transformation: The REmap Transition Pathway (Background Report to 2019 Edition). International Renewable Energy Agency, Abu Dhabi.

Köberle, A.C., Gernaat, D.E.H.J., van Vuuren, D.P., 2015. Assessing current and future techno-economic potential of concentrated solar power and photovoltaic electricity generation. Energy 89, 739–756.

Lilliestam, J., Labordena, M., Patt, A., Pfenninger, S., 2017. Empirically observed learning rates for concentrating solar power and their responses to regime change. Nat. Energy 2, 17094.

Lilliestam, J., Labordena, M., Patt, A., Pfenninger, S., 2019. Author correction: empirically observed learning rates for concentrating solar power and their responses to regime change. Nat. Energy 4, 424.

Lovegrove, K., Csiro, W.S., 2012. Introduction to concentrating solar power (CSP) technology, Chapter 1. In: Lovegrove, K., Stein, W. (Eds.), Concentrating Solar Power Technology, 2012. Woodhead Publishing, pp. 3–15.

Lunz, B., Stöcker, P., Pitz-Paal, R., Sauer, D.U., 2016. Evaluating the value of concentrated solar power in electricity systems with fluctuating energy sources. AIP Conf. Proc. 1734, 160010.

Neij, L., 2008. Cost development of future technologies for power generation—a study based on experience curves and complementary bottom-up assessments. Energy Policy 36, 2200–2211.

Pietzcker, R.C., Stetter, D., Manger, S., Luderer, G., 2014. Using the sun to decarbonize the power sector: the economic potential of photovoltaics and concentrating solar power. Appl. Energy 135, 704–720.

Pitz-Paal, R., Dersch, J., Milow, B., 2004. European Concentrated Solar Thermal Roadmapping. ECOSTAR, DLR, Germany.

Platzer, W.J., Dinter, F., 2016. A learning curve for solar thermal power. AIP Conf. Proc. 1734, 160013.

Pitz-Päal, R., 2017. Concentrating solar power, still small but learning fast. Nat. Energy 2, 17095.

Samadi, S., 2018. The experience curve theory and its application in the field of electricity generation technologies – a literature review. Renew. Sustain. Energy Rev. 82, 2346–2364.

San Miguel, G., Corona, B., 2018. Economic viability of concentrated solar power under different regulatory frameworks in Spain. Renew. Sustain. Energy Rev. 91, 205–218.

Sargent & Lundy LLC Consulting Group, 2003. Assessment of Parabolic Trough and Power Tower Solar Technology Cost and Performance Forecasts, Chicago, IL, NREL Report SR-550-34440.

SolarPACES, 2019. Available from: <https://www.solarpaces.org/csp-technologies/csp-projects-around-the-world/> (last accessed 23.04.19.).

Tasbirul Islam, M., Huda, N., Abdullah, A.B., Saidur, R., 2018. A comprehensive review of state-of-the-art concentrating solar power (CSP) technologies: current status and research trends. Renew. Sustain. Energy Rev. 91, 987–1018.

Teske, S., Leung, J., Crespo, L. Bial, M., Dufour, E., Richter, C., 2016. Solar Thermal Electricity – Global Outlook. ESTELA/Greenpeace/SolarPACES. Available from: <http://www.solarpaces.org/wp-content/uploads/gp-estela-solarpaces_solar-thermal-electricity-global-outlook-2016_full-report.pdf> (last accessed 23.04.19.).

Twidell, J., Weir, T., 2015. Renewable Energy Resources, third ed. Routledge, London.

Viebahn, P., Kronshage, S., Trieb, F., Lechon, Y., 2008. Final report on technical data, costs, and life cycle inventories of solar thermal power plants. Project No. 502687, New Energy Externalities Developments for Sustainability. Available from: <https://www.solarthermalworld.org/sites/gstec/files/concentrating%20solar%20thermal%20power%20plants.pdf> (last accessed 21.06.19.).

WBG, 2016. Global Solar Atlas. World Bank Group. Available from: <https://globalsolaratlas.info/downloads/world> (last accessed 21.06.19.).

Further reading

Zhao, Z.-Y., Chen, Y.-L., Thomson, J.D., 2017. Levelized cost of energy modeling for concentrated solar power projects: a China study. Energy 120, 117–127.

CHAPTER 13

Light-emitting diode lighting products

Brian F. Gerke

Lawrence Berkeley National Laboratory, Berkeley, CA, United States

Abstract

Over a span of only a few years in the early to mid-2010s the lighting industry was revolutionized by the arrival on the market of products using light-emitting diodes (LEDs) for general lighting applications. During that period, LED lighting products underwent both a dramatic growth in sales along with a precipitous decline in price. In the United States alone, sales of LED-based standard household light bulbs rose by more than two orders of magnitude, while falling nearly 10-fold in price. This rapid rate of change in both price and production represented an excellent opportunity for tracking the effects of technological learning in real time. Measuring the learning effect directly on manufacturing costs is challenging, owing to a lack of available data on cost, as well as very rapid evolution in product design. Because LED products are sold directly to consumers through online retail channels, it was possible to use web-crawling techniques to track retail prices at high frequency and track the price decline in great detail. A variety of studies using web crawling and other retail tracking approaches pointed consistently to a steady 20%–30% annual rate of price decline for household LED light bulbs from 2011 to 2018, in both the United States and elsewhere. Coupling this with a public sales index for the US market, a picture emerges of a technological learning curve characterized by an 18% price decline for each doubling of cumulative sales. Projecting these trends forward implies that substantial price declines are still to come, with prices expected to drop by more than a factor of four between 2015 and 2030.

Chapter Outline

Technological Learning in the Transition to a Low-Carbon Energy System.
DOI: https://doi.org/10.1016/B978-0-12-818762-3.00013-3

13.1 Light-emitting diode lighting technology in the 2010s

In the past decade, lighting products utilizing light-emitting diodes (LEDs) have radically transformed the global lighting market. LEDs are semiconductor devices that emit light over a narrow wavelength band via electroluminescence when an electrical current is applied. Since their invention in the 1960s, LEDs have been valued for their highly efficient conversion of electrical energy into light, but for most of their history, they have been low-intensity sources with emission in the red (long-wavelength) end of the visible spectrum, limiting their range of uses to indicator lighting and other low-output applications. In 1993 the breakthrough invention of blue-light LEDs (see Feezell and Nakamura, 2018, for a review) created the possibility of using LEDs to produce white light via phosphor down-conversion processes similar to those historically used in fluorescent lighting or via color-mixing. Within little more than a decade, LED devices suitable for general lighting applications were available commercially and being used in lighting products ranging from replacement light bulbs (referred to in the lighting industry as *lamps*), to novel luminaire designs with directly integrated LEDs. Owing to these products' highly efficient production of visible light compared to traditional technologies, the potential was widely recognized for LED lighting products to drive a major reduction in global energy consumption.

Since then, LED lighting products have undergone rapid adoption in the lighting market. In the United States, for example, a study of the 2010 lighting market for the United States Department of Energy (US DOE) estimated that there were 67 million total LED lighting installations in the United States, representing 0.8% of the national lighting stock (Ashe et al, 2012). By 2015, this value had increased 10-fold, to an estimated 701 million total installations, representing 8% of the US lighting stock (Buccitelli et al., 2017). In terms of market share in the standard household light bulbs (technically referred to as *A-line lamps*), LED products have grown from a negligible presence in 2010 to capture more than 50% of the market by late 2017 in the United States, according to a sales index[1] reported periodically by the National Electrical Manufacturing Association (NEMA, 2018b). Fig. 13.1 shows the relative growth in sales from 2011 to 2018, for LED A-line lamps and A-line lamps using more traditional technologies. Traditional incandescent lamp sales show a sharp decline, driven by national efficiency standards that effectively phased out this technology starting in 2012. Over the same period, LED sales have grown by more than

[1] See https://www.nema.org/Intelligence/Pages/Lamp-Indices.aspx.

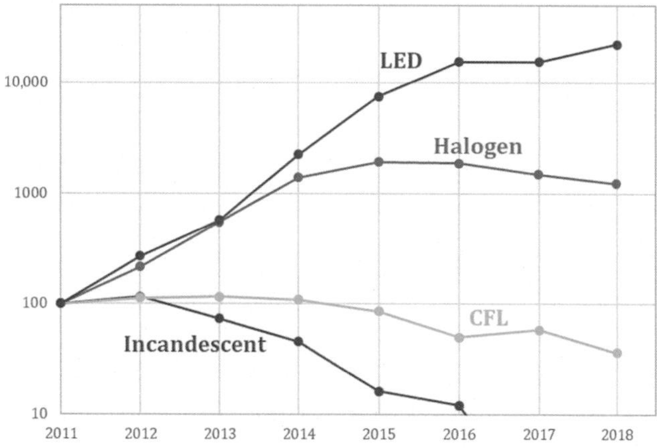

Figure 13.1
Relative annual sales of A-line lamps utilizing different technologies, compared to sales of each technology in 2011. LED lamp sales have grown by more than two orders of magnitude, while sales of incandescent, CFL, and halogen technologies have fallen in response. *CFL*, Compact fluorescent; *LED*, light-emitting diode. *Source: Author's compilation of historical data from the NEMA A-line lamp indices.*

two orders of magnitude, while sales of halogen and compact fluorescent lamps have begun to decline.

This dramatic market growth for the LED technology has been observed worldwide across a broad range of lighting products. In India, for instance, a recent report on LED adoption observes a fivefold increase in sales of LED lighting products between 2014 and 2016, with a concurrent sales decline in other lighting technologies (Chunekar et al., 2017). Globally, the International Energy Agency (IEA) estimates that LED lighting products grew to make up fully one quarter of lighting installations in 2017 (see Fig. 13.2), despite less than a decade's significant presence in the market [International Energy Agency (IEA), 2018].

The rapid adoption of LED technology has significant implications for global energy consumption, both now and in the future. The IEA estimates that an annual worldwide electricity savings from LED lighting adoption in grid-connected applications grew from 20 TWh in 2010 to more than 140 TWh by 2016 [International Energy Agency (IEA), 2016]. In the EU alone, annual energy savings from LED lighting technologies are projected to rise to more than 200 TWh (De Almeida et al., 2014) by 2030, while in the United States, annual savings are projected to reach 3.9 EJ (3.7 quadrillion BTUs), corresponding to over 300 TWh, by 2035 (Penning et al., 2016). At the same time, ultraefficient LED lighting technologies have also dramatically increased access to off-grid electric lighting in the developing world, with substantial positive implications for human-health and economic development (Alstone and Jacobsen, 2018).

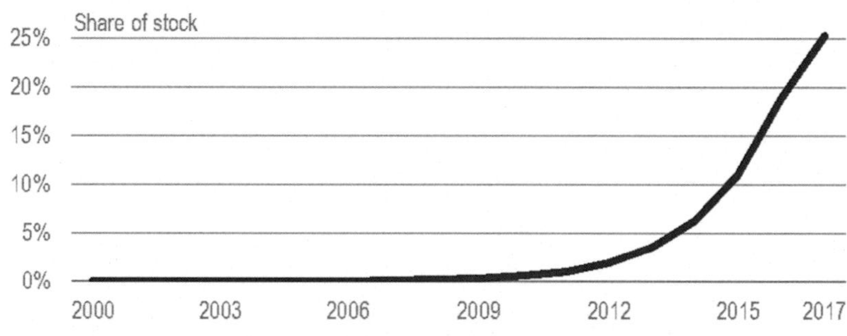

Figure 13.2
Evolution in the share of LED products in the installed stock of lighting products worldwide, as compiled by the IEA. *IEA*, International Energy Agency; *LED*, light-emitting diode. *Source: (IEA, 2018).*

In light of their potential for energy savings the development and adoption of LED lighting products have received considerable support from government policies aimed at reducing energy consumption and greenhouse gas emissions. Minimum efficiency standards that effectively banned traditional incandescent lamps, such as those implemented in the United States (US Congress, 2007) and the European Union (European Parliament, 2009), created a market for LED products to capture. Concurrently, other programs were implemented to support the rapid deployment of LED lighting technology, such as US DOE's Solid State Lighting (SSL) program (US DOE, 2018), which worked with the industry to develop a roadmap, technological targets, and testing protocols to support LED development. In a highly visible example of such support the US DOE's SSL program held the "L-Prize" competition, which helped to bring the first viable mass-market LED A-line lamp to the US market in late 2011 (US DOE, 2016a). Additionally, a wide variety of local and regional programs, typically managed by electrical utilities, encouraged adoption of LED products via subsidies, direct installation, or other means. In addition to the policy support, LED adoption was also encouraged by attractive features specific to LED lighting products, such as their capacity for dimming, directional control, and color tuning.

However, the most important driver of rapid adoption for LED lighting products was surely the precipitous decline in price that followed their early market entry. In 2008 the early market-entrant A-line LED lamps purchased for testing under the US DOE's SSL program had prices in excess of USD 100/klm of light output, falling to prices around USD 50/klm in 2012 following the introduction of the L-prize lamp (Tuenge, 2013). By 2015 typical observed prices had fallen to USD 16/klm (Penning et al., 2016). Similar declines were seen worldwide. Again using the example of India, a fivefold drop in retail price was observed for LED lighting products from 2014 to 2018 [International Energy Agency (IEA), 2018].

A rapid price decline of this nature was widely anticipated when LED lighting products first entered the market at prohibitively high price points late in the first decade of this century. Part of the reason for this expectation was Haitz's law (Haitz et al., 1999), which is the observation that the price of LEDs, per unit of light output, has declined 10-fold every decade since their invention in the 1960s, corresponding to a decline of roughly 25% per year—a phenomenon that has also been observed for white-light LEDs since their introduction (Haitz and Tsao, 2011). Haitz's Law alone is insufficient to account for the dramatic price drop for LED lighting products, however, since these products also consist of other componentry besides the LED package itself, such as drivers and other electronics, heat sinks, electrical components, and optics. Moreover, combining all of the components into an omnidirectional lighting product required substantial investments in research and development, with significant additional opportunities for cost reduction associated with the assembly process. Industry estimates gathered by the US DOE's SSL program (Bardsley et al., 2014, 2016) indicate that, although the LED packages have seen the most dramatic decline, significant cost reductions are ongoing among most or all of these categories, and these are expected to drive a 40% reduction in manufacturing cost between 2015 and 2020.

Thus the rapid mass-market adoption of LED lighting products that took place immediately following their market entry made these products an excellent laboratory for measuring the effects of technological learning, with limited confounding influence from inflation or other long-term economic trends. The rapid growth in demand for these products early in their existence yielded very rapid doublings in production—sometimes several times per year—which enabled learning analyses to be performed with only a few years of price-tracking data.

In this chapter, we survey efforts to measure learning curves and other price trends for LED lighting products in the early to mid-2010s, during the period of their initial US market adoption. To focus on a concrete example, we devote most of our discussion to LED A lamps in the United States lighting market, since this market has seen the broadest range of studies relevant to technological learning. In Section 13.2, we discuss the approaches taken to collect data on price and production for LED lighting products, and we summarize key issues related to data collection for a mass-market product in a competitive market environment. In Section 13.3, we summarize the results from various price-trend studies for LED lighting products, which, taken as a whole, find that LED lighting products declined in price by some 20%–30% per year in the early 2010s and that, in the United States, these declines corresponded to a learning curve with an 18% price decline for each doubling of cumulative shipments to the US market. Section 13.4 presents the results of various price forecasts for LED lighting products, based on the measured price trends and expected future production. In Section 13.5, we conclude and discuss implications of the observed price declines for industry and policy.

13.2 Methodological issues and data availability

13.2.1 Price data for light-emitting diode lighting products

Because lamps and luminaires are marketed directly to consumers and are widely available via conventional retail channels, data on the retail price of LED lighting products are relatively easy to find and collect. Retail prices present a conceptual challenge, in principle, for learning-curve analysis of LED lighting products, since these products consist of a new technology (the LEDs and associated electronics), whose cost may be declining rapidly, in a familiar package (the light bulb) whose componentry (such as the enclosure, electrical connectors) are largely technologically mature and likely declining only slowly in price. A multifactor learning approach could be considered in this situation; however, comprehensive component-level cost and production data for LED lighting products are extremely difficult to obtain,[2] making such an analysis infeasible.

Moreover, the unique challenges of using LED technology in general lighting applications, such as heat dissipation or overcoming LEDs' fundamentally directional nature, led to a diverse and fast-evolving set of product designs throughout the 2010s. This rapid churn in product design suggests that significant technological learning was occurring throughout the manufacturing and assembly process of LED lighting products, not just in the fabrication of the LEDs themselves, so that a learning analysis based on total product price is likely the most appropriate approach during this early phase of market adoption.

However, the rapid evolution in LED lighting products also introduced a wide variety of new product features that also affect price, posing a challenge for determining a single, well-defined price at any given point in time, and confounding efforts to measure the underlying learning dynamics for the base technology. For instance, the earliest LED lighting products intended for general illumination had relatively low light output, similar to 40 or 60 W traditional incandescent bulbs. Over time, products with higher output (e.g., replacements for 75 and 100 W bulbs) were introduced to the market at a substantial price premium that eased with time. It is thus essential for any price-trend analysis to control for lumen output, in order to account for the varying maturity and market penetration of bulbs having different output levels. Additional features that can impact the price of LED lighting products, and whose relative market penetration varied significantly during the 2010s, include lifetime; color temperature (the perceived "warmth" or "coolness" of the light); color rendering (the ability of the light to reflect the true color of illuminated objects); dimmability; color tunability; remote controllability

[2] The US DOE's SSL program does compile estimates, based on industry input, for the relative manufacturing costs of LED lamps, broken down into broad categories (Bardsley et al., 2014, 2016). However, these are likely insufficiently precise and detailed to support a component-based learning analysis; and in any case, they do not include estimates of cumulative production for each component, which would also be required.

(e.g., via a smart-phone application); and the esthetic appearance of the light bulb itself. In a highly competitive market, such as the ones that generally exists for new technologies, one would naturally expect a degree of market differentiation on these features, with some products minimizing features in the interest of price, while others add desirable features at a price premium. It is important that a learning-curve analysis for LED products accounts for such market dynamics, either via a multifactor learning approach or by other means.

A common method of tracking the evolution of LED products since their introduction has been *web crawling*, using automated software tools to collect data on price and product attributes from online retail outlets (Gerke et al., 2014, 2015; McGaraghan 2015; Penning et al., 2016). This approach allows data to be collected repeatedly, on a regular cadence and at a high frequency (for instance, Gerke et al., 2014, 2015, collected LED light bulb prices on a weekly or biweekly basis for more than 3 years), with little additional effort required beyond the initial development of the web-crawling software. Thus even very rapid price declines can be tracked in detail.

One downside to using retail web crawling in the context of learning-curve analysis is that the market consists of a wide variety of products at different price points, and it may be challenging to aggregate these prices to estimate a typical price in each time period for which cumulative production has been measured. On the positive side, however, in addition to price, retail sites usually also display information about product specifications, so web-crawling software can also collect data on the diverse product features that may serve as confounding factors in a price trend analysis, enabling these features to be controlled for via multivariate regression (see Gerke et al., 2015). Web crawling also allows data collection to be easily extended to a broad range of different products, so that prices can be tracked separately for different LED lighting products, ranging from A-line and reflector lamps to replacements for fluorescent tubes, to integrated LED luminaires. For instance, Penning et al. (2016) used web crawling to track prices for 24 different categories of LED lighting products over a period of 7 years.

The confounding influence of product features on price-trend analyses applies to a wide variety of consumer products, but for LED lighting products, there are two additional concerns that could obscure the effects of technological learning. First, early in the decade, there was substantial concern about price volatility in the market for rare earth elements (which would impact the price of LED phosphors), so that analyses of cost-effectiveness for lighting efficiency policy took the impact of such volatility into effect (US DOE, 2014). However, in recent years, this volatility has eased, and it is not expected to significantly impact price trends for LEDs (US DOE, 2016b). Also, a large number of utilities and local policymaking bodies have provided subsidies to encourage the adoption of LED products, and many of these programs are structured as so-called upstream or mid-stream incentives, in which the subsidy is paid directly to the manufacturer, distributor, or retailer, to produce a lower consumer-facing retail price (rather than providing a rebate to the consumer after purchase). These programs could obscure or enhance the underlying price decline if they are not accounted for. Fortunately, the web-crawling approach can overcome this issue, since it is possible to crawl prices as displayed to a customer outside of the geographic region covered by a subsidy (e.g., by crawling from a server location outside of the relevant country).

13.2.2 Approach to inferring cumulative production

While price data for LED lighting products are relatively easy to obtain, it is more difficult to find data that can inform estimates of cumulative production for these products. The market for LED lighting is a competitive one, and actual data on manufacturing output and sales tend to be very closely held by firms. A key source of information on this front for the US market are the sales indices for lighting products published roughly twice yearly by NEMA (2018a). These indices provide the relative quarterly sales of various lighting products, including household LED light bulbs (A-type lamps), compared to a selected baseline quarter.

From the perspective of determining a learning rate, having an index of relative sales such as the NEMA indices, rather than absolute sales, presents no obstacle. A relative sales index is given by $I_p = q_p/q_0$, where q_p is the sales in time period p, and q_0 represents the sales in a selected reference period. Then the cumulative production Q_p can be computed in units of q_o by summing up from the period of introduction p_i: $\tilde{Q}_p \equiv Q_p/q_0 = \sum_{p'=p_i}^{p} q_{p'}/q_0 = \sum_{p'=p_i}^{p} I_{p'}$. Although the absolute sales multiplier q_0 remains unknown, this value cancels out of the learning-curve equation when we write it as

$$P = \left(\frac{Q}{Q_0}\right)^{-b} = \left(\frac{\tilde{Q}}{\tilde{Q}_0}\right)^{-b}.$$

In this case, $\tilde{Q}$ is the independent variable, while $\tilde{Q}_0$ and b are the parameters being estimated. A residual challenge in developing a cumulative production estimate is that the NEMA lamp indices only began tracking LED sales figures as of 2011; several years after the first LED lamps were introduced to the market. Fortunately, the scale of LED shipments at this stage was sufficiently small that it is reasonable to backcast the shipments to the year of introduction by assuming a simple trend, without introducing significant error to the estimate of cumulative production. (For more detail on this procedure, see Section 13.3.2.)

A bigger challenge posed by the NEMA indices (or other sources of market tracking data) is that they represent sales in a limited geographic region (the United States), whereas the rapid adoption of LED lighting products is a worldwide phenomenon, and one would expect

Table 13.1: Data issues related to light-emitting diode lighting learning-curve analysis, and resolutions applied.

Issue	Resolution applied	Applicability
Data are not for cost but for price	Use price data as indicator for costs	☑
Data not available for desired cost unit		
Data are valid for limited geographical scope	Assume regional production tracks global production	☑
Cumulative production figures not available	Use relative sales index	☑
Data are in incorrect currency or currency year	Correct for inflation	☑
Early cumulative production figures are not clear or available	Backcast early cumulative production from trend in available data	☑
Supply/Demand affecting costs significantly		
Lack of empirical (commercial scale) data		

technological learning to be driven by the growth in cumulative global production. To perform a learning-curve analysis using sales data from only one geographical region, one assumes that the market being analyzed represents an approximately constant fraction of the global market over the period being analyzed. Since the market growth may proceed at substantially different rates in different regions, this assumption, though necessary, may be a significant source of error in the estimated learning rate. Similarly, most market-tracking data will focus on specific types of LED lighting product (e.g., household A-type lamps in the case of the NEMA indices), and there is a risk that these do not represent a constant fraction of total production for all LED products. Fortunately, because the United States represents a significant fraction of the global lighting market, and since household A-type lamps are a dominant product category, the fraction of global production represented by the NEMA A-lamp indices is likely to be fairly stable, and so these can be used as a reasonably reliable proxy for global LED production, at least over a period of a few years.

13.2.3 Summary of data and methodological issues

Table 13.1 summarizes the data issues encountered in performing learning-curve analysis for LED lighting products, as described in this section, and the approaches that have been taken to address them.

13.3 Results

13.3.1 Time-trend analyses

Owing to the ready availability of price data for LED lighting products, alongside the challenges described in the previous section in estimating cumulative production, numerous

Table 13.2: Estimated annual rates of price decline for light-emitting diode (LED) A-line lamps from various studies of LED price trends.

Study	Region	Period of data collection	Estimated annual price decline rate (%)	Note
Tuenge (2013)	United States	2008–12	65	Estimated from the results of a power-law fit
Gerke et al. (2014)	United States	2011–14	28	Based on 25th percentile observed lamp price for low-lumen lamps only
Gerke et al. (2015)	United States	2011–15	28	Based on multivariate regression, including controls for various product features
McGaraghan (2015)	California	2013–15	21–35	Trend range represents separate estimates for low and high CRI products
Penning et al. (2016)	United States	2010–16	28	Estimated from the early-adoption period of a learning-based model
International Energy Agency (IEA) (2018)	India	2014–18	33	Estimated from a reported 80% price decline over the period, for bulk procurement (not consumer price)
Heidari et al. (2018)	Switzerland	2010–16	22	Based on exponential fit to reported data

CRI, Color-rendering index.

studies have estimated price trends for different categories of LED lighting products assuming an exponential decline with time (or some other time trend), rather than a learning-curve model based on cumulative production. In this section, we summarize the various time trend analyses and compare the various annual rates of price decline obtained using A-line lamps as a common technology for comparison. Table 13.2 presents a summary of the studies considered and the estimated rates of price decline.

In an early example a report from the US DOE's SSL program (Tuenge, 2013) used the purchase price of lamps acquired for performance testing within the program to track the price of various LED lighting products, ranging from household LED lamps to integrated LED fixtures, to street lights, during the period from 2008 to 2012, during the early years of market entry for the technology. As may be expected for a very new technology, the observed price declines were quite extreme, ranging from a factor of 2 to 6 in average price over the 4-year period of observation. Although the report used a power-law trend with price and did not report fit parameters explicitly, we can estimate the annual rate of price decline that would be obtained from an exponential fit, based on the fractional price decline over the observation period. For A-line lamps the fit suggests a decline rate of 65% per year during the period of observation.

Several years later, two reports from Lawrence Berkeley National Laboratory (LBNL) (Gerke et al., 2014, 2015) used web-crawling data to estimate learning curves for LED

A-line lamps. As a prelude to the learning analysis, both studies also estimated a declining exponential trend of price with time. The 2014 analysis divided the data sample into ranges of lumen output corresponding to the standard traditional incandescent wattages. Finding a faster price decline as lumen output increases, and noting that more luminous lamps had entered the market more recently than dimmer lamps, the authors focus on the lowest lumen range (310–749 lm, corresponding to a 40 W incandescent lamp) as the best estimate of the baseline decline rate, at 28% per year. The 2015 report undertook a more thorough regression analysis, including lumen output and other features, as explicit regression variables impacting the lamp price. The result was a more rigorous estimate of the underlying price decline rate, which happened to be unchanged from the 2014 report at 28% per year.

In a formal comment on proposed energy-consumption standards for LED lighting products in California (McGaraghan, 2015), prepared by the consultancy Energy Solutions on behalf of the California Investor Owned Utilities, web crawling data were used to monitor price declines for LED A-line lamps categorized by their color-rendering index (CRI) into typical and high-CRI ranges. The authors found a price decline rate of approximately 21% per year for the typical products, while the high-CRI lamps, which had entered the market later, were falling at a faster rate of roughly 35% per year. This difference in price decline rates highlights a challenge, mentioned in Section 13.2, in using web crawling data to estimate a learning rate: lamps with the high-CRI feature were a growing fraction of the market and had a more rapidly declining price, which could confound efforts to estimate a learning rate for the underlying technology.

A 2016 forecast of energy savings potential from LED products from the US DOE's SSL program (Penning et al., 2016) developed price forecasts for numerous different categories of LED lighting products by using web crawling data to estimate learning-based price trends, using the same regression model as Gerke et al. (2015). Although it did not report fit parameters explicitly, for the purpose of comparing to other studies, we can infer the measured rate of price decline from the forecasted price declines early in the forecast period, when the trend would still be expected to be approximately exponential. For LED A-line lamps the report forecasts a price drop of a factor of four from 2015 to 2020, which would correspond to an exponential decline at a rate of 28% per year.

The picture that emerges from these studies is of a steady 20%–30% annual price decline for LED lamps in the US market. Relatively fewer studies have been published on LED lamp price trends outside the US market, so it is difficult to know how well these results scale to the rest of the world. However, in 2018, the IEA referenced an 80% price decline between 2014 and 2018 for India's LED bulk procurement program, Unnat Jyoti by Affordable LEDs for All (UJALA) [International Energy Agency (IEA), 2018], which translates to a roughly 33% annual price decline, broadly in line with (or even slightly faster than) trends observed in the United States. Most recently, a study from the University of Geneva (Heidari et al., 2018) presented data on the price of LED lamps (as well as other technologies) in Switzerland, between 2010 and 2016, based on a compilation of data from a variety of online and other sources. They fitted an exponential model of price decline, with an additional constant term based on an assumed price floor. Based on the data reported in this study, the price of LED lamps on the Swiss market fell by a factor of four over the period considered, from CHF 35.3 in 2010 to CHF 8.7 in 2016. An exponential fit to the reported data implies a 22% annual rate of price decline over the period, which is similar to trends observed in the United States.

13.3.2 Learning-curve analyses

If an emerging technology follows a learning curve, one would naturally expect an exponential price decline with time in the early adoption period, based on commonly used models for market adoption. Growth in the adoption of new technologies is often observed to obey an *S*-shaped curve (Bass, 1969) that is well approximated by an exponential growth curve in the early adoption period. By the properties of exponential functions, exponential growth in production implies that the *cumulative* production Q is also growing exponentially: $Q \propto e^{\beta t}$. If we insert this relation into the learning-curve equation, we find that the price is expected to fall exponentially when production is growing exponentially:

$$P \propto Q^{-b} \propto e^{-b\beta t} = e^{-\alpha t}. \tag{13.1}$$

Since LED lighting products are still early in their market-penetration curve, the exponential price-decline models in the studies mentioned above are consistent with an assumption that prices followed a technological learning curve. If overlapping production data become available in the future, a learning parameter could be estimated from the exponential price trends by combining the rate of price decline with the growth rate of production: $b = \alpha/\beta$.

More explicit attempts to estimate a learning curve for standard A-type LED light bulbs were undertaken in the two reports from LBNL (Gerke et al., 2014, 2015). In both studies the authors used web-crawling data that had been collected on a weekly or biweekly basis from online retailers in the US market, starting in late 2011 and continuing through the year of publication. The authors then combined the web-crawled price data with the quarterly NEMA shipments index to estimate a learning curve. Because the NEMA indices did not include LED lamps prior to 2011, the authors extrapolate an exponential growth curve backward to an assumed introduction year of 2004, prior to which shipments were assumed to be zero. With this backcast, it was then possible to compute an index of cumulative shipments.

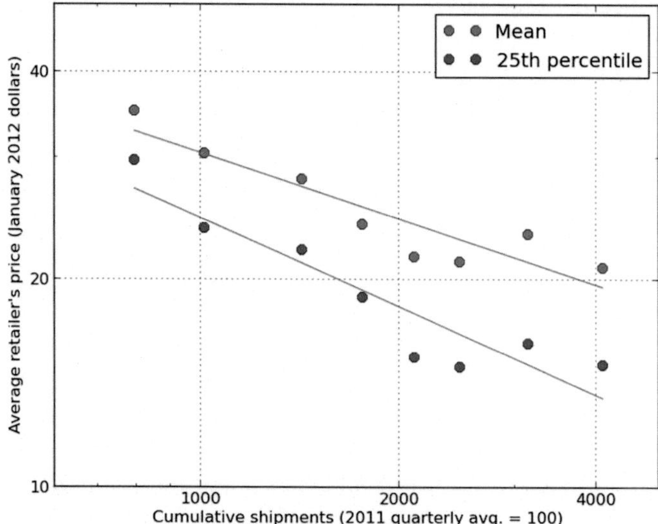

Figure 13.3
Learning curves fitted to the mean and 25th percentile in price for A-type LED light bulbs near 500 lm, offered for sale at the US online retailers in the period from late 2011 to 2014. Cumulative shipments are shown in units where the average quarterly shipment volume in 2011 is equal to 100. The mean and 25th percentile learning curve fits imply a 20% and a 26% drop in price for each doubling of cumulative shipments, respectively. *LED*, Light-emitting diodes.
Source: Reproduced from Gerke et al. (2014).

As discussed in Section 13.3, the broad diversity in product features and prices captured in high-frequency web-crawling data presented a challenge for estimating a typical price value that can be fitted against the cumulative shipments index. The 2014 report estimates a learning curve based on the price of LED light bulbs having relatively low lumen output, comparable to a traditional 40-W incandescent bulb (approximately 500 lm), to avoid the confounding effects of higher lumen bulbs that had entered the market at a significant price premium during the analysis period. The report aggregates the prices of different light bulbs offered on the market by selecting either the mean price (to represent a typical overall price) or the 25th percentile (to represent a typical price for a purchased item) from each retailer, and then averaging this statistic across retailers to smooth out differences in pricing strategy. To convert the resulting weekly prices into a quarterly price, the authors took a 6-week average about the end date of each quarter. Fig. 13.3 shows the result of fitting these aggregated quarterly prices against the cumulative NEMA shipments index to estimate a learning curve. The study found learning parameters of $b = 0.32 \pm 0.05$ for the mean price and $b = 0.43 \pm 0.07$ for the 25th percentile price (corresponding to a 20% and 26% price decline per doubling of cumulative shipments, respectively).

The growing difference between the mean and 25th percentile price, observed in Gerke et al. (2014), is suggestive of a market whose price structure is undergoing differentiation on product features. To better account for these market dynamics the same authors took a different approach in their 2015 study. Sidestepping the issue of aggregating weekly price data to a quarterly value, they instead fitted price and cumulative shipments as exponential functions of time, then estimated a learning parameter via Eq. (13.1). This approach allowed the use of a more thorough regression model for price, utilizing all of the web-crawled data and including as variables additional product features, such as lumen output, color temperature, and brand name. They found that including the product features—especially brand name—resulted in a slower estimated price trend, suggesting that an evolving market landscape, with brands competing on price and other features, was driving a steeper price decline than the baseline trend from technological learning. Using a regression model that accounts for these effects results in an estimated learning parameter $b = 0.30$, corresponding to an 18% price decline per doubling of cumulative production.

Within the reported uncertainty the results of the 2015 study on learning rate are consistent with the results for the learning rate estimated for the mean price statistic in the 2014 report. This suggests that a value of approximately 0.30 is a reasonably a robust estimate of the learning parameter b for A-type LED light bulbs in the US market, and that steeper observed price declines may be driven in part by product differentiation on features, or other competitive effects.

In addition to the LBNL studies, as discussed in Section 13.3.1, an LED energy-savings forecast from US DOE (Penning et al., 2016) also used web-crawling data to fit price trend regression models for LED lighting products. Although that report did not estimate learning parameters, it does provide forecasts out to 2030 for an impressively broad array of products, which we discuss in Section 13.4.

13.3.3 Main drivers of the price decline

As discussed previously, actual cost data are difficult to obtain for the componentry and manufacturing processes of LED lighting products. However, the US DOE's SSL program periodically polls manufacturers to build a picture of the relative cost for different components of the manufacturing process and their evolution over time. The main manufacturing cost components for an LED lighting product can be subdivided into several categories: the LED packages themselves; the driver and other electronics; thermal, mechanical, and electrical components; optics; assembly; and overhead costs, including research and development, engineering, regulatory compliance, packaging, and distribution (Bardsley et al., 2014, 2016). The relative costs in each of these categories varies substantially among different lighting products, but as of 2016, the dominant cost category was generally the thermal, mechanical, and electrical components (such as heat sinks,

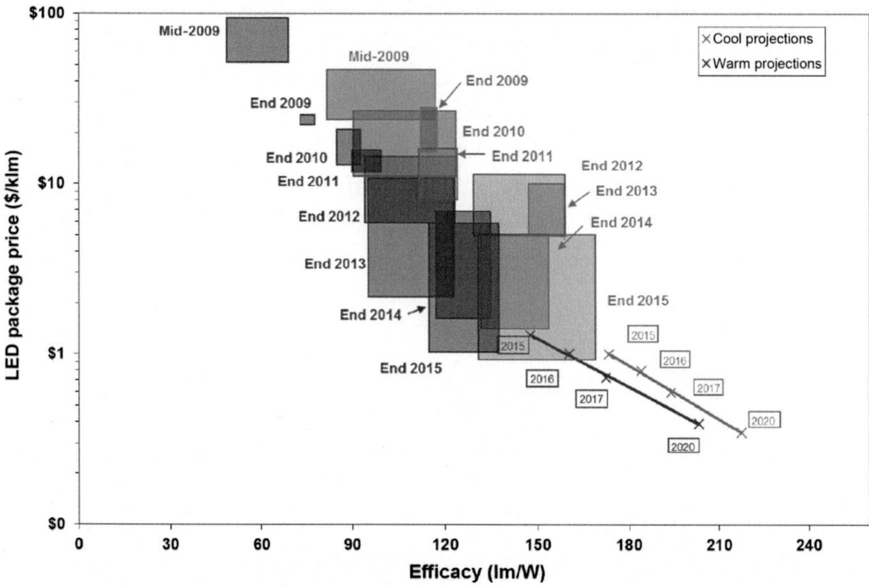

Figure 13.4

Typical ranges of retail price and luminous efficacy for LED packages from 2009 to 2015, with projections to 2020. As noted in the original source, efficacies are as obtained when operating the packages with power density of 1 W/mm^2 at an operating temperature of 25°C; cool white packages are assumed to have CCT of 5700 K and CRI of 70, while warm white packages assume CCT = 3000 K and CRI = 80, and while rectangles represent the full region mapped by maximum efficacy and lowest price for each time period, the maximum efficacy may not have been available for purchase at the lowest price. *CCT*, Coordinated color temperature; *CRI*, color-rendering index; *LED*, light-emitting diode. *Source: (Bardsley et al., 2016).*

electrical connectors, fasteners, and housing), which made up more than one quarter of costs. These were followed in importance by the driver electronics, the LED packages, assembly, overhead, and optics (Bardsley et al., 2016).

In the early to the mid-2010s the LED package category underwent the most dramatic cost decline: LED packages were by far the dominant cost component at the start of the decade (Bardsley et al., 2014), but by 2016, they had fallen to less than 20% of total manufacturing cost across a range of product types (Bardsley et al., 2016). This decline was driven by a more than 10-fold decline in the price of LED packages between 2009 and 2016, as shown in Fig. 13.4. Over the same period the luminous efficacy of LED packages (lumens of output per Watt of electricity) roughly doubled, facilitating new product designs utilizing fewer packages per lamp or luminaire, further reducing the overall costs in the LED package category. Costs in the other main categories declined as well, though at a slower pace. Notably, there were numerous changes to the overall design of LED lighting products as their market adoption increased (Bardsley et al., 2016). New system designs can significantly impact overall costs owing to changes in the bill of materials and assembly process, even if the per-unit cost of components remains unchanged. Because of this, it can be difficult to account for the full price trend using a reductive, purely component-based approach. In this situation a more holistic learning analysis at the product level, such as those described earlier in this section, may be more effective at capturing the full effect of learning.

13.4 Future outlook

13.4.1 Product-level price projections

Two different 2016 studies from US DOE projected future prices for LED lighting products, using learning curves estimated from historical data combined with forecasts of shipments to the US market. The agency published a Notice of Proposed Rulemaking (NOPR) (US DOE, 2016b) proposing new energy efficiency standards for general service lamps (GSLs), along with a detailed analysis of the expected national impacts of such standards (US DOE, 2016c). The analysis included a projection of the shipments of LED A-line lamps under different policy scenarios, based on a model of stock turnover and consumer adoption, and it used the learning curve estimated by Gerke et al. (2015) to project the price declines expected to occur in each case. In addition, as discussed in Section 13.3, a broader energy savings forecast from the US DOE's SSL program used a learning-based model, along with a shipments projection to forecast price declines for a variety of different LED lighting products (Penning et al., 2016). In both the cases the forecasts were based on analysis of trends in the total product retail price, rather than a bottom-up component-based analysis of manufacturing costs.

The supporting analysis for the GSL NOPR included a detailed projection of annual US shipments and price points for LED A-line lamps having different lumen outputs and efficiency levels, from 2015 to 2049 (US DOE, 2016c). Fig. 13.5 displays the results of this projection for the scenario with no new efficiency standards.[3] Shown on a semilog plot are the projected cumulative shipments of LED A-line lamps to the US market (*solid black curve*); the projected lamp price for lamps having four different lumen outputs, corresponding approximately (from bottom to top) to the output of standard 40, 60, 75, and 100 W traditional incandescent lamps (colored curves); and the per-kilolumen price averaged across all four lumen outputs (*dashed black curve*), for ease of comparison to

[3] Since neither the proposed standard nor any other new efficiency standard covering LED A-line lamps had been finalized at the time of writing, the no-new-standards scenario represents the current status quo.

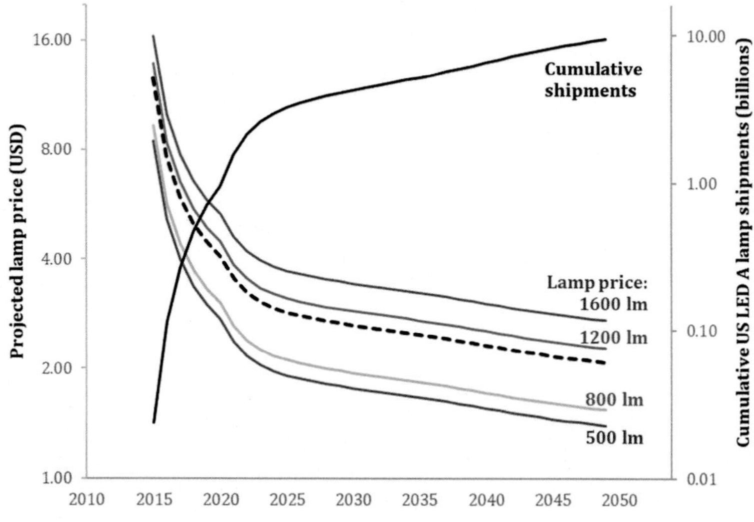

Figure 13.5

Projected cumulative shipments and price for LED A-line lamps in the US market, from an analysis performed in support of proposed energy-efficiency standards for general service lamps. The price trends indicate price projections for lamps with four different lumen outputs, corresponding to the standard wattages of (from bottom to top) 40, 60, 75, and 100 historically used for incandescent lamps. Also shown is the normalized price per kilolumen of light output, averaged across all four lamp types shown (dashed curve). *Source: Developed from data in US DOE (2016d).*

other results.[4] The projection indicates a rapid growth of cumulative shipments through the early 2020s, as LED market-adoption rates increase, followed by a slower rate of growth due to natural stock replacement and market growth after market saturation is reached. The lamp prices mirror the cumulative shipments growth, undergoing a dramatic drop of approximately a factor of four from 2015 to the early 2020s,[5] followed by a slower, yet steady, price decline. The ultimate result is a projected sixfold price decline from 2016 to 2049, with the per-kilolumen price falling to USD 2.60 by 2030 and to approximately USD 2 by the end of the period.

The forecasts from the US DOE SSL program (Penning et al., 2016) cover a much broader range of LED lighting products, but with less detail than the standards analysis. Table 13.3 shows price forecasts out to 2035 for a selection of different commonly used lamp types;

[4] Averaging across the lumen ranges is necessary, since lamp luminous efficacy (lumens per Watt) increases with increasing lumen output, so that the higher output lamps tend to be less expensive per kilolumen than the lower output ones.

[5] A particularly sharp feature is visible in Fig. 13.5 around 2020, where there is a sudden sharp increase in cumulative shipments and a corresponding acceleration in the price decline due to learning impacts. This represents the projected impact of a provision in the US law that would effectively eliminate most A-line halogen incandescent lamps in 2020, if certain conditions are met (US Congress, 2007).

Table 13.3: Price forecasts from Penning et al. (2016) is based on learning-curve analysis for various types of light-emitting diode lamps.

Product category	2015	2020	2025	2030	2035
A-line lamps	16	4	3	3	3
Large directional	21	12	9	8	7
Small directional	47	13	10	9	9
Linear tube	20	7	5	4	3
Low and high bay	30	17	13	11	10
Decorative	28	8	6	6	5
Area and roadway	23	15	12	11	10

All values are forecast lamp price in USD per kilolumen.

the full report covers a wider range of products, including integrated LED luminaires. The results overall indicate a substantial price drop for all LED lighting products, ranging from a factor of two to a factor of more than six out to 2030. As expected for an emerging technology, the bulk of the price decrease occurs in the early part of the forecast period, with a flatter trend for all products in the later years. The forecast for A-line lamps is broadly in line with the standards analysis, indicating a roughly fivefold price decline from 2015 to 2030 (although the standards analysis has marginally lower absolute price estimates throughout).

13.4.2 Component-based manufacturing cost projections

As discussed in Section 13.3.3, although specific component-level price data for LED lighting products are difficult to obtain, the US DOE SSL program periodically polls manufacturers for cost information to develop an overview of relative manufacturing costs for LED lighting products. The information gathered through that effort includes the best estimates of manufacturers for near-term cost declines. Fig. 13.6 shows the resulting estimates for the case of A-line LED lamps, as of 2016 (Bardsley et al., 2016). Overall, manufacturing costs were expected to decline by 40% by 2020, led by a continuing rapid decline in the cost of LED packages, with a more modest decline across all other categories, with the exception of thermal, mechanical, and electrical components, which were expected to undergo a modest cost increase.

13.4.3 Discussion

Both the product-level price forecasts and the component-level cost projections indicate that fairly dramatic price declines are expected to continue for LED lighting products. Nevertheless, there is a clear tension between the two approaches, with manufacturers projecting a 40% decline in costs from 2015 to 2020, while the price-based forecasts point to a fourfold price drop over the same period. Some discrepancy is perhaps to be expected,

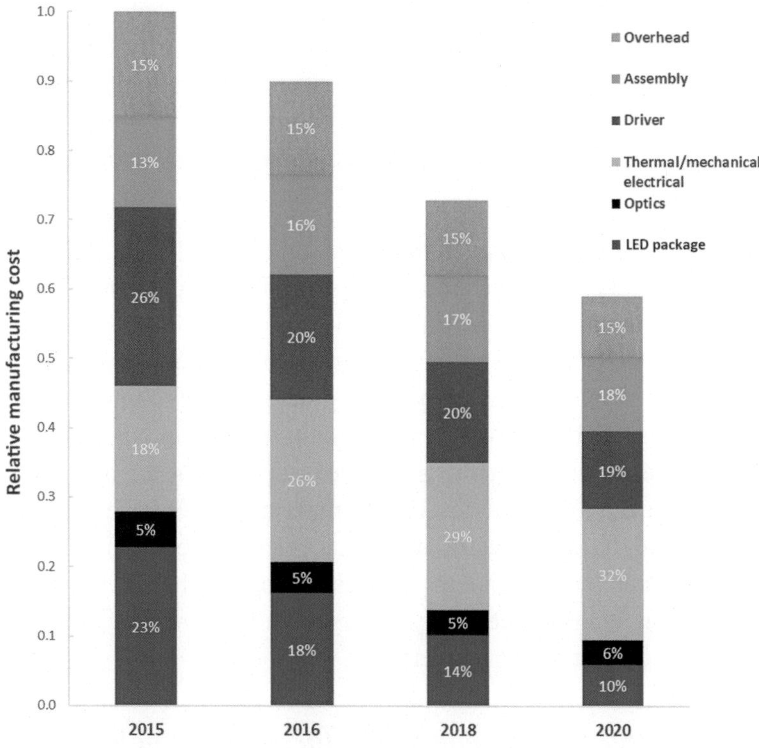

Figure 13.6
Projected evolution of relative manufacturing costs in different cost categories for LED A-type lamps, based on manufacturer input gathered by the US DOE SSL Program. *LED*; Light-emitting diodes; *US DOE*, United States Department of Energy. *Source: (Bardsley et al., 2016).*

given that the price forecasts are based on historical trends, whereas the cost projections reflect manufacturer expectations for componentry and labor, as informed by current manufacturing practice. One explanation for this difference could involve changing margins for manufacturers. Indeed, prices for emerging technologies have been observed, in certain cases, to fall at different rates from manufacturing costs, when an "umbrella" pricing period maintains prices at an elevated level, relative to the declining manufacturing costs, followed by a competitive "shakeout" period that drives prices down at a faster rate than costs, before the price decline settles to the same rate as the cost trend (Hedley, 1976).

Alternatively, or in addition, future evolution in product designs could reduce the required bill of materials for manufacturing or simplify the assembly process, leading to faster declines in total manufacturing cost (and ultimately product price) than would be expected based only on the component-level projections (see Bardsley et al., 2016, for further discussion of this point). For example, as LEDs have become more efficient, the need for thermal management has become less acute, reducing the amount of material needed for heat sinks, so the contribution of these components to total cost may fall more rapidly than the material cost evolution would imply. This represents a technological learning effect that is not captured by a cost forecast for the individual components of current manufacturing practice; in this sense the more holistic price-based approach may lead to more accurate forecasts.

Indeed, there is some indication that even the more rapid price-based forecasts may underestimate the true rate of price decline. There have been recent anecdotal observations of LED A lamp prices that have *already* fallen below the USD 3/klm price that was forecast for 2030 in the c.2016 price-based forecasts. Observations of lower than expected prices, occurring earlier than expected, may simply stem from evolving utility or governmental incentives, or they may arise from changes to lamp quality and features, as competition for customers drives firms to pursue lower prices, at the expense of reduced product quality or features (e.g., lifetimes, color rendering, or dimmability). It is also possible that the learning-based forecasts of A-lamp prices underestimate the price decline because they use the cumulative shipments of A-lamps only, whereas the cumulative shipments of all LED lighting products may have grown more rapidly (since LED A lamps were a relatively early entrant to the market).

13.5 Conclusions and recommendations for science, policy, and business

The 2010s decade saw a steady and rapid decline in price for LED lighting products, with prices falling several fold from the high initial market-entry prices observed near the beginning of the decade. For LED A-line lamps sold in the US market, a steady decline of 20%–30% per year was observed through the first half of the decade. At the same time, LED lighting products were also experiencing fast growth in consumer uptake, resulting in a many-fold increase in cumulative production over a period of only a few years. This situation presented an unusual opportunity to observe significant technological-learning effects in near real time as they occurred over a period of only a few years. In the instance of A-line lamps in the United States, researchers were able to measure a learning curve robustly using only 2–3 years of data on price and lamp sales, finding a price decline of 18% for every doubling in cumulative production (Gerke et al., 2015).

Despite the unmistakable price declines and production growth, there are nevertheless challenges to measuring the effects of technological learning for a consumer product such as LED lighting products in the context of a rapidly evolving and highly competitive market. In such a market, data on absolute component costs and production output tend to be closely held by manufacturing firms, rendering relevant data difficult to obtain. Because LED lighting products are distributed through mass market and online consumer-facing channels, gathering information on retail price was straightforward using web-crawling approaches, permitting price trends to be easily tracked. However, during the early years of market adoption for LED lighting products, there was also dramatic evolution in the mix of products and product features available in the market, such as the introduction of lamps with increasingly higher lumen output. This churn in the mix of product features creates a significant confounding factor for measuring the price effects of learning; to isolate learning effects, it was essential to control for features either by restricting the sample under consideration (Gerke et al., 2014) or by using multivariate regression (Gerke et al., 2015). Generally speaking, evolving product features should be expected to pose challenges when undertaking any learning analysis based on the retail prices of consumer products. Thus for researchers analyzing learning effects based on retail price, it will be important, in addition to collecting price and production data, to also collect data on any product features that might significantly impact price and control for these in any analysis of learning.

The dramatic price drop for LED lighting products observed in the past decade, and the future declines still projected to occur, are also a reminder that it is important to account for the effects of technological learning when making strategic business or policy decisions regarding new and emerging technologies. Early LED lighting products entered the market at price points on the order of $100/klm, and, despite their energy efficiency and lifetime advantages, they were not cost competitive with the most efficient incumbent technologies (see Garbesi et al., 2018, for a cost-effectiveness comparison between early LED products and incumbent technologies). Nevertheless, rapid price declines meant that LED lamps had captured a majority of the US A-line lamp market by the late 2017 (NEMA, 2018b) which likely conferred a significant market share advantage on firms that entered the LED market early, despite the prohibitive early price points.

In the context of policy the LED price decline was a strong reminder of the importance of including the effects of technological learning when considering the economic impacts of energy efficiency policy: a projected fourfold decline in the price of LED lighting products by 2050 (US DOE, 2016b) is likely to have substantial implications for the expected impacts of policies that impact these products. (For a broader discussion of technological learning in the context of developing energy efficiency policy, see Desroches et al., 2013.) Moreover, forward-looking approaches for policy development may have interacted positively with technological learning effects to help encourage the transition to LEDs. Few or no viable LED lighting products existed that could replace incumbent technologies when the United States and EU announced the phase-out of less efficient traditional lighting technologies (in 2007 and 2009, respectively), but these announcements themselves helped to create a market for LED lighting products, by effectively displacing an incumbent technology. In concert with directly supportive policies, such as state and utility efficiency programs and the US DOE's SSL program, the phaseout announcements may themselves have helped to spur research and development, driving a faster price decline for LED lighting products, and a faster technological transition, that would have occurred in the absence of the policies. (For a broader discussion of the potential interactive effects between energy efficiency policy and price trends, see Van Buskirk et al., 2014.)

Thus a key lesson from the market for LED lighting products in the 2010s is that an appreciation for the effects of technological learning is essential for sound decision-making with regard to emerging technologies, both for market actors and for policymakers. Decisions that may seem bold, or even foolhardy, in the context of status-quo market conditions may in fact appear wise and beneficial once the full effects of technological learning are considered.

Acknowledgments

Lawrence Berkeley National Laboratory is supported by the US Department of Energy under Contract No. DE-AC02-05CH1131. I would like to thank Allison Ngo, Andrea Alstone, and Liat Zavodivker for assistance in compiling the lamp sales index data from NEMA. I also thank Joe Ritchey and his colleagues at the IEA for providing the data underlying Fig. 13.2; and Stephane de la Rue du Can, Michael McGaraghan, and Peter Alstone for providing background references that contributed to the information in this chapter.

References

Alstone, P., Jacobsen, A., 2018. LED advances accelerate universal access to electric lighting. C. R. Phys. 19 (3), 146.

Ashe, M., Chwastyk, D., de Monasterio, C., Gupta, M., Pegors, M., 2012. 2010 U.S. lighting market characterization. In: Prepared for United States Department of Energy Office of Energy Efficiency and Renewable Energy. Available from: <https://www1.eere.energy.gov/buildings/publications/pdfs/ssl/2010-lmc-final-jan-2012.pdf>.

Bardsley, N., Bland, S., Chwastyk, D., Pattison, L., Pattison, M., Stober, K., et al., 2014. Manufacturing roadmap: solid state lighting research and development. In: Prepared for United States Department of Energy Office of Energy Efficiency and Renewable Energy. Available from: <https://www1.eere.energy.gov/buildings/publications/pdfs/ssl/ssl_mfg_roadmap_aug2014.pdf>.

Bardsley, N., Hansen, M., Pattison, L., Pattison, M., Stober, K., Taylor, V., et al., 2016. Solid state lighting program R&D plan. In: Brodrick, J. (Ed.), Prepared for United States Department of Energy Office of Energy Efficiency and Renewable Energy. Available from: <https://www.energy.gov/sites/prod/files/2018/09/f56/ssl_rd-plan_jun2016.pdf>.

Bass, F.M., 1969. A new product growth model for consumer durables. Manage. Sci. 15 (5), 215–227.

Buccitelli, N., Elliott, C., Schober, S., Yamada, M., 2017. 2015 U.S. lighting market characterization. In: Prepared for United States Department of Energy Office of Energy Efficiency and Renewable Energy. Available from: <https://www.energy.gov/sites/prod/files/2017/12/f46/lmc2015_nov17.pdf>.

Chunekar, A., Mulay, S., Kelkar, M., 2017. Understanding the Impacts of India's LED Bulb Programme, 'UJALA'. Prayas Energy Group Report. Available from: <http://www.prayaspune.org/peg/publications/item/354-understanding-the-impacts-of-india-s-led-bulb-programme-ujala.html>.

De Almeida, A., Santos, B., Bertoldi, P., Quicheron, M., 2014. Solid state lighting review—potential and challenges in Europe. Renew. Sustain. Energy Rev. 34, 30–48.

Desroches, et al., 2013. Incorporating experience curves in appliance standards analysis. Energy Policy 52, 402–416.

European Parliament, 2009. Directive 2009/125/EC of the European Parliament and of the Council of 21 October 2009 establishing a framework for the setting of ecodesign requirements for energy-related products. Off. J. L 285, 10, 31.10.2009.

Feezell, D., Nakamura, S., 2018. Invention, development, and status of the blue light-emitting diode, the enabler of solid-state lighting. C. R. Phys. 19 (3), 113.

Garbesi, K., Gerke, B.F., Alstone, A.L., Atkinson, B., Valenti, A., Vossos, V., 2018. Energy-efficient lighting technologies and their applications in the residential and commercial sectors, Energy Efficiency and Renewable Energy Handbook, second ed. CRC Press, Boca Raton, FL.

Gerke, B.F., A.T. Ngo, A.L. Alstone, and K.S. Fisseha. 2014. The Evolving Price of Household LED Lamps: Recent Trends and Historical Comparisons for the US Market. Lawrence Berkeley National Laboratory report number LBNL-6854e. Available from: <http://eta-publications.lbl.gov/sites/default/files/lbnl-6854e.pdf>

Gerke, B.F., A.T. Ngo, and K.S. Fisseha. 2015. Recent Price Trends and Learning Curves for Household LED Lamps from a Regression Analysis of Internet Retail Data. Lawrence Berkeley National Laboratory report number LBNL-184075. Available from: <http://eta-publications.lbl.gov/sites/default/files/lbnl-1006273.pdf>

Haitz, R., F. Kish, J. Tsao, J. Nelson, 1999. The case for a national research program on semiconductor lighting. In: Annual Forum of the Optoelectronics Industry Development Association, Washington, DC.

Haitz, R., Tsao, J.Y., 2011. Solid-state lighting: 'The Case' 10 years after and future prospects. Phys. Status Solidi (A) 208 (1), 17–29.

Hedley, B., 1976. A fundamental approach to strategy development. Long Range Plann. 9 (6), 2–11.

Heidari, M., Majcen, D., van der Lans, N., Floret, I., Patel, M.K., 2018. Analysis of the energy efficiency potential of household lighting in Switzerland using a stock model. Energy Build. 158, 536–548.

International Energy Agency (IEA), 2016. Energy Efficiency Market Report. 2016. Available from: <https://www.iea.org/eemr16/files/medium-term-energy-efficiency-2016_WEB.PDF>.

International Energy Agency (IEA), 2018. Energy Efficiency 2018: Analysis and Outlooks to 2040. Available from: <https://webstore.iea.org/download/direct/2369?fileName = Market_Report_Series_Energy_Efficiency_2018.pdf>.

McGaraghan, M., 2015. CA IOU comments on LED lamps. In: California Energy Commission Docket Number 15-AAER-06, TN# 206868.

NEMA, 2018a. Lamp Indices. Available from: <https://www.nema.org/Intelligence/pages/lamp-indices.aspx> (accessed 10.02.18.).

NEMA, 2018b. LED A-Line and Halogen Lamp Shipments Increase in Third Quarter 2018. Available from: <https://www.nema.org/Intelligence/Indices/Pages/LED-A-line-and-Halogen-Lamp-Shipments-Increase-in-Third-Quarter-2018.aspx> (accessed 10.02.18.).

Penning, J., Stober, K., Taylor, V., Yamada, M., 2016, Energy savings forecast of solid-state lighting in general illumination applications. In: Prepared for United States Department of Energy Office of Energy Efficiency and Renewable Energy. Available from: <https://www.energy.gov/sites/prod/files/2016/09/f33/energysavingsforecast16_2.pdf>.

Tuenge, J.R., 2013. SSL pricing and efficacy trend analysis for utility program planning. In: Prepared for United States Department of Energy Office of Energy Efficiency and Renewable Energy. Available from: <http://apps1.eere.energy.gov/buildings/publications/pdfs/ssl/ssl_trend-analysis_2013.pdf>.

US Congress, 2007. Energy Independence and Security Act of 2007. 42 U.S. Code 152 § 17001.

US DOE, 2014. Final Rule Technical Support Document, Energy Conservation Program for Consumer Products and Certain Commercial and Industrial Equipment: General Service Fluorescent Lamps and Incandescent Reflector Lamps. Available from: <https://www.regulations.gov/contentStreamer?documentId = EERE-2011-BT-STD-0006-0066&contentType = pdf>.

US DOE, 2016a. L Prize® Competition Drives Technology Innovation, Energy Savings. Available from: <https://www.energy.gov/sites/prod/files/2017/10/f38/rdimpactsummary_lprize.pdf>.

US DOE, 2016b. Energy Conservation Standards for General Service Lamps: notice of proposed rulemaking (NOPR) and announcement of public meeting. In: Federal Register of the United States, 81 FR 14528.

US DOE, 2016c. Notice of Proposed Rulemaking Technical Support Document: Energy Efficiency Program for Consumer Products and Commercial and Industrial Equipment: General Service Lamps. Available from: <https://www.regulations.gov/document?D = EERE-2013-BT-STD-0051-0042>.

US DOE, 2016d. Analytical Spreadsheet: National Impacts Analysis and Shipments Analysis. Available from: <https://www.regulations.gov/document?D = EERE-2013-BT-STD-0051-0040>.

US DOE, 2018. Available from: <https://www.energy.gov/eere/ssl/solid-state-lighting> (accessed 10.12.18.).

Van Buskirk, R.D., Kantner, C.L.S., Gerke, B.F., Chu, S., 2014. A retrospective investigation of energy efficiency standards: policies may have accelerated long term declines in appliance costs. Environ. Res. Lett. 9 (11), 114010.

PART III

Application of experience curves in modeling

CHAPTER 14

Experience curves in energy models—lessons learned from the REFLEX project

Steffi Schreiber[1], Christoph Zöphel[1], Christoph Fraunholz[2], Ulrich Reiter[3], Andrea Herbst[4], Tobias Fleiter[4] and Dominik Möst[1]

[1]Energy Economics, TU Dresden, Dresden, Germany, [2]Institute for Industrial Production (IIP), Karlsruhe Institute of Technology (KIT), Karlsruhe, Germany, [3]TEP Energy GmbH, Zurich, Switzerland, [4]Fraunhofer Institute for Systems and Innovation Research ISI, Karlsruhe, Germany

Abstract

The consideration of technological learning in energy system models is of crucial importance as modeling future energy pathways across several sectors by considering cost developments influences model results significantly. Implementing experience curves in energy system models is one of few methods to consider the relation between cumulative installed capacity deployment and unit cost reductions of a technology. In this chapter the endogenous and exogenous approach of implementing technological learning in different energy system models are compared and analyzed in detail. Therefore the corresponding strengths and limitations of these approaches are encountered as well as possible solutions to overcome these constraints are estimated. To determine the influence of uncertainty in experience curves, sensitivity analyses with three different bottom-up models are conducted. The analysis of the literature and the lessons learned from the REFLEX project reveal that the endogenous approach is feasible, especially for top-down models but related to several challenges. Thus a balance between modeling accuracy and increasing complexity needs to be maintained while interpreting modeling results carefully.

Chapter Outline

Technological Learning in the Transition to a Low-Carbon Energy System.
DOI: https://doi.org/10.1016/B978-0-12-818762-3.00014-5

14.1 Technological progress and experience curves in energy system modeling

As discussed in Chapters 2 and 3, one of the key applications of experience curves lies in energy system modeling. By using experience curves in modeling, future technology cost trajectories resulting from technological progress can be taken into account. The objective of this chapter is to estimate the importance of experience curve implementation in energy system models to derive realistic scenario pathways for long-term perspectives. In this context the chapter contributes to the existing literature by answering the following research question: What are the chances and barriers of endogenous and exogenous incorporation of experience curves in energy system models?

To answer the research question the modeling efforts and the experiences gained from the REFLEX project are analyzed. The project received funding from the European Union's (EU) Horizon 2020 research and innovation program.[1] The core objective of REFLEX is to analyze and evaluate the development toward a low-carbon energy system with focus on flexibility options in the EU to support the implementation of the Strategic–Energy–Technology Plan. The analyses are based on a modeling environment that considers the full extent to which current and future energy technologies and policies interfere and how they affect the environment and society while considering technological learning of low-carbon and flexibility technologies. Miscellaneous technologies exist, which affect the European energy system in different ways. Within the REFLEX project, experience curves for most relevant technologies are developed and implemented in sector-based energy system models. Therefore special attention is given to determination of uncertainty ranges of progress ratios, i.e., the slopes of the experience curves, as these can have a major impact on modeling results, especially for long-term modeling until 2050. In an early state of the project, it was planned to incorporate the derived experience curves in the various energy models to model technological learning endogenously. However, the endogenous implementation of experience curves is challenging. Therefore the exogenous incorporation of experience curves is applied in most of the REFLEX models. The chapter gives an overview of the advantages and disadvantages of the endogenous and exogenous implementation of experience curves in energy system models.

14.2 Description of applied energy models with implemented experience curves

In REFLEX, several energy system models with miscellaneous modeling approaches are soft-linked with both one- and bidirectional data exchanges. The basis for this linkage

[1] GA-No. 691685, http://reflex-project.eu/.

builds a scenario framework and a common database with the result that all models are based on identical assumptions. The soft-linked model coupling approach is used since the features and high techno-economical details of the models to couple can be integrated without excessively increasing the computational costs. This way, REFLEX takes advantage of the individual strengths of each model while covering all aspects and sectors of the European energy system. In addition an important disadvantage of the integrated approach can be avoided since relying on extremely complex single models may decrease the understanding, and thus acceptance of model results particularly when communicated to a broader public. Fig. 14.1 gives an overview of the energy model system within REFLEX and shows which models incorporate the developed learning curves (Kunze, 2018; Zöphel et al., 2019). In the following the focus is given on three models: the demand projection model FORECAST as well as the electricity market models PowerACE and ELTRAMOD.

Fig. 14.2 gives an overview of different energy modeling approaches that are used to project the future energy demand and supply of countries or regions. In general energy system models can be distinguished between top-down models (macroeconomic models)

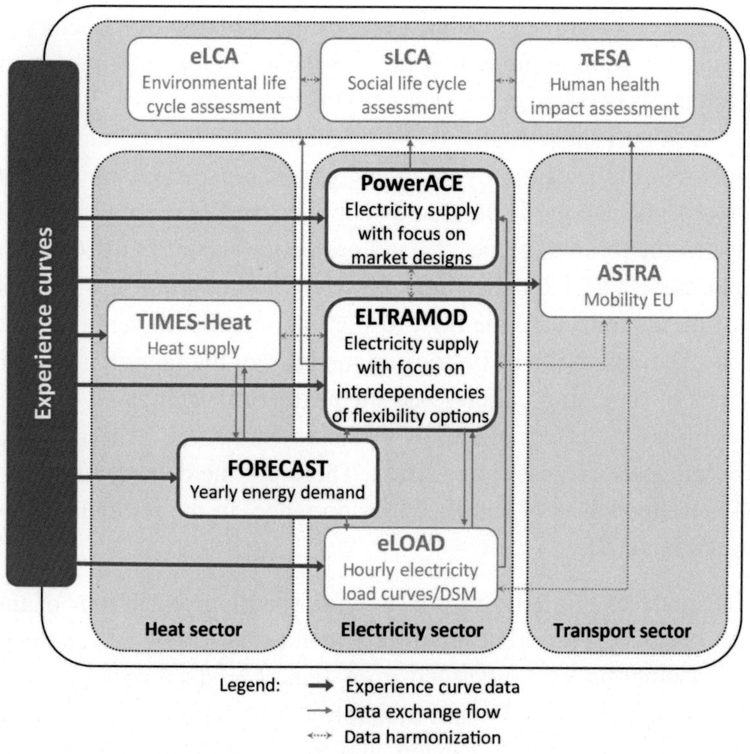

Figure 14.1
Applied models and their interlinkage within the REFLEX project. *Source: Own illustration based on the REFLEX project.*

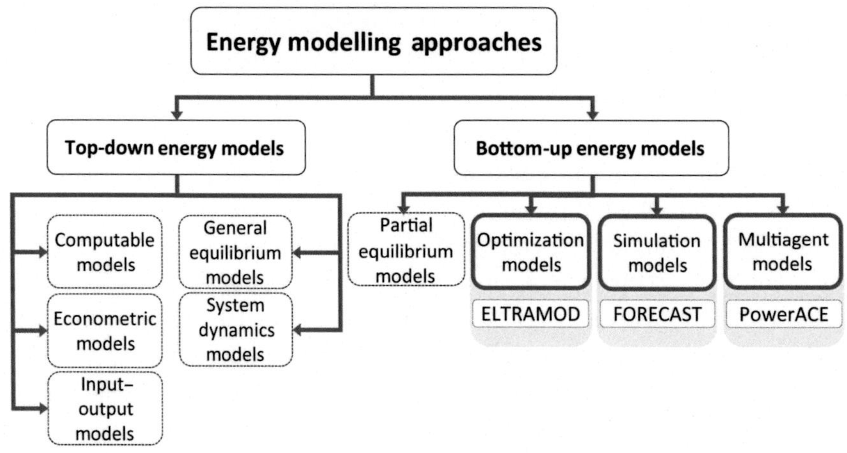

Figure 14.2
Overview of energy system modeling approaches. *Source: Based on Herbst et al. (2012).*

and bottom-up models (detailed techno-economic or process-oriented models). Top-down models are applied to depict the whole economy on a national or regional level. Therefore effects of energy as well as climate change policies are assessed in monetary units. Further, macroeconomic models equilibrate market developments by maximizing consumer welfare applying feedback loops between economic growth, employment, and welfare as well as by using production factors (Herbst et al., 2012). In contrast, bottom-up models are applied to depict energy sectors and the economy in an aggregated perspective by simulating economic developments, energy demand and supply as well as employment. The REFLEX models presented in the following, the demand projection model FORECAST as well as the electricity market models ELTRAMOD and PowerACE, are bottom-up models. The main characteristics of these models are the high resolution of technological details compared to top-down models. The objectives of bottom-up models are the identification of best technology choices by assessing policies and their effects as well as investment costs and benefits by determining the effect of energy efficiency measures, synergy effects across sectors, and sectoral costs (Herbst et al., 2012). Therefore the consideration of technological learning within these models is of significant importance. In the following the three models are described more in detail.

The FORECAST model is primarily responsible in providing projections of the future energy demand, considering different types of demand-related policies and consumer prices, among others. Consumer prices are derived from market models' output (e.g., wholesale electricity prices from ELTRAMOD) and transformed into retail prices within FORECAST. Demand projections cover all energy carriers and are available on an annual basis.

The FORECAST model is designed as a tool that can be used to support strategic decisions. Its main objective is to support the scenario design and analysis for the long-term development of energy demand and greenhouse gas emissions for the industry, residential, and tertiary sectors on country level. FORECAST considers a broad range of mitigation options to reduce CO_2 emissions, combined with a high level of technological detail. It is based on a bottom-up modeling approach, considering the dynamics of technologies and socioeconomic drivers. Technology diffusion and stock turnover are explicitly considered to allow insights into transition pathways. The model further aims to integrate different energy efficiency and decarbonization policy options (Herbst, 2017; Fleiter et al., 2018; Elsland et al., 2014; Braungardt et al., 2014).

The output of the FORECAST model, i.e., the annual electricity demand, is used as input into the hourly load curve model eLOAD, which is taking into account demand side management (DSM) potentials as a function of consumer price tariffs.

ELTRAMOD is a bottom-up electricity market model and a linear optimization model which calculates the cost-minimal generation investments and dispatch in power plant capacities, storage facilities, and power-to-x technologies [i.e., power-to-heat (p2h), power-to-gas (p2g)] while considering specific constraints. Further, it allows fundamental analyses of the European electricity market. The net transfer capacities between regions are considered, while the electricity grid within one country is neglected. Each country is treated as one node with country-specific hourly time series of electricity and heat demand as well as renewable feed-in. All technologies are represented by different technological characteristics, such as efficiency, emission factors, ramp rates, and availability. Technology-specific economic parameters are annualized capacity-specific overnight investment costs, operation and maintenance costs, fixed costs, as well as costs for ramping up and down the generation. In addition, hourly prices for CO_2 allowances, as well as hourly wholesale fuel prices are considered in ELTRAMOD. The geographical scope covers the member states of EU28, Norway, Switzerland, and the Balkan countries. In REFLEX, ELTRAMOD is used to analyze the penetration of different flexibility options and their contribution to renewable energy source (RES) integration as well as the interdependencies among various flexibility options in the European electricity system, taking existing regulatory frameworks into account (Ladwig, 2018; Schubert, 2016; Müller et al., 2013; Gunkel et al., 2012).

PowerACE analyzes the impact of different market designs (energy-only market, strategic reserve, capacity markets) and policy measures on investments in low-carbon technologies with focus on flexibility options and their contribution to the security of supply on a national and European level. PowerACE is an agent-based, bottom-up simulation model for wholesale electricity markets. The main generation companies in the modeled system are represented by individual agents. Other agents are modeled to bid electricity demand and generation from renewable energy sources as well as to operate markets. Concerning the short-term markets, the focus is on the day-ahead market. The bidding behavior of generation agents is generally based on variable generation costs, startup costs, and potential strategic considerations (e.g., scarcity markup). Regarding the geographical scope, different market areas can be coupled with a day-ahead market-coupling algorithm based on net transfer capacities. The short-term clearing of the markets is performed on an hourly basis. Furthermore, the model contains an investment planning module executed on an annual basis, which aims to determine a Nash equilibrium of investment decisions in all market areas. Potential revenues for power plants can be generated from selling electricity on energy markets as well as from different capacity remuneration schemes (e.g., central capacity market, strategic reserve) depending on the respective configuration (Fraunholz and Keles, 2019; Keles et al., 2016; Ringler et al., 2017).

Within the REFLEX project, the generated insights about learning curves, in particular with respect to new energy efficiency technologies, are integrated into the three presented models to further enhance the accuracy and consistency of the modeling approach and consider technological cost reduction in the underlying investment decisions.

14.2.1 Key new and incumbent technologies

As introduced, the bottom-up models include detailed technology options to derive energy demand and supply projections. For the long-term impact and high installation rate of the different technologies, cost development is of high relevance. Therefore for a wide set of the represented technologies, a review was conducted to estimate the relevance for model impact as well as future cost reduction potentials (in terms of investment costs), differentiated for the considered demand sectors such as industry, residential, and tertiary as well as supply sectors as electricity and heat (see Table 14.1 for summary of reviewed technologies). Besides prospective technologies for decarbonizing the energy demand sectors (e.g., heat pumps) with currently low installation rates, established technologies with high market shares (e.g., oil-based heating systems) are also considered in the review. However, the potential cost reductions of such systems are rather low, as a doubling of the installed capacity

Table 14.1: Summary of reviewed technologies regarding relevance of cost reduction and related impact on model results.

Category	Technology	Cumulative data unit	Model
Electricity generation	CCS (postcombustion CO_2 capture)	(M)W installed	ELTRAMOD, PowerACE
Electricity storage	Utility lithium-ion battery	(M)Wh + (M)W	ELTRAMOD, PowerACE
	Utility redox-flow battery	(M)Wh + (M)W	ELTRAMOD, PowerACE
Energy demand technologies	Heat pumps (power-to-heat)	(M)W	ELTRAMOD, FORECAST
	Electrolysis alk. (power-to-gas)	(M)W	ELTRAMOD, FORECAST

CCS, Carbon capture and storage.

especially in respect of climate mitigation scenarios seems to be unlikely. Therefore results and discussion of technologies mainly focuses on key new and incumbent technologies, which are expected to have growth potential in context of a decarbonized energy system.

Depending on the state of the art of the technology and the actual market share, the uncertainty of investment costs can have a relevant impact on model results. In the following sections, we describe some of the considered technologies and their impact in more detail (based on Louwen et al., 2018a).

14.2.1.1 Electricity generation technologies

To achieve the CO_2 emission reduction targets, carbon capture and storage (CCS) will play a crucial role by generating low-carbon electricity for power and by providing process energy for the industry sector. Various types of CCS processes exist, which can be distinguished between the postcombustion capture, precombustion capture, and the oxy-fuel combustion capture (OECD/IEA, 2016). For the following analyses, only the postcombustion capture process is considered, as this process is assumed to be more suitable for most of the power plants. Within this process, CO_2 is captured from the flue gas of a power plant or from the industry process after combustion. Seventeen large-scale CCS projects were operating globally in 2017 and capturing over 30 $MtCO_2$ annually. The projects are mainly located in the United States, Canada, Norway, Brazil, Saudi Arabia, and United Arab Emirates. CCS in industrial processes with natural gas processing are dominating and representing a niche market where CCS pilot projects are possible to realize. In the electricity sector, CCS is less developed as 2 out of 17 CCS projects are applied in the power sector (Global CCS Institute, 2017). Thus due to the lack of existing CCS pilot projects, only few empirical data exist. Therefore experience curves from various technologies related to or used as a proxy for CO_2 capture at natural gas and coal-fired power plants are applied to derive learning rates. For instance, Rubin et al. (2007) determine future cost reductions that can be achieved by CCS plants based on rates of cost reductions of process technologies that are similar with capture plant components. In the REFLEX project, learning rates of 2.1%–2.2% for CCS technologies are developed based on these assumptions from Rubin et al. (2007). Technological breakthroughs that would result in further cost reductions and higher learning rates are neglected. As existing CCS projects are insufficient and provide no reasonable basis for empirical data, the learning rates are of high uncertainty and should be analyzed by sensitivity analyses.

14.2.1.2 Electricity storages

Lithium-ion batteries are the dominating grid-connected electrochemical storages providing electricity (69% market share of electrochemical storages in 2016). However, redox-flow batteries are of increasing importance (5% market share of electrochemical storages in 2016) (EASE/EERA, 2017). To push the deployment of redox-flow batteries, issues, such as the long-term performance and reliability, need to be solved. Electrochemical storages represent only 1% of the global grid-connected energy storages, while pumped hydro storages are presenting 96%. Leading countries in the electrochemical storage market are the United States, Korea, Japan, Germany, Italy, and Chile (REN21, 2017). In consideration of increasing renewable energy generation, energy system flexibility is needed, and therefore the scale-up of the energy storage market will continue (EASE/EERA, 2017). To derive experience curves for residential and utility-scale lithium-ion as well as for redox-flow batteries, data was taken from Schmidt et al. (2017) and own data collection (Louwen et al., 2018a). Learning rates devised are in the range of 12.5%–15.5% depending on the storage type. Spillover effects are neglected to avoid increasing modeling complexity. However, in reality, spillover effects will probably occur, for example, between lithium-ion batteries and electric vehicles.

14.2.1.3 Energy demand technologies

Heat pumps are one of several energy demand technologies with growing importance for an energy system with high shares of intermittent renewable energy sources. The p2h technology can not only satisfy the district and residential heating as well as cooling demand but can also reduce the electricity surplus (negative residual load) in times where the feed-in of renewables exceeds the electricity demanded. The heat pump market is difficult to assess due to the lack of data and the inconsistency of available information (Louwen et al., 2018a). Countries with growing sales in heat pumps are Lithuania, Ireland, France, and Poland (EHPA, 2015). Heat pumps are providing less than 1% of the total heating and cooling demand in Europe in 2015 (Fleiter et al., 2017). However, the market size in coming decades remains uncertain. Cost reductions are depending on technological learning of manufacturing, installation of heat pumps, and related system components (e.g., vapor compressor, heat exchanger) as well as economies of scale (Weiss et al., 2008). Within the REFLEX project, learning rates of ca. 11% are observed (Louwen et al., 2018a).

Another important energy demand technology is power-to-gas (hydrogen) via electrolysis. Hydrogen is used in many industries such as the chemical, refining, food, and electronic industry as well as fuel for transportation. Alkaline electrolysis is the most economical and mature electrolyzer technology. However, alkaline electrolysis is not cost-competitive yet compared to large steam methane reforming. Further technological improvements are needed for widespread commercialization. Moreover, the uptake of electrolysis is dependent on developments in different sectors. Within the REFLEX project, derived experience curves for alkaline electrolysis showed a learning rate of 17.8% ($\pm$ 5.3%) (Louwen et al., 2018a).

14.2.2 Overview of energy markets involved

Fossil-based energy markets have developed in alignment to the different energy technology diffusions and vice versa in the past (Fouquet, 2016). Therefore established market systems and platforms exist, which are supported and used by major market players and actors. This is of high relevance as these market players have large investments and interests placed in this system.

However, to decarbonize the current energy system, an energy market transition has to take place (Fouquet, 2016) and new markets have to be developed as well as other markets have to be transformed or even closed. With this transition, new opportunities arise for established players as well as for market entrants while other market players will be forced out of the market.

The electricity market in Europe will continue to play a dominant role in achieving climate mitigation targets (Green and Staffell, 2016), while other markets, due to declining fuel demand, will rather decrease (e.g., coal and oil). However, new markets need to evolve within the market transition, as new fuels and energy technologies are needed for the transition. It is expected that, for example, p2g, p2h as well as new storage markets will appear. Given the transition theory of Fouquet (2016), the full deployment and market shares of beyond 80% will take at least decades.

14.3 Model results with exogenous technological learning

In this chapter the influence of technological learning on the overall modeling results is discussed. Table 14.2 represents the main characteristics of each model and opposes them comparatively.

The analyses illustrated next are conducted under the REFLEX scenario framework that assumes a centralized energy system with high shares of renewable energy sources (High-RES centralized). More ambitious climate policies are reflected in this scenario. A major objective is to limit the global temperature increase by 2°C, by more drastically decreasing the greenhouse gas emissions. Therefore higher contribution from experience curves and the need for flexibility options due to large share of renewable energy sources (80% RES share on power generation in 2050) as well as increasing CO_2 prices (ca. 150 EUR/tCO_2 in 2050) are assumed. The centralized system is characterized by stronger deployment of wind onshore and offshore, large-scale centralized energy sources, heat supply on a centralized level (e.g., higher share of district heating), and hydrogen production in large-scale electrolyzer with distribution by trailers and pipelines (Zöphel et al., 2019).

14.3.1 Developments of novel technologies and concurrent price reductions

Most of the large-scale energy models are incorporating experience curves rather exogenously than endogenously to specify technology performance improvements and cost trajectories (Rubin et al., 2015). The considered REFLEX models employ learning curves

Table 14.2: Overview of modeling characteristics.

	FORECAST	ELTRAMOD	PowerACE
Demand side/supply side	Demand-side projection models	Supply-side models	
Sector	Industry, residential, tertiary sector	Electricity sector	
Model type	Bottom-up energy system model		
	Simulation model	Optimization model (linear problem)	Agent-based simulation model
Model target	Simulated yearly electricity and heat demand	Optimal investment and dispatch decision at minimal system costs	Nash equilibrium for investment decision and welfare-maximizing coupling of day-ahead markets
Perfect foresight/ competition	No	Yes	No
Time resolution	2014–50	2014, 2020, 2030, 2040, 2050	2015–50
	Yearly basis	Invest yearly basis Dispatch hourly basis	Invest yearly basis Dispatch hourly basis
Geographical resolution	EU27 + NO + CH	EU28 + NO + CH + Balkan	AT, BE, CH, CZ, DE, DK, FR, IT, NL, PL
Scenario	Centralized scenario with 80% RES share on power generation in 2050		
Technological learning	Exogenous implementation of experience curves		
Technologies with technological learning	• Heat pumps • Other heating technologies	• Gas, coal, lignite CCS • Utility lithium-ion battery • Utility redox-flow battery • Electrolysis • Heat pumps	• Gas, coal, lignite CCS • Utility lithium-ion battery • Utility redox-flow battery

CCS, Carbon capture and storage.

exogenously to reduce the modeling complexity and computational time. However, the experience curves are derived from empirical data and are converted to specific investment costs for selected technologies. Table 14.3 illustrates the learning rates and their estimated error values for CCS technologies, different types of battery storages, and energy demand technologies as heat pumps and electrolysis. A learning rate of, for example, 10% means that the cost of a technology decreases with 10% for every doubling of cumulative production.

The experience curves are all showing production or price decline, whereas none of the selected technologies are identified with constant or increasing costs (not over several cumulative doublings of deployment). *CCS-related technologies* are expected to deploy widely in future decades (Heuberger et al., 2017; Capros et al., 2016). Due to the lack of CCS projects in the past, there exist only few data and studies that anticipate the potential

Table 14.3: Overview of learning rates and errors of selected technologies.

Category	Technology	Learning rate (%)	Error	Cumulative data unit	Model
Low-carbon fossil fuel technologies	Coal, lignite CCS	2.1	n.a.[a]	MW installed	ELTRAMOD, PowerACE
	Gas CCS	2.2	n.a.[a]	MW installed	ELTRAMOD, PowerACE
Electricity storages	Utility lithium-ion battery	12.9[b]	3.7%	GWh installed	ELTRAMOD, PowerACE
	Utility redox-flow battery	14.3[b]	6.1%	GWh installed	ELTRAMOD, PowerACE
Energy demand technologies	Heat pumps (power-to-heat)	10.0[c]	n.a.[c]	Units sold	ELTRAMOD, FORECAST
	Electrolysis alk. (power-to-gas)	17.7	5.3%	GW installed	ELTRAMOD, FORECAST

CCS, Carbon capture and storage.

[a]Due to limited availability of empirical data, proxy technologies are used for CCS.

[b]Since the experience curve data was delivered for implementation in the models, updates have been made to these values (see also Chapter 8).

[c]For heat pumps, we observed no clear trends at the time of model implementation, hence the learning rate is an expert estimate.

Source: Based on Louwen et al. (2018a).

cost reductions by using proxy technologies (Louwen et al., 2018a). This might be the only feasible solution until empirical data become available. Therefore the learning rates produced (2.1%–2.2%) to project the future cost reductions of CCS should be used carefully. An overview of the implemented experience curves is given in Table 14.3.

Battery storages, such as *lithium-ion* and *redox-flow batteries*, are increasingly developing on a global level. Recent studies and own analyses indicate learning rates in the range of 12%-18%, whereby most values are around 15%. It is expected that especially batteries at utility level and for electric vehicles will become more important in the next decades.

Important energy demand technologies are *heat pumps*, which are expected to play a major role in future heating applications. Heat pumps have been applied for decades; however, only few studies exist, which are presenting data sets on their technological learning (Weiss et al., 2008). The literature shows that the learning rates for heat pumps can differ significantly between countries. More data for heat pumps has been gathered, but in principle, for the application in the energy models, the learning rate of 10% was applied (Louwen et al., 2018a).

Storage technologies are playing a crucial role with the increasing electricity generation by intermittent renewable energy sources. Power-to-gas with the production of hydrogen (and other power-to-x technologies) can use the electricity surplus to produce green fuels such as hydrogen and other chemicals. *Alkaline electrolysis* has been developed since the 1980s. Therefore achieving learning rates of about 18% is feasible. Nevertheless, the experience curves of power-to-gas are uncertain because of the limited amount of data, especially for the early phase of production (Louwen et al., 2018a).

14.3.2 Effect of experience curve implementation on market developments including uncertainty analysis of experience curve parameters

Depending on the technology, the uncertainties of learning rates are different and will have miscellaneous effects on the modeling results. Therefore sensitivity analyses for each model are conducted to assess the combined effects of uncertain learning rates of selected technologies on the model outcomes.

14.3.2.1 Uncertainty of experience curve parameters in FORECAST

The FORECAST group of models is currently split into three submodules, which are generally run in separate instances. This means that exogenous learning is applied to consider investment and growth of certain technologies and the interactions with respective learning rates within all sectors. In other words, currently, the installation of, for example, heat pumps in the residential sector cannot have direct influence on costs for the same technology in other sectors, such as tertiary or industry. To overcome this issue within the project an iterative approach is applied summing up all capacity values of the different submodules for the first iteration and then applying the cost reductions based on the learning effects, given higher installation rates of this technology in the consecutive iteration. This iterative process can be repeated until further additional installations have no further representative effect on cost reductions. Given the applied learning rate of approx. 11%, specific cost reductions of 20% until 2050 are observed based on the installed cumulative capacities for air-source heat pumps (see Fig. 14.3) compared to the originally applied minimal generic cost reductions.

Although the cost effects of the learning mechanism for certain technologies in the demand sectors are to be considered, the impact on the overall installation rate of these technologies was found to be limited. Besides, the implementation of technology preferences independent of pure cost parameters but rather political frameworks (e.g., centralized vs decentralized), the overall demand reduction for, e.g., heating on building level, reduces the cost impact of the heating system on overall system cost.

14.3.2.2 Uncertainty of experience curve parameters in PowerACE

In the electricity market model PowerACE the future technology mix emerges from the endogenously simulated investment decisions of the different agents. These decisions, in turn, are largely affected by future price expectations on the day-ahead market as well as

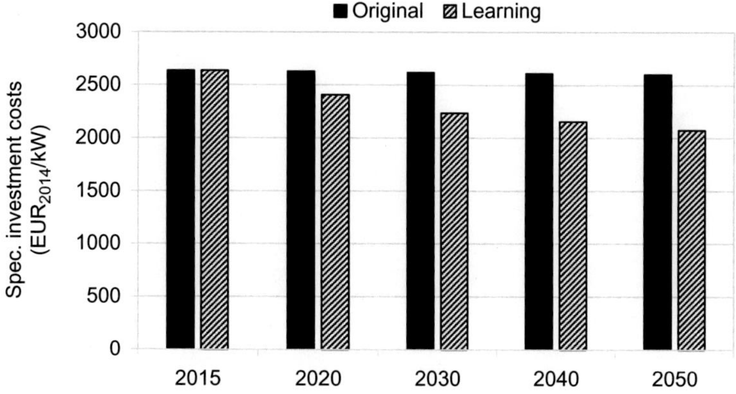

Figure 14.3
Development of specific investment costs of air-source heat pumps with learning as implemented in FORECAST. *Source: Own illustration based on REFLEX data.*

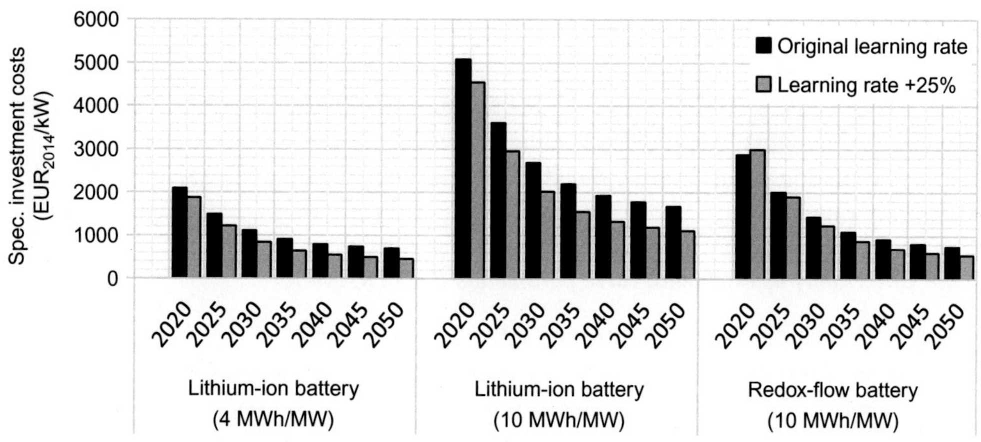

Figure 14.4
Development of specific investment costs of lithium-ion and redox-flow batteries considering technological learning as implemented in PowerACE. *Source: Own illustration based on REFLEX data.*

assumed investment and operating costs of the different investment options. For this reason, apart from the High-RES centralized scenario as defined previously, an additional sensitivity analysis is run, in which the learning rates for lithium-ion and redox-flow batteries are increased by 25%, respectively. Fig. 14.4 presents the original specific investment costs of the two battery types as well as the resulting values under the assumed higher learning rates.

Since the profitability of electricity storage increases under higher shares of renewables, the scenario results only differ between 2040 and 2050. Fig. 14.5 therefore shows the model-endogenous investment decisions for two exemplary countries (France and Italy) in this

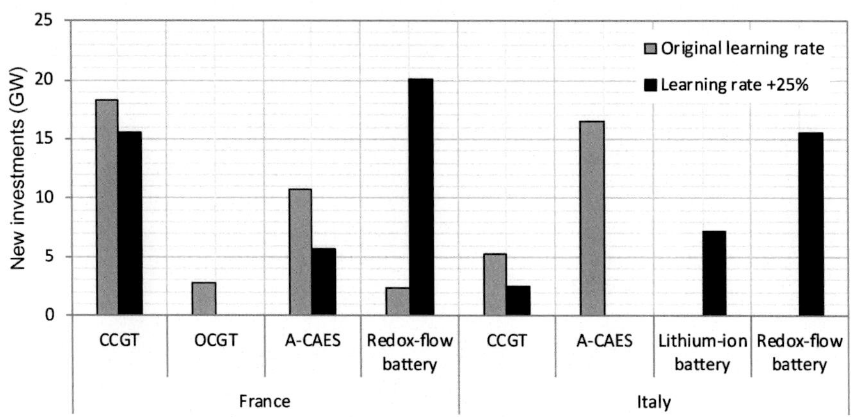

Figure 14.5
New-generation and storage investments between 2040 and 2050 in PowerACE for two selected countries. *Source: Own illustration based on REFLEX data.*

time frame. In both countries the higher learning rates lead to a technology switch and especially investments in A-CAES are replaced by lithium-ion and redox-flow batteries. However, investments in combined-cycle gas turbines (CCGTs) and open-cycle gas turbines are also affected to some extent. This finding shows that when implementing technological learning in simulation models, it is crucial to carefully analyze the impact of higher or lower learning rates.

14.3.2.3 Uncertainty of experience curve parameters in ELTRAMOD

The linear optimization model ELectricity TRAnsshipment MODel (ELTRAMOD) estimates the cost-minimal and optimal technology investments and dispatch under the given the scenario framework. The scenario assumptions that are mainly influencing the modeling results are the increasing deployment of intermittent renewable energy sources, the growing electricity demand as well as the increasing CO_2 prices (ca. 8 EUR/tCO_2 in 2014 to 150 EUR/tCO_2 in 2050). Two sensitivity analyses are conducted to estimate the impact of exogenous implemented experience curves on the overall modeling results.

The first sensitivity assumes that the specific investment costs of lithium-ion and redox-flow batteries are reduced by -50% of their original investment costs, considering learning rates of 12.9% ($\pm$3.7%) for lithium-ion and 15.1% ($\pm$6.1%) for redox-flow batteries. The second sensitivity assumes no technological learning of CCS technologies compared to original modeling results with learning rates of 2.1% for coal and lignite CCS and 2.2% for gas CCS power plants. Fig. 14.6 represents the development of specific investment costs for batteries and CCS resulting from the original assumed learning rates that are considered for the scenario assessment of the High-RES centralized world. The lithium-ion batteries are distinguished by their storage capacity in small- (1 MWh/MW), medium- (4 MWh/MW),

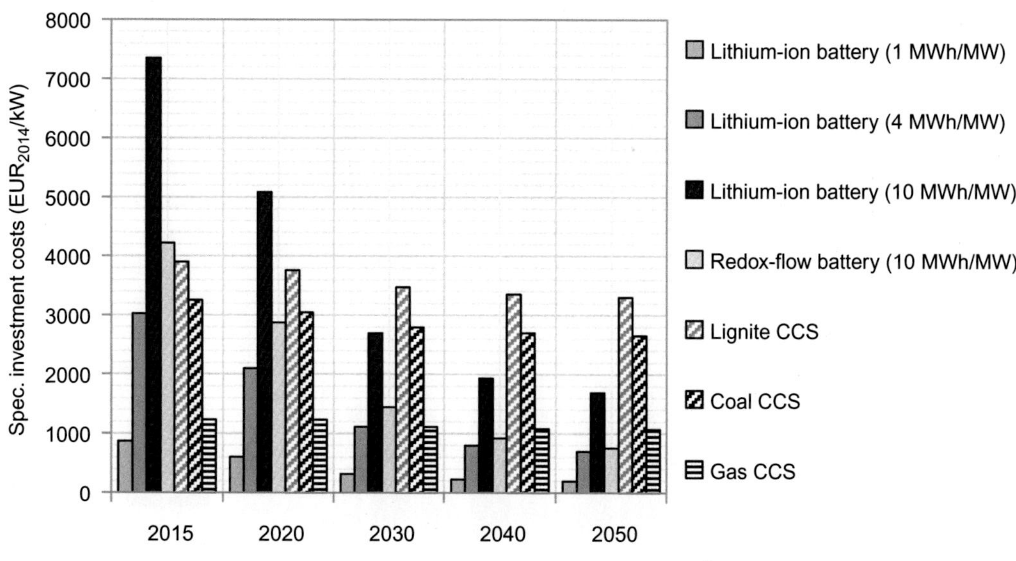

Figure 14.6

Development of specific investment costs of batteries and CCS technologies considering technological learning. *CCS*, Carbon capture and storage. *Source: Own illustration based on REFLEX data.*

and large- (10 MWh/MW) scale utility batteries. The specific investment costs are representing the whole battery system costs including the investments for converter (17% for lithium-ion, 61% for redox-flow batteries) and storage (83% for lithium-ion, 39% for redox-flow batteries).

Fig. 14.7 compares the original scenario results with the modeling output of the two sensitivity analyses concerning reduced battery costs and no technological learning for CCS technologies in the EU28 countries, Norway, Switzerland, and the Balkan countries. All three modeling outputs from 2020 to 2050 are comparatively similar. Only the years 2040 and 2050 show noticeable differences in the power plant portfolio, which results from the changes in investment costs for batteries and CCS. As expected, in the sensitivity with reduced battery costs by −50%, more redox-flow batteries in 2040 and 2050 are deployed compared to the original scenario results. Due to the high shares of intermittent renewable energies, battery storages with large storage capacity are needed. As redox-flow batteries are modeled with a storage capacity of 10 MWh/MW and with comparatively low investment costs compared to large-scale lithium-ion batteries, redox-flow batteries are the preferred and cost-optimal option. In the sensitivity without technological learning of CCS technologies, no CCS is installed in 2040, and only few capacities can be observed in 2050 compared to the original scenario results. As the specific investment costs for lignite and coal CCS are almost three times as high as for gas CCS, only the latter ones are installed to provide the needed low-carbon conventional electricity generation.

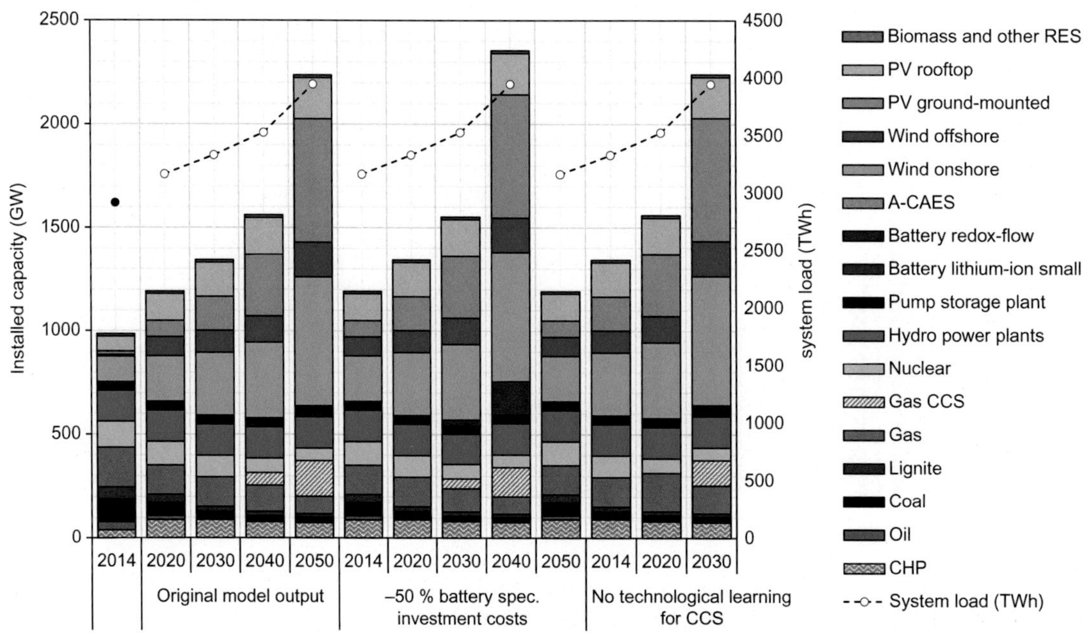

Figure 14.7

Comparison of installed cumulative capacities in the different REFLEX modeling scenarios, under consideration of uncertainties in technological learning. *Source: Own illustration based on REFLEX data.*

Fig. 14.8 gives a detailed overview of the main changes in the capacity mix across the sensitivities compared to the original scenario.

Sensitivity of −50% of specific investment costs for batteries

With increasing technological improvements and costs reductions of batteries, the amount of installed redox-flow batteries increases to 30 GW in 2040 and 160 GW in 2050, whereby in the original scenario, results consist only of 10 GW in 2050. Few capacities of small lithium-ion batteries are deployed (4 MW). In general, battery capacities are not installed as much as expected in ELTRAMOD due to the model coupling with eLOAD, which provides the optimized hourly system load curve through the dispatch of demand-side flexibility options that are mainly applied in the residential sector (e.g., p2h and electric vehicles). Due to the enforced electrification in other sectors, for example, by increasing deployment of power-to-x technologies, the value of electricity storages from a market perspective is decreasing. In addition, other studies, such as Gils et al. (2017), have shown that less storages are required in wind-dominated energy systems as the High-RES centralized scenario. Furthermore, due to the higher installed capacity of batteries, the need for additional conventional generation

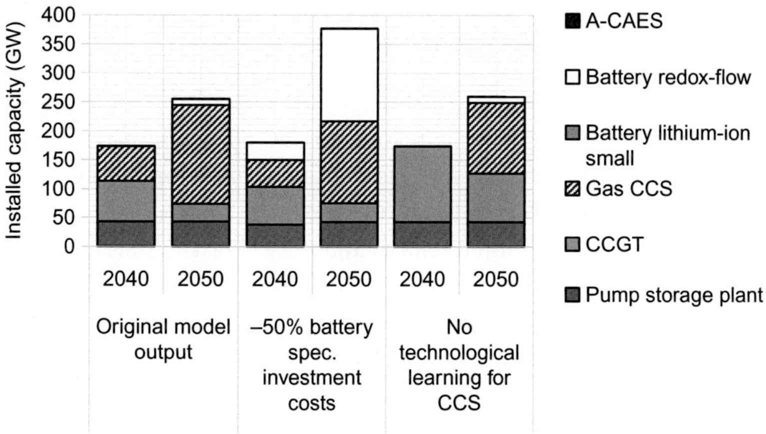

Figure 14.8
Development of storage, CCS and, CCGT capacities in the different REFLEX modeling scenarios, under consideration of uncertainties in technological learning. *CCGT*, Combined-cycle gas turbine; *CCS*, carbon capture and storage. *Source: Own illustration based on REFLEX data.*

technologies, such as gas CCS, is declining in 2040 (−13 GW) and 2050 (−30 GW) in the sensitivity results.

Sensitivity of no technological learning for carbon capture and storage technologies

If the CCS technology does not experience technological improvements and thus no cost reductions, power plants with CCGTs will play a crucial role in 2040 and 2050. CCGT replaces the needed conventional capacity provided by gas CCS in the original scenario results. Without technological learning of CCS, additional 60 GW of CCGT is deployed in 2040 and 50 GW in 2050 compared to the original scenario results. Further, no CCS is installed in 2040, and in 2050, the CCS capacity is reduced by ca. 21 GW in the sensitivity results in contrast to the original scenario results.

14.4 Lessons learned and conclusion

Large efforts have been taken to derive learning curves for selected technologies within the REFLEX project. Experience curves have been implemented in various energy models to analyze the future of the European energy system in different sectors, including the electricity and heat supply as well as demand for the industry, residential, tertiary, and transport sector. Although experience curves are usually based on historical data, it is only one of the few methods to derive evidence-based cost trajectories for the future (Louwen et al., 2018b). However, for miscellaneous reasons, explained in the chapters before, the endogenous approach is not as practical as expected and faces a number of challenges. This chapter summarizes the lessons learned from the REFLEX project for modelers and policy makers regarding the exogenous and endogenous implementation of experience curves in energy system models.

For many technologies the availability of consistent time series of empirical data for cost developments is not given by far. Especially for new and upcoming technologies that are key technologies to achieve the greenhouse gas emission reduction targets, such as CCS, electricity storages, demand-side technologies (heat pumps, electrolysis), or electric transport vehicles are defined by limited data availability. For CCS and electricity storages, almost a complete lack of empirical and commercial data exists (Louwen et al., 2018b). Thus evidence-based cost projections for these technologies are difficult and require approximations by using similar technologies. Such decarbonization technologies are supported by policy makers across different countries. Therefore research should be pushed to estimate data on accurate cost estimations. Further, more pilot projects, for example, on industrial CCS, should be supported while ensuring the availability of transparent cost data (Louwen et al., 2018b).

As experienced in the REFLEX project, the endogenous approach of implementing learning curves cannot be applied in all energy models. As the experience curve function is characterized by its nonlinearity and nonconvexity, it cannot be implemented endogenously in optimization models without a piecewise linear approximation of the curve to guarantee a global optima. Further, for perfect foresight or myopic models, the endogenous incorporation can result in lock-in effects. Technological learning is defined as a global process, and thus experience curves should be derived on global cost developments. However, the REFLEX models among others are modeling smaller geographical scales at the EU or at country level. Hence, the complete endogenous modeling of technological progress is not feasible for those models and requires assumptions on global capacity deployments (Louwen et al., 2018b). Therefore using the exogenous approach can be less complex and justified in the case of the REFLEX project. For future projects, it should be taken into account that well-aligned global models could take care of global developments and are therefore highly appreciated in projects where technological learning is considered.

Market dynamics and raw material prices are nontechnological learning—related factors; however, they can significantly affect the technology cost developments and thus also experience curve parameters. Multifactor learning curves can partially replicate these effects but are simultaneously increasing the modeling complexity and data requirements. However, as multifactor learning curves are not widely used yet, further research is required in the context of model implementation (Rubin et al., 2015). Moreover, research, accurate data, and better econometric models are needed to estimate underlying factors that influence technological innovations and their diffusion.

Technology cost developments can influence each other. This is usually the case if applications share common components, for example, technologies based on lithium-ion batteries (electricity storages, electric vehicles, and residential applications) benefit from each other's developments (Seebregts et al., 2000). These spillover effects are difficult to account in experience curves, as data on technology cost components is even less transparently available and modeling complexity will drastically increase.

Experiences of the REFLEX project show that further research is needed to investigate what modeling accuracy improvements can be reasonably expected with a more detailed approach and evaluate this against higher input data requirements as well as increasing modeling complexity (e.g., through multifactor or component-based learning curves). Further, the use of learning curves to project future cost developments is affected by high uncertainties. Therefore an improved characterization of uncertainties and the identification of robust conclusions are playing a crucial role (Rubin et al., 2015). The lack of these analyses could increase the credulity in inaccurate modeling outputs and their inappropriate policy recommendations.

While considering technological progress in energy system models, modelers should consider the strengths and limitations of the endogenous and exogenous approach and choose the approach that is more conforming to the model structure and purpose by having the characteristics of each approach in mind when interpreting the modeling results (Gillingham et al., 2008).

References

Braungardt, S., Elsland, R., Dehler, J., 2014. Modelling the effect of the ecodesign and labelling directives – bottom-up analysis of EU-27 residential electricity use. In: International Energy Policy & Programme Evaluation Conference. Berlin.

Capros, P., Vita, A., De, Tasios, A., Siskos, P., Kannavou, M., et al., 2016. EU Reference Scenario 2016 – Energy, transport and GHG emissions – Trends to 2050. European Commission, Brussels. Available from: https://doi.org/10.2833/9127.

EASE/EERA, 2017. European Energy Storage Technology Development Roadmap Towards 2030. Update 2017. The European Association for Storage of Energy (EASE)/European Energy Research Alliance (EERA), Brussels, Belgium. Available from: https://doi.org/10.1287/inte.31.4.109.9664.

EHPA, 2015. European Heat Pump Market and Statistics Report 2015. Executive Summary. European Heat Pump Association (EHPA).

Elsland, R., Peksen, I., Wietschel, M., 2014. Are internal heat gains underestimated in thermal performance evaluation of buildings? Energy Procedia 62, 32–41. Available from: https://doi.org/10.1016/j.egypro.2014.12.364.

Fleiter, T., Elsland, R., Rehfeldt, M., Steinbach, J., Reiter, U., Catenazzi, G., Stabat, P., 2017. EU Profile of heating and cooling demand in 2015. Deliverable of the project Heat Roadmap Europe, Retrieved from http://heatroadmap.eu/output.php.

Fleiter, T., Rehfeldt, M., Herbst, A., Elsland, R., Klingler, A.-L., Manz, P., 2018. A methodology for bottom-up modelling of energy transitions in the industry sector: the FORECAST model. Energy Strategy Rev. 22 (2018), 237–254. Available from: https://doi.org/10.1016/j.esr.2018.09.005.

Fouquet, R., 2016. Historical energy transitions: speed, prices and system transformation. Energy Res. Soc. Sci. 22, 7–12. Available from: https://doi.org/10.1016/j.erss.2016.08.014.

Fraunholz, C., Keles, D., 2019. D5.2 Report on investments in flexibility options considering different market designs. In: Report for REFLEX Project.

Gillingham, K., Newell, R.G., Pizer, W.A., 2008. Modeling endogenous technological change for climate policy analysis. Energy Econ. 30 (6), 2734–2753. Available from: https://doi.org/10.1016/j.eneco.2008.03.001.

Gils, C.H., Scholz, Y., Pregger, T., de Tena, D.L., Heide, D., 2017. Integrated modelling of variable renewable energy-based power supply in Europe. Energy 123 (2017), 173–188. Available from: https://doi.org/10.1016/j.energy.2017.01.115.

Global CCS Institute, 2017. The Global Status of CCS: 2017. Global Carbon Capture and Storage Institute Ltd, Australia. Retrieved from: <http://www.globalccsinstitute.com/sites/www.globalccsinstitute.com/files/uploads/globalstatus/1-0_4529_CCS_Global_Status_Book_layout-WAW_spreads.pdf>.

Green, R., Staffell, I., 2016. Electricity in Europe: exiting fossil fuels? Oxford Rev. Econ. Pol. 32 (2), 282–303. Available from: https://doi.org/10.1093/oxrep/grw003.

Gunkel, D., Kunz, F., Müller, T., von Selasinsky, A., Möst, D., 2012. Storage Investment or Transmission Expansion: How to Facilitate Renewable Energy Integration in Europe? Tagungsband VDE-Kongress Smart Grid – Intelligente Energieversorgung der Zukunft.

Herbst, A., 2017. Kopplung eines makroökonomischen Modells mit einem "bottom-up" Energienachfrage-Modell für die Industrie (dissertation). Europa-Universität Flensburg, Flensburg.

Herbst, A., Toro, F., Reitze, F., Jochem, E., 2012. Introduction to energy systems modelling. Swiss Soc. Econ. Stat. 148 (2), 111–135. Available from: https://doi.org/10.1007/bf03399363.

Heuberger, C.F., Rubin, E.S., Staffell, I., Shah, N., Mac Dowell, N., 2017. Power capacity expansion planning considering endogenous technology cost learning. Appl. Energy 204 (2017), 831–845. Available from: https://doi.org/10.1016/j.apenergy.2017.07.075.

Keles, D., Bublitz, A., Zimmermann, F., Genoese, M., Fichtner, W., 2016. Analysis of design options for the electricity market: the German case. Appl. Energy 183, 884–901. Available from: https://doi.org/10.1016/j.apenergy.2016.08.189.

Kunze, R., 2018, D2.4 Updated Data Management Plan (DMP). Report for the REFLEX Project.

Ladwig, T., 2018. Demand Side Management in Deutschland zur Systemintegration erneuerbarer Energien (dissertation). Technische Universität Dresden, Dresden. http://nbn-resolving.de/urn:nbn:de:bsz:14-qucosa-236074.

Louwen, A., Krishnan, S., Derks, M., Junginger, M., 2018a. D3.2 comprehensive report on experience curves. In: Report for REFLEX Project.

Louwen, A., Junginger, M., Krishnan, S., 2018b. Technological learning in energy modelling: experience curves. In: Policy Brief for REFLEX Project.

Müller, T., Gunkel, D., Möst, D., 2013. How does renewable curtailment influence the need of transmission and storage capacities in Europe? In: 13th European IAEE Conference. Düsseldorf.

OECD/IEA, 2016. 20 Years of Carbon Capture and Storage – Accelerating Future Deployment. International Energy Agency (IEA), Paris. Available from: https://doi.org/10.1787/9789264267800-en.

REN21, 2017. Renewables 2017 Global Status Report. REN21 Secretariat, Paris.

Ringler, P., Keles, D., Fichtner, W., 2017. How to benefit from a common European electricity market design. Energy Policy 101, 629–643. Available from: https://doi.org/10.1016/j.enpol.2016.11.011.

Rubin, E.S., Azevedo, I.M.L., Jaramillo, P., Yeh, S., 2015. A review of learning rates for electricity supply technologies. Energy Policy 2015 (86), 198–218. Available from: https://doi.org/10.1016/j.enpol.2015.06.011.

Rubin, E.S., Yeh, S., Antes, M., Berkenpas, M., Davison, J., 2007. Use of experience curves to estimate the future cost of power plants with CO_2 capture. Int. J. Greenhouse Gas Control 1 (2), 188–197. Available from: https://doi.org/10.1016/S1750-5836(07)00016-3.

Schmidt, O., Hawkes, A., Gambhir, A., Staffell, I., 2017. The future cost of electrical energy storage based on experience rates. Nat. Energy 2 (8), 17110. Available from: https://doi.org/10.1038/nenergy.2017.110.

Schubert, D.K.J., 2016. Bewertung von Szenarien für Energiesysteme: Potenziale, Grenzen und Akzeptanz (dissertation). Technische Universität Dresden, Dresden.

Seebregts, A., Kram, T., Schaeffer, G.J., Bos, A., 2000. Endogenous learning and technology clustering: analysis with MARKAL model of the Western European energy system. Int. J. Global Energy Issues 14 (1–4), 289–319.

Weiss, M., Junginger, H.M., Patel, M.K., 2008. Learning Energy Efficiency: Experience Curves for Household Appliances and Space Heating, Cooling and Lighting Technologies. Utrecht University, Utrecht, The Netherlands.

Zöphel, C., Schreiber, S., Herbst, A., Klingler, A.-L., Manz, P., Heitel, S., et al., 2019. D4.3 Report on cost optimal energy technology portfolios for system flexibility in the sectors heat, electricity and mobility. In: Report for the REFLEX Project.

CHAPTER 15

Global electric car market deployment considering endogenous battery price development

Stephanie Heitel[1], Katrin Seddig[2], Jonatan J. Gómez Vilchez[3] and Patrick Jochem[2]

[1]*Fraunhofer Institute for Systems and Innovation Research ISI, Karlsruhe, Germany,* [2]*Institute for Industrial Production (IIP), Karlsruhe Institute of Technology (KIT), Karlsruhe, Germany,* [3]*European Commission*, Joint Research Centre, Ispra, Italy*

Abstract

Electric road vehicles are a promising strategy to reduce CO_2 emissions and air pollutants. Their market penetration depends on several factors including the vehicle price relative to conventional technologies and the electric range. This chapter analyses the impact of battery price developments on global electric car sales. To simulate battery prices based on global learning, the two system dynamics models ASTRA (Europe) and TE3 (key non-European markets) are coupled, covering together all major car markets. Experience curves are endogenously implemented for battery, plug-in hybrid, and fuel cell electric cars. Sensitivity analyses of learning rates for batteries and of battery capacities are conducted to investigate their impact on the market share of electric vehicles. As a core result, it turns out that battery prices, which rely strongly on the learning rate, have a strong impact on global electric car sales. This effect is even stronger than many market inducing policy measures.

Chapter Outline

* The views expressed are purely those of the authors and may not in any circumstances be regarded as stating an official position of the European Commission.

Technological Learning in the Transition to a Low-Carbon Energy System.
DOI: https://doi.org/10.1016/B978-0-12-818762-3.00015-7

15.1 Introduction

15.1.1 Motivation

Despite today's image as a new innovative technology to tackle climate change and air pollution, electric vehicles (EVs) have a long history. Already in the early 19th century, EVs were invented and tested (Guarnieri, 2012), but the early market deployment was disappointing against the alternative drivetrains horse and steam engine. Towards the end of the 19th century, EVs became leaders in some markets (Abt, 1998). In those days, they outperformed their drivetrain competitors, for example, by becoming the first vehicle traveling at a speed greater than 100 km/h and the first four-wheel-drive vehicle (Abt, 1998). However, the boom of EVs was stopped by the invention of the electric starter motor for the internal combustion engine vehicles (ICEVs) in 1912 (Abt, 1998). In the second half of the 1910s the market share of EVs collapsed because the complicated and long-lasting charging process as well as the bulky, huge, and costly battery could not cope with the properties of the ICEV. Nevertheless, in those days some users still preferred the silent locomotion. Since then, EVs were doomed to a shadowy existence. Only in the 1970s during the first oil crises and in the following years, the interest in EVs underwent a first small renaissance. Market share and the public interest still starve at low levels until the beginning of the 21st century. This time, environmental concerns and considerable developments in battery technology (driven by the market success of mobile devices) led to a promising market introduction of several dozen car models by the automotive industry (Kieckhäfer, 2013). Today, the strong demand for batteries from consumer electronics as well as the developments in lithium-ion batteries has led to fast decreasing prices (similar to photovoltaics before) and high reliability of batteries (Schmidt et al., 2017). In addition, further significant developments in battery technology are highly likely (Freunberger, 2017). This has changed the situation for EVs considerably, as their costs are now in the range of conventional vehicles.

Furthermore, the still fast-increasing greenhouse gas (GHG) emissions in transport not only on the global level, but also on the European scale, make a significant change in the transportation system unavoidable (Creutzig et al., 2015). If no substantial change in the modal split away from motorized road transport towards rail transport, shipping, cycling, and walking can be achieved, EVs might become a necessary cornerstone for the transport transition—especially if economic growth in developing and emerging countries continues.

Consequently, for a sustainable energy system, which also includes transportation, EVs play a leading part. Their market success depends, among other factors, on decreasing battery costs, which are mainly driven by global learning curves as in the photovoltaic market (Nelson et al., 2012). Therefore a global perspective is important if the market diffusion of EVs is analyzed.

In the following, we couple two car market models: one focusing on the European market (ASTRA) and the other covering the main non-European car markets (TE3). The coupling of the two models allows considering the EV market development dependent on a model-endogenous decrease of battery costs that is mainly based on global EV sales volumes and experience curves. This consideration leads to a reinforcing feedback process: the lower the battery costs, the more EVs are sold, the faster the battery costs are falling.

The structure of the chapter is the following: first, literature reviews are conducted to provide the status quo of EV market penetration studies and to introduce the system dynamics (SD) modeling approach and its suitability for the transport sector. In the next two sections the main characteristics of the two models, the implementation of the experience curves, and the model coupling procedure are described. Then, the assumptions for the Reference scenario and the sensitivity analyses are explained. Finally, simulation results are presented and discussed.

15.1.2 Status quo of electric vehicle market diffusion studies

Besides the battery price, mainly the (perceived) availability of charging stations as well as other preconceptions of potential EV users are decisive for their successful market take-up (Lieven et al., 2011; Rezvani et al., 2015). Many market penetration studies of EVs not only focus on their impact on the environment but also have a geographical focus limited to one specific country [e.g., on Greece, Chatzikomis et al. (2014); on China, Shen et al. (2014); on Germany, Jochem et al. (2015); and on New Zealand, Shafiei et al. (2017)]. Most of them consider only the emissions during the usage phase instead of analyzing the whole life cycle of the vehicle. But also the focus on the life cycle assessment is an evolving branch in literature [e.g., Kim et al. (2016) or Hawkins et al. (2012)]. Other authors look at the market potential alone (Gnann et al., 2015; Jochem et al., 2018), the barriers to widespread adoption of EVs (She et al., 2017; Wang et al., 2017b), or the impact of government incentives to promote EVs (Hardman et al., 2018; Ma et al., 2017; Wang et al., 2017a).

The huge differences in future market shares resulting from these studies indicate the high uncertainty in this market. To further contribute to the discussion a model coupling approach of two SD transport models was applied, thus enabling the simulation of global experience curve effects and consequently an endogenous-driven market penetration of EVs.

15.1.3 System dynamics transport modeling

The SD method was developed by Jay W. Forrester in the late 1950s (Forrester, 1995). SD focuses on dynamic problems and the feedback structure underlying complex systems, with the purpose of identifying the leverage points that may be used to steer the system toward a (desired) state (Forrester, 1997; Sterman, 2000). The result of formally applying systems thinking (Meadows and Wright, 2008) is a computer simulation model that captures a system's feedback structure and its resulting behavior over time.

Twenty-five years ago, Abbas and Bell (1994) pointed out the advantages of applying SD to transport modeling. Among these, the explicit consideration of feedback interactions between demand and supply (in our exercise, between EV sales and battery prices) was mentioned. Since then, knowledge on the application of the method to analyze transport problems has accumulated, as reviewed by Shepherd (2014). In the context of EV market development, SD competes with alternative methods such as agent-based modeling (ABM) [see the review by Jochem et al. (2018)]. Few transport applications (e.g., Kieckhäfer et al., 2014; Shafiei et al., 2013) integrate SD and ABM in a single model. Furthermore, the use of discrete choice frameworks to simulate powertrain market shares [as pioneered in an SD model by Ford (1995)] remains common practice.

In addition to the two SD models described in the next sections, there is an increasing number of SD models focusing on the market development of alternative vehicle powertrain technologies. Broadly speaking, these models attempt to capture the dynamic problem that EVs face to become a mainstream technology, by considering the factors that influence their market diffusion [see, for example, Shepherd et al. (2012)]. In general, there is a tendency to model these factors at the country level [e.g., Wansart (2012) for Germany and Keith (2012) for the United States (US)] or at the EU level (Harrison et al., 2016). As claimed by Gómez Vilchez and Jochem (2019) a truly global perspective is still missing. This situation is not ideal, as the battery manufacturing and automotive sectors have a global nature. Specifically, this calls for the endogenization of the battery cost, which affects EV sales in each market but is, in a feedback loop, also affected by global EV sales.

In this study, we focused on the endogenization of the battery price for battery EVs (BEVs) and plug-in hybrid EVs (PHEVs). The two simulation models used for this purpose are described in the following sections.

15.2 Model description

15.2.1 ASTRA

15.2.1.1 Overview

ASTRA (ASsessment of TRAnsport Strategies) is an integrated assessment model applied for more than 20 years for strategic policy assessment in the transport and energy field. The model is based on the SD approach and built in Vensim. ASTRA covers all EU28 member states plus Norway and Switzerland and all transport modes for passenger and freight transport including road transport, rail transport, inland waterways, navigation, aviation, and the active modes cycling and walking. A strong feature of ASTRA is the ability to simulate and test integrated policy packages, to analyze modal shifts and to provide indicators for the indirect effects of transport on the economic system. The ASTRA model covers the time period from 1995 until 2050. Results in terms of main indicators are available on an annual basis.

ASTRA simulates the development of passenger and freight transport per mode with an adapted classical four-stage modeling approach (traffic generation, traffic distribution, mode choice, route selection). The model is calibrated to reproduce major indicators until 2016 such as transport performance, fleet composition, fuel consumption, CO_2 emissions, and the gross domestic product (GDP) according to the main European reference sources such as Eurostat. For future trends until 2050 the EU Reference Scenario (Capros et al., 2016) provides parameters such as the development of GDP, population, and energy prices and serves for validation of the ASTRA model behavior (e.g., transport performance and fleet development) in the Reference scenario.

15.2.1.2 Modular structure of the ASTRA model

ASTRA consists of six different modules, each related to one specific aspect, such as the economy, the transport demand or the vehicle fleet. The modules cover the following main features:

- *Population:* The development of the population with its demographic structure and income groups is simulated based on factors such as fertility and death rates, immigration, and employment. Module outputs such as the number of persons in the working age and of persons in age-classes that permit to acquire a driving license are provided to other modules.
- *Economy:* This module simulates the linkages of the transport sector with the whole economic system, covering the estimation of GDP, input–output matrices, employment, and the demand side with consumption and investment.
- *Trade:* In this module, foreign trade within Europe and to regions in the rest of the world is simulated via bilateral relationships between country pairs and by sector.

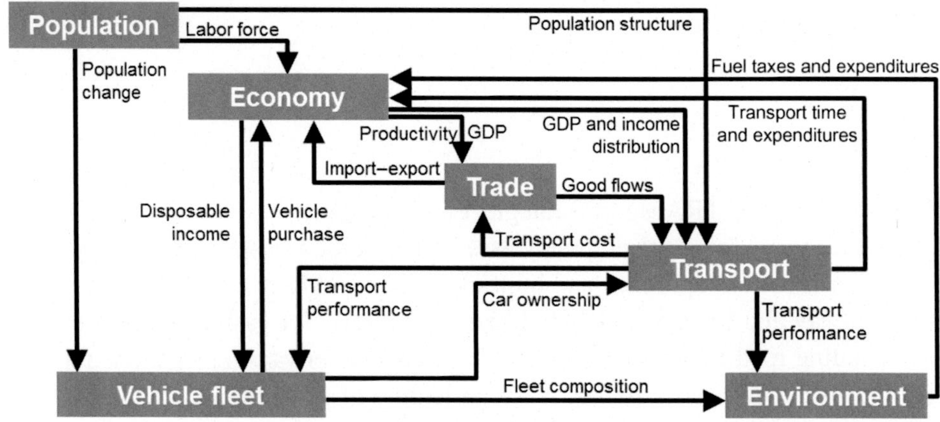

Figure 15.1
Overview of the ASTRA model structure with main linkages between modules. Source: *Adapted based on Fermi et al. (2014).*

- *Transport:* This module covers the transport demand estimation, the distribution and the modal split of passenger and freight movements. The number of passenger trips is driven by the employment situation, car ownership, and the number of people in different age-classes. Cost and time, as major decision criteria for mode choice, are influenced by infrastructure capacity and travel speeds, transport charges, and fuel prices.
- *Vehicle fleet:* This module provides the vehicle fleet composition for all road modes by age-class, emission standard, and drive technology. The number of new vehicle purchases is simulated based on the development of country-specific ownership and scrappage rates, and the diffusion of technologies is based on a discrete choice approach (see more details below).
- *Environment:* Using information provided by the vehicle fleet and the transport module, air pollutant and CO_2 emissions, fuel and energy consumption, fuel taxes and expenditures as well as accidents are calculated.

A key feature of ASTRA as an integrated assessment model is that the modules are linked together. Fig. 15.1 provides an overview of the ASTRA model structure visualizing the main linkages between modules. A more detailed description of the ASTRA model can be found in Fermi et al. (2014) or on the ASTRA website www.astra-model.eu.

15.2.1.3 Vehicle fleet and technology diffusion in ASTRA

The vehicle fleet module is the main module for our analyses. The module can be differentiated into three submodels, which simulate (1) the aging of the vehicle stock, (2) the number of newly registered vehicles, and (3) the choice of the drive technology for these new vehicle purchases. The number of newly registered cars is influenced by several factors, in particular, population growth, GDP development, income distribution, average car prices, and the urban trend to use car-sharing that can reduce the car-ownership rate. The car-ownership rate is part of a negative feedback loop that dampens the number of new cars and represents a saturation effect of the market when a certain level of motorization is reached.

The diffusion of alternative drive technologies in the road vehicle fleet is simulated separately for different vehicle categories. These categories comprise private and commercial cars, light-duty vehicles, heavy-duty vehicles in four gross vehicle weight categories, urban buses, and coaches. Based on the technical characteristics of available fuel options today and in the future and the heterogeneous requirements of the different users, a set of technologies is available for each vehicle category. For cars, available technologies are gasoline, diesel, liquefied petroleum gas (LPG), compressed natural gas (CNG), BEVs, PHEVs, and fuel-cell EVs (FCEVs).

The technology choice is based on an adapted total cost of ownership (TCO) approach that considers consumer prices for vehicles, the cost for energy consumption, maintenance costs, taxes, insurance, and road charges. Apart from the associated costs, important factors for the diffusion of new vehicle technologies are the deployment of the charging and filling station infrastructure and the ranges of the vehicles. Both factors are covered in a component named fuel procurement cost that is calculated based on the density of the filling or charging station infrastructure for each country and the average range per charge or filling. Experience curves are implemented for the development of prices of major technology components as part of the total vehicle consumer prices for BEVs, PHEVs, and FCEVs. The total vehicle price calculation consists of the following three parts:

- A technology-independent vehicle base price with development of prices over time due to increasing safety, efficiency, and convenience of the vehicles.
- A price for major technology-dependent components of new powertrain vehicles, in particular batteries and fuel-cell stacks that are implemented via experience curves.
- A technology-dependent part representing diverse components related to the specific technology, which either do not show additional learning effects anymore like for internal combustion engines or that are assumed to decrease slightly over time.

The probability of the choice of a certain fuel option is finally estimated with a discrete choice approach using logit, separately for private and commercial cars for each country. New vehicles diffuse with this resulting technology share into the vehicle stock. In the logit probability formula the calculated TCO are entered as negative benefits according to the following equation:

$$P_{cc,ct,i} = \frac{exp\left(-\lambda_i \times C_{cc,ct,i} + \mathrm{LC}_{cc,ct,i}\right)}{\sum_{ct} exp\left(-\lambda_i \times C_{cc,ct,i} + \mathrm{LC}_{cc,ct,i}\right)} \tag{15.1}$$

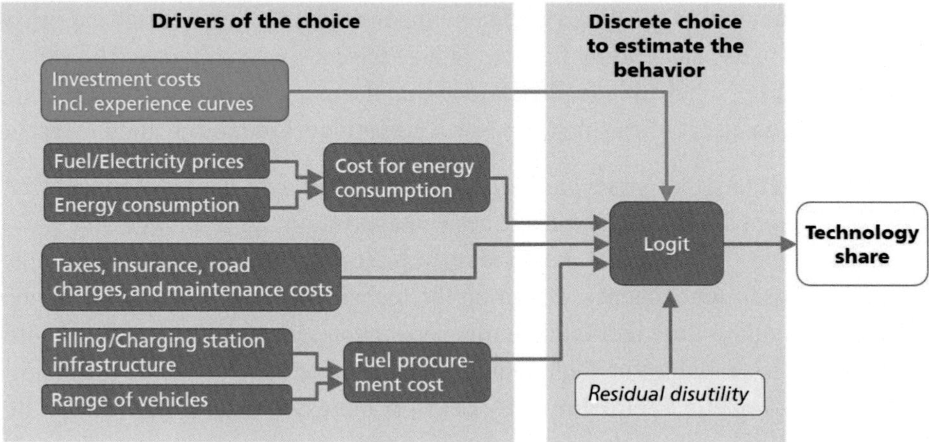

Figure 15.2
Technology share calculation for new vehicles in the ASTRA vehicle fleet module.

where P is the share of purchased cars per car technology ct, C is the total cost of ownership per vehicle-km, λ is the multiplier lambda per country i, LC is the logit constant representing the residual disutility, cc is the index for the two car categories private and commercial cars, ct is the index for seven car drive technologies, and i is the index for countries.

Fig. 15.2 provides a schematic representation of the calculation of the technology share for new vehicles in the ASTRA vehicle fleet module.

15.2.2 TE3

15.2.2.1 Overview

The TE3 (Transport, Energy, Economics, and Environment) model is a simplified representation of the road passenger transport system, with a focus on car powertrain technologies and their impact on energy demand and GHG emissions. TE3 is also based on the SD approach and implemented in Vensim. TE3 is capable of generating scenarios and suitable for policy analysis [see an application in Haasz et al. (2018)]. On the model validation, see Gómez Vilchez et al. (2015).

The initial version of the model includes China, France, Germany, India, Japan, and the US and runs from 2000 to 2030. This version of TE3 is available at www.te3modelling.eu. In this study an updated version of TE3 has been used. The main differences with respect to the initial version of the model are as follows: (1) France and Germany have been excluded, as these countries are considered in ASTRA and (2) the model time horizon has been extended to 2050.

15.2.2.2 Modular structure of the TE3 model

The TE3 model consists of nine modules (see Fig. 15.3). A detailed description of each of them can be found in Gómez Vilchez (2019).

- *Population–GDP*: This module incorporates the external projections on population and GDP, which is used to derive the income per capita. Additional macroeconomic variables translate economic values from nominal to real.
- *Car stock*: The aim of this module is to determine the resulting aggregate car sales derived from the country-specific car ownership projections as well as to simulate the market shares by car technology.
- *Travel demand by car*: The variable "average annual vehicle-km traveled by car" is determined in this module, which is then used for the estimation of energy use.
- *Infrastructure*: The deployment of public refueling and recharging infrastructure is analyzed in this module as a result of investment.
- *Technology choice*: This module comprises the model's main behavioral assumptions (see more details below).
- *Production costs*: In this module, three broad classes of car attributes–technical features, production costs, and consumer costs–are considered. In particular, the production costs submodule is of interest as therein the experience curves are implemented.
- *Energy*: This module consists of three main parts: energy prices, electricity mix, and energy use. Each submodule respectively aims at (1) generating the price evolution of the various fuels, (2) determining the share of power generation by energy source, and (3) estimating energy use.
- *Emission*: This module is divided into six submodules: emission factors, new car emissions, manufacturing and scrappage, tank-to-wheel, well-to-tank, and lifecycle. These are used to calculate the corresponding GHG emissions.
- *Policy*: In this core module, policy analysis is undertaken. The different policy measures (see more details below) affect several of the remaining modules.

15.2.2.3 Vehicle fleet and technology diffusion in TE3

In the TE3 model the vehicle fleet is disaggregated by age and powertrain technology, for each country. An aging chain (see Sterman, 2000) with three stocks is implemented: new cars, medium(-aged) cars, and old cars. In terms of car technology the current model version includes nine powertrains: gasoline (G), diesel (D), flexible fuel (FF), LPG, natural gas (NG), hybrid EV (HEV), PHEV, BEV, and FCEV. These are powered by seven energy sources in the model: gasoline, diesel, ethanol (E85), autogas, CNG, electricity, and hydrogen (H_2).

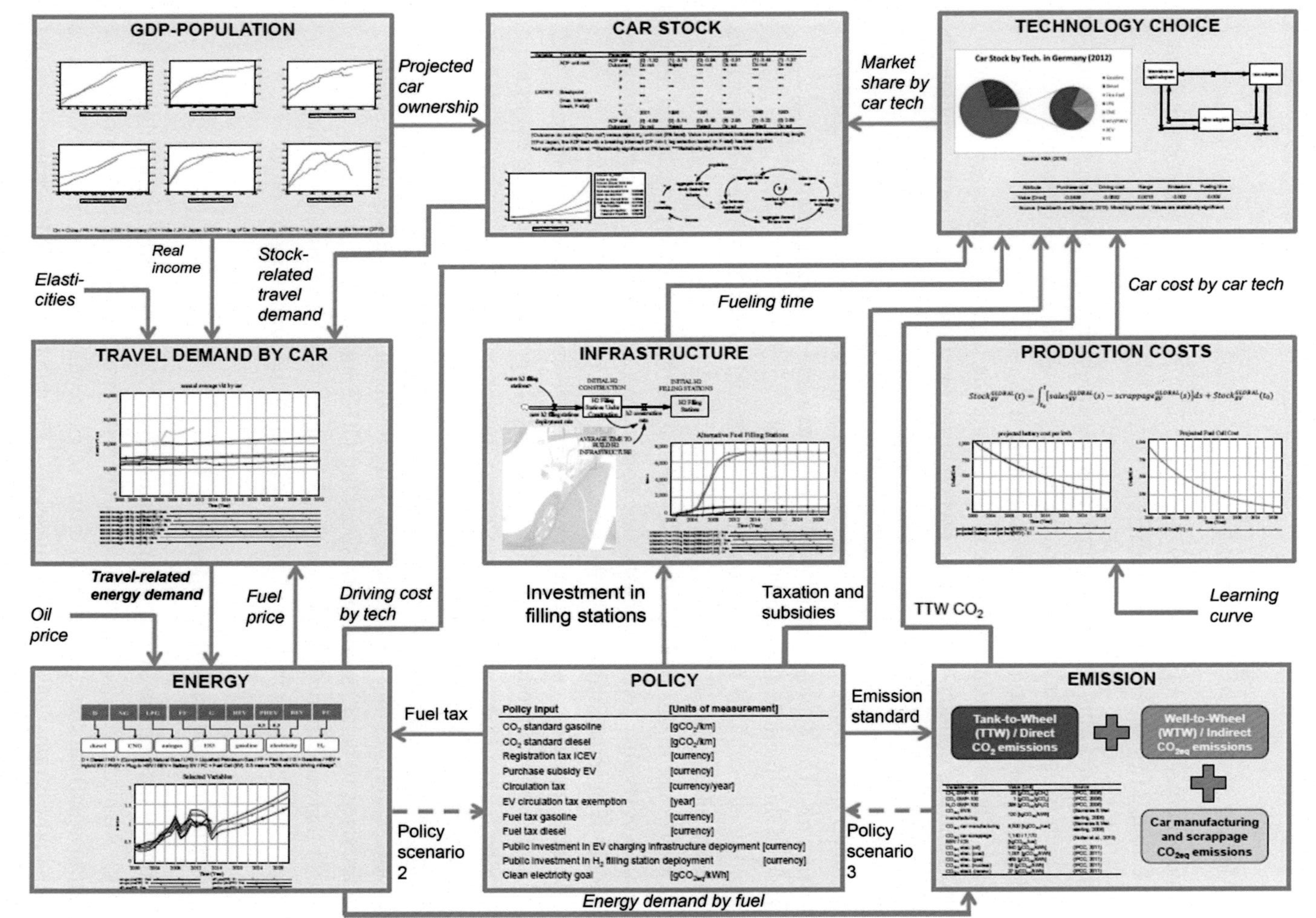

Figure 15.3

Overview of the TE3 model structure with main linkages between modules. Source: *From Gómez Vilchez et al. (2016).*

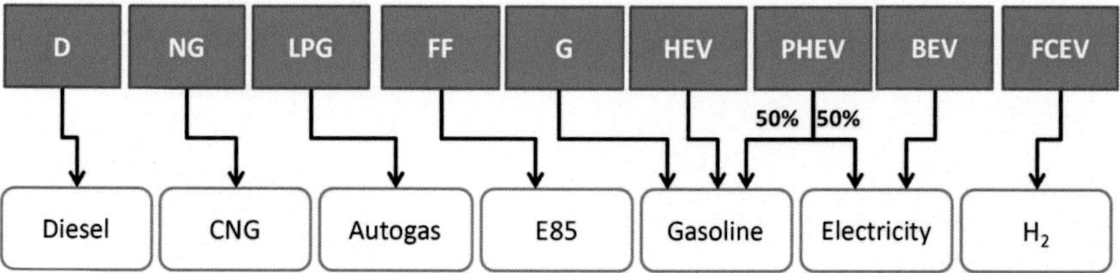

Figure 15.4
Linkages between car technologies and energy sources in the TE3 model. Source: *Adapted from Gómez Vilchez (2016).*

Fig. 15.4 shows the mapping of powertrains to energy sources. As can be seen, PHEVs can be powered by gasoline and electricity (a 50% split is assumed by default).

Projection of car ownership is accomplished through econometric estimation, and an adjustment mechanism is employed to derive annual aggregate demand for new cars [see Gómez Vilchez (2019) for details].

Technology diffusion fundamentally depends on the assumed behavior of consumers and the evolving attributes of the various car options. Concerning consumer behavior, TE3 reflects to some extent the ideas by de Wolff (1938) on first versus replacement demand as well as by Kotler et al. (2008) on consumer segmentation. For the latter, four consumer segments are explicitly modeled in TE3: habit(-oriented), innovators, low-cost buyers, and utility maximizers.

Habit-oriented consumers are those who make repeating car purchases, with the assumption that they are satisfied with a certain technology and stick to their current powertrain. It is implicitly assumed that innovators have a high level of income and, following Rogers (2003), represent a relatively small fraction of the market. Low-cost buyers dominate first-time sales. The implicit assumption is that people buy their first car when they are young and their level of income low. As a result, the car purchase price is the key factor influencing the choice. The segment of utility maximizers is the one where stronger economic rationality is assumed. This group evaluates the six attributes purchase price, usage cost, driving range, station coverage, recharging time, and level of car emissions. A logit probability formulation is introduced to calculate choice, which can be constrained by technology availability and the degree of powertrain popularity. These ideas are captured in a stylized manner in Fig. 15.5. In the model, there is the possibility that the proportion of consumers within a certain segment may dynamically change. For instance, as first-time car purchasers age and gain experience with the car, they accumulate sufficient work experience (which probably but not necessarily leads to higher income) and become more aware of the various car attributes by the time they make their next car purchase (i.e.,

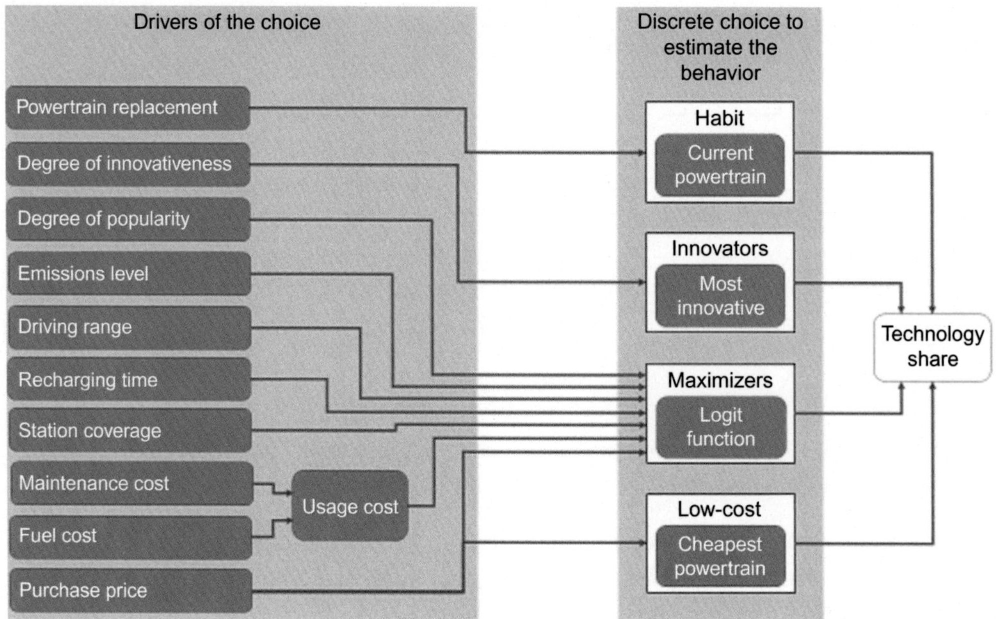

Figure 15.5
Technology share calculation for new cars in TE3.

repeating purchase). This hypothesis is captured in the model by a low number of low-cost buyers making repeating purchases. Furthermore, as the market introduction of electric cars brings attributes such as driving range and recharging time to the forefront of choice, this may result in a larger proportion of consumers behaving with stronger economic rationality. In TE3, this possibility is represented by a flow variable from habit-oriented consumers to utility maximizers. This, together with the evolution of the car powertrain attributes, is crucial for technology development over time.

15.2.3 Comparison of the two models

Table 15.1 provides an overview of the main features of the two models that are relevant for the simulation of the EV diffusion. The differences between the two models can more clearly be seen in Table 15.2, which shows the assumed choice behavior as previously described.

In both models the drivers of choice (cf. Fig. 15.2 for ASTRA and Fig. 15.5 for TE3) can be influenced by policy measures. The list of such measures can be seen in Table 15.3.

Table 15.1: Main characteristics of the two models.

	ASTRA	TE3
Method	System dynamics	
Software	Vensim	
Countries	EU28, Norway, Switzerland	China, India, Japan, the United States
Time horizon	1995–2050	2000–2050
Car technologies	7	9
Gasoline	x	x
Diesel	x	x
Flexible fuel		x
LPG	x	x
CNG	x	x
HEV		x
PHEV	x	x
BEV	x	x
FCEV	x	x

BEV, battery electric vehicle; *CNG*, compressed natural gas; *FCEV*, fuell-cell electric vehicle; *HEV*, hybrid electric vehicle; *LPG*, liquefied petroleum gas; *PHEV*, plug-in hybrid electric vehicle.

Table 15.2: Choice behavior assumed in the two models.

	ASTRA	TE3
Type of new car sales	All	First-time and repeating
Consumer segments	Utility maximizers	Innovators, habit-oriented, low-cost, and utility maximizers
Decision rule	Logit probability	Deterministic choice and logit probability

Table 15.3: Policy measures in the two models.

	ASTRA	TE3
CO_2 emission standards for new cars	X	X
Sales ban of conventional ICEVs	X	
Fuel taxation	X	X
Purchase subsidy	X	X
Investment in infrastructure[a]	X	X
Scrappage scheme		X

ICEV, Internal combustion engine vehicle.
[a]Public slow and fast recharging as well as H_2 refueling stations.

15.3 Experience curves and model coupling

15.3.1 Implementation of experience curves

In SD models, experience curves can be implemented endogenously using the stock and flow concept [see, e.g., Sterman (2000)]. The cumulative production of batteries as the input value for the experience curve function is implemented by a stock in the model.

Battery capacities belonging to the new EV purchases flow into this stock. Thus the demand of new BEVs and PHEVs is used as a proxy for the development of battery production assuming average battery capacities. A similar approach is used for fuel cells.

Experience curves were implemented in both models according to the following equation:

$$C(X) = C_0 \cdot X^{\log_2(1-l)} \tag{15.2}$$

The battery cost per kWh C is a function of the cumulative battery production X. Values for the cost of the first unit C_0 and the learning rate l were derived in Louwen et al. (2018). The learning rate l was set to 15.2% for batteries (in kWh) and to 18.0% for fuel cells (in kW) for the reference case, in other words, together with the base 2 a doubling of market volumes leads to a price decrease of 15.2% for batteries and of 18.0% for fuel cells.

15.3.2 Description of the interface and feedback loops between models

As the learning parameters refer to global learning, both models are coupled by exchanging their EV sales numbers. First, each model simulates the cumulative battery production endogenously for its inherent countries. Then, battery capacities of the vehicle purchases in countries covered by the other model are added to the cumulative production, and the model is rerun. Iterations of data exchange and reruns continue until battery prices, and sales numbers, respectively, get stable in both models. Fig. 15.6 visualizes this approach that simulates the global EV diffusion using the experience curves in an endogenously implemented way by establishing this soft link between the two models. For the non-European countries [abbreviated as rest of the world (RoW)], TE3 provides sales numbers for China, India, Japan, and the US. In order to scale up to the cumulative production of global car purchases covering all non-European countries, the EV sales numbers of these countries are multiplied with a small factor that was assumed based on the global EV outlook (International Energy Agency, 2018).

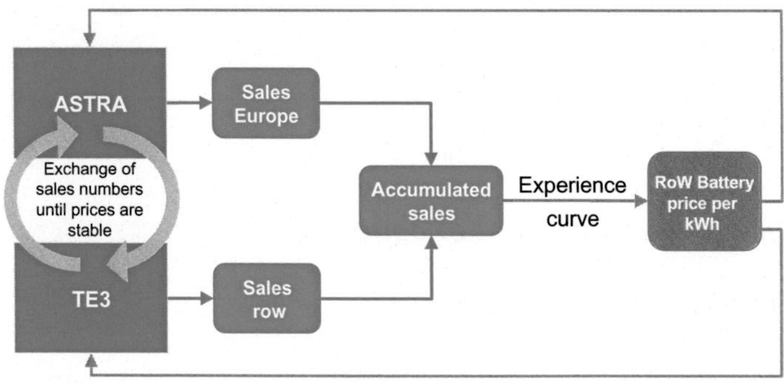

Figure 15.6
Model coupling of ASTRA and TE3 for the simulation of global learning.

15.4 Scenario and sensitivity analyses

First, we simulate a Reference scenario that reflects the effects of current policies. Based on its results, we investigate sensitivities and effects of global learning on battery prices and EV market share by varying learning rates and average battery capacities per car.

The Reference scenario assumes that all policy measures related to the transport sector existing at the beginning of 2019 are implemented in their defined form and continue to be valid until the year 2050. This comprises in particular regulations on CO_2 standards for cars and national plans for the deployment of alternative fuel and charging infrastructure. Sensitivity analyses for the learning rates and the average size of the battery will be performed, as both comprise essential uncertainties. The learning rate values cover an uncertainty range in form of the standard error. Thus, the learning rate will be varied around the mean value using the positive standard deviation for a maximum learning rate and the negative standard error as the minimum value. The development of the average battery capacity in a vehicle is also quite uncertain and will depend on several factors, such as driving profiles of EV owners, the size and weight of future cars, the deployment of fast-charging infrastructure, perceived range reliability, and the development of battery prices. While battery prices decrease, range anxiety is still a big barrier to the purchase of EVs (Funke et al., 2019). Therefore we investigate two different developments of the average battery capacity: in a first sensitivity analysis, battery capacities will be kept constant over time. A second analysis assumes strongly increasing battery capacities that could occur if the trend to larger and heavier cars such as sports utility vehicles (SUVs). (Munoz, 2019) continues and because reduced battery costs over time represent an opportunity to increase battery capacity. Table 15.4 summarizes the assumptions for the main parameters in the base scenario and for the sensitivity analyses.

Table 15.4: Parameters defined for scenario and sensitivity analysis.

			Battery capacity in kWh			
	Short designation	**Learning rate**		**2020**	**2030**	**2050**
Reference scenario	REF	15.2%	BEV	24	36	36
			PHEV	10	10	10
Sensitivity—Learning rate	LR high	18.1%	Like REF			
	LR low	12.3%				
Sensitivity—Battery capacity	BC constant	Like REF	BEV	24	24	24
			PHEV	10	10	10
	BC strong increase		BEV	24	60	90
			PHEV	10	20	30

15.5 Results and discussion

15.5.1 Results of the Reference scenario

In the Reference scenario, battery prices decline to 102 USD/kWh in 2030 and 68 USD/kWh in 2050. Only 11% of accumulated battery capacity until 2050 is produced for European cars. Thus the main driver of battery price decline is the sales in RoW, mainly for cars in China and India. The developments of accumulated produced battery capacity and resulting battery prices are visualized in Fig. 15.7.

The EV share in European car sales is consistently growing, achieving 15% in 2030 and 39% in 2050. The PHEV share in EV car stock reaches around 50%, however, it is hard to predict. While PHEV numbers were negligibly small in the first years of EV sales in some countries, their share increased in several instances between 2013 and 2017 (International Energy Agency, 2018). This increase was taken into account in the ASTRA calibration process and, thus, is reflected also in the development towards 2050. FCEVs rarely diffuse in the reference case with the defined learning rates, being too expensive and only interesting for niche applications.

The EV share for the four key non-European countries is increasing until 2050. In particular, China and India will have around 60% sales share of EV by 2050. Both countries have introduced policy targets that aim at quite high EV sales and an increased number of EVs in the overall stock. This is the reason why the sales share for both countries is already increased by 30% before 2030. The high stock of BEVs arises as again China and India have introduced higher subsidies for BEVs than for PHEVs. This trend for

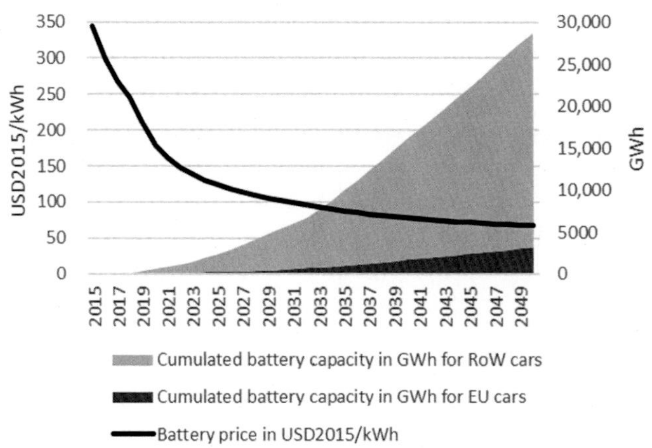

Figure 15.7
Battery price and produced battery capacity for the Reference scenario.

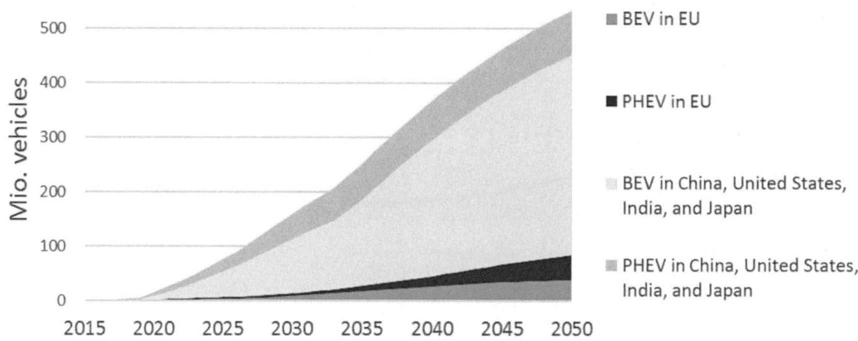

Figure 15.8
Electric car stock in the Reference scenario.

BEVs is assumed to remain until 2050. The development of the EV car stock is depicted in Fig. 15.8. The EV car stock in the simulated countries reaches 532 million vehicles in 2050.

For the TE3 model a specific model behavior was observed. Due to the fact that a significant proportion of consumers in China and India are assumed to be low-cost buyers purchasing the cheapest powertrain, there is the possibility that a drastic market adjustment occurs if electric cars become less expensive than other technologies in the simulation. We observed this effect mainly for the Chinese (e.g., between 2020 and 2027) and the Indian market, resulting in demand jumps. The market uptake remains, however, within realistic limits, for example, sales volumes are not expected to reach the 7 million BEVs before 2025 in China. In 2018, we observed already 1.2 million registrations of BEVs and PHEVs together (ZSW, 2019). By 2025 China plans that at least 20% of car sales are electric (Xinhua, 2017).

15.5.2 Results of the sensitivity analyses

The variation of the learning rate has the highest impact on the battery price development (see Fig. 15.9). Assuming a low learning rate of 12.3% the battery price lies still at 128 USD/kWh in 2050 compared to 40 USD/kWh for a high learning rate of 18.1%. In contrast, the assumption on different developments of the battery capacities in a car has only a minor impact on the battery price, being still in the range of the Reference scenario of 68 USD/kWh in 2050 with −5 USD/kWh for the assumption of a strong increase of the battery capacity and +4 USD/kWh for constant small capacities. The mechanisms that lead to these respective battery price developments are various and will be further explained in the following.

For the learning rates the mechanism is simple: higher learning rates lead to a faster decline of battery prices making EVs less costly and thus more attractive. The mechanism is partly

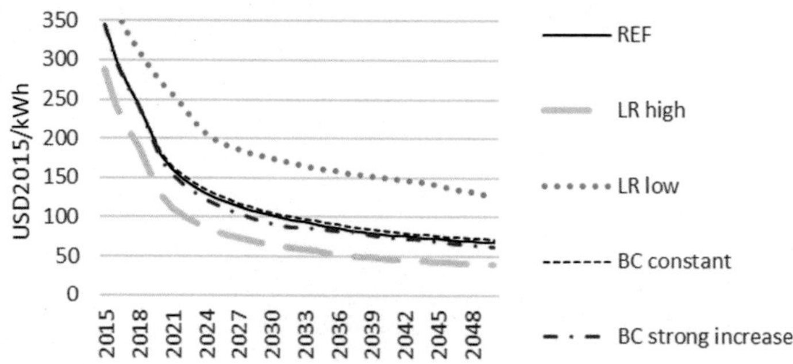

Figure 15.9
Development of battery prices in USD_{2015} in the Reference scenario and for the sensitivity analyses.

reinforcing as the lower battery prices caused by higher learning rates lead to higher car sales, which means a larger volume of produced battery capacity, which leads to higher experience curve effects and thus lower battery prices. Low learning rates have the opposite effect with higher EV prices leading to lower EV shares.

The models lead to different market reactions on EV sales due to the specific market characteristics and the described underlying choice models. First of all, the TE3 model shows stronger effects of variation of parameter values on EV sales compared to the ASTRA model for Europe where besides initial investment costs for a vehicle also available public charging options play an important role for a higher diffusion of EV sales. Furthermore, while constant and thus relatively small battery capacities lead to a strong uptake of EV sales in all investigated RoW countries as car prices are lower with smaller batteries, EV sales decrease for Europe compared to the Reference scenario because of lower ranges and higher efforts for charging. This disadvantage of lower battery capacities would need to be compensated by a far larger deployment of charging infrastructure including fast chargers than assumed in current policies in the Reference scenario. In Europe the highest EV sales share is achieved when assuming either a high learning rate or a strong increase of the battery capacity (see Fig. 15.10). Furthermore, the PHEV share in EV stock in 2050 is lowest in these two cases with 52%. For LR low, this share is higher by +7% as the total vehicle price of PHEVs is less affected by high battery costs due to the smaller battery capacity compared to BEVs. For BC constant, the PHEV share in EVs is even 61% (+9%) because the lower battery capacities result in shorter ranges thus making BEVs less attractive.

For all the four key non-European countries the battery price shows a major impact as it is reflected in the purchase price of BEVs and PHEVs. Therefore in particular when the learning rate is low and the battery price is relatively high, the growth of the EV sales is

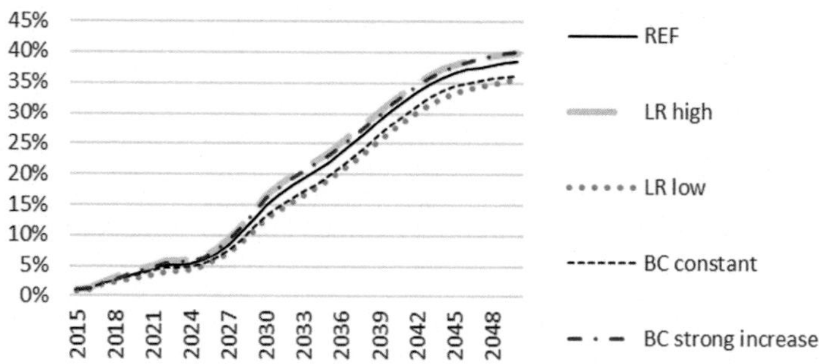

Figure 15.10
Electric vehicle market shares for Reference scenario and sensitivity analyses in Europe.

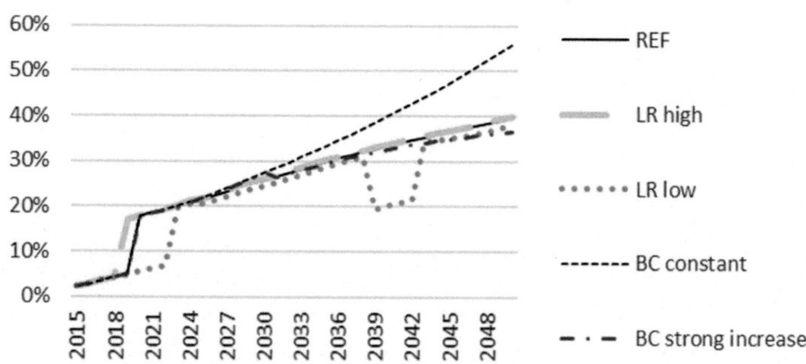

Figure 15.11
Electric vehicle market shares for Reference scenario and sensitivity analyses in the United States.

lower than with other sensitivity analyses. In some cases the sales share of specific countries jumps (cf. LR low in Fig. 15.11) because a turning point for another technology (e.g., car price for BEVs drops below the one of ICEVs) is achieved in one group segment. Hence, this segment decides in favor of the new technology. Quite often there is a significant shift from the PHEV to the BEV share or vice versa depending on whether the battery price is high or low. The high learning rate and the constant battery capacity implies a steady increase in EV shares for all countries. The sensitivity for an increase in the battery capacity leads to jumps in the market shares as well. Again, this is due to modeling reason as an overall market segment chooses one or the other powertrain technology. In particular, when the battery capacity increases, the overall battery price for the EV increases, which leads to an increased purchase price. This often leads to a decrease in sales in the short run as the attributes such as higher range cannot or only little offset this effect. Figs. 15.10 and 15.11 show the EV share in car sales in Europe and the US, respectively.

15.5.3 Effects of global learning via model coupling

For the sensitivity analyses the Reference scenario was used as the starting point. The first run with changed parameters is simulated by ASTRA. The sales numbers for RoW still represent the Reference scenario in this run. Then, several iterations of data exchange with TE3 and model runs are performed until battery prices and EV sales get stable. Fig. 15.12 illustrates the effects that occurred by coupling the two models showing the resulting values in the ASTRA model for each run until stable values were achieved. The impact of the model coupling differs between the analyzed parameter variations.

There is a strong impact on battery prices for the low learning rate because EV sales drop significantly in RoW with this assumption. Thus the distance between the curve from run 1 and run 2 for LR low is clearly visible in the left part of Fig. 15.12. In the same figure, we can see that for the assumptions of a high learning rate, the effect is negligible as the battery price curve changes only slightly from the first to the second and third run. When assuming a strong battery capacity increase, the battery price is substantially lower compared to the Reference scenario in the first run but then increases with each iteration of EV sales exchange with TE3 as EV sales drop in RoW for larger battery capacities (see right part of Fig. 15.12).

These detailed results show that model coupling can be valuable compared to assuming exogenously defined sales numbers because the impact on battery price development can differ substantially when feedback mechanisms in the models take effect, and the result might not be obvious beforehand.

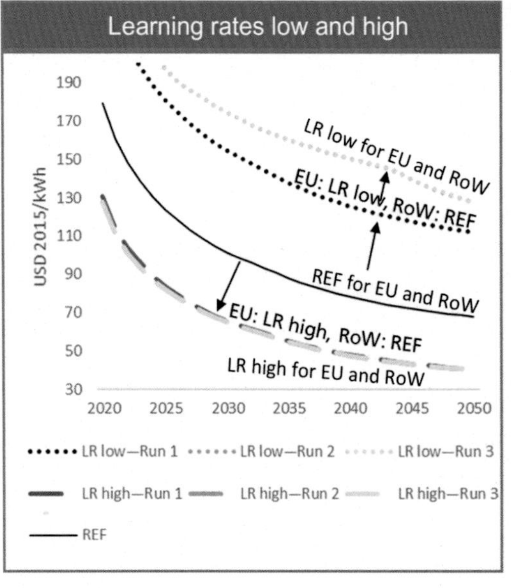

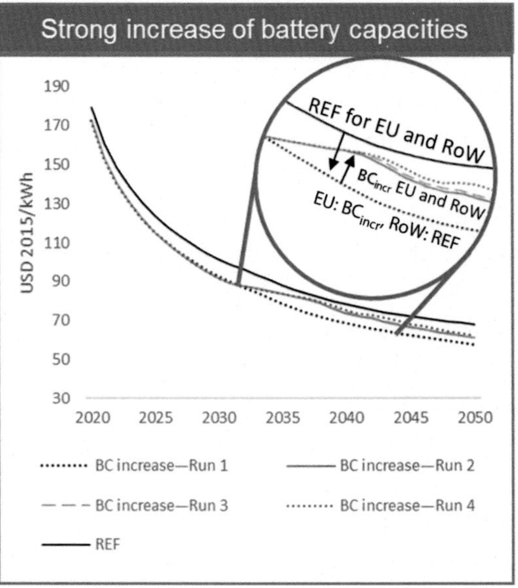

Figure 15.12
Development of the battery price in ASTRA due to the exchange of electric vehicle sales numbers for RoW with TE3 for each run (run 1 to X, X with stable values). *RoW*, rest of the world.

15.5.4 Comparison with other studies and limitations

Our results indicate the significant impact of vehicle prices and battery capacities on the market share. Compared to other studies our results stand out neither in terms of overall stock development nor national market shares (cf. Tsiropoulos et al. (2018) and Fig. 15.13).

Not surprisingly, the learning rate has a significant influence on these numbers. Also the development of ranges (i.e., battery capacity of cars) has a strong influence on our modeling results. Increasing the battery capacity will increase—*ceteris paribus*—the electric range, the average recharging time, and the vehicle purchase price (see Fig. 15.14). As opposed to the range, the last two negatively influence technology diffusion. This, in turn, influences the cumulative battery manufacturing experience, consequently affecting battery cost and price. This process is captured by a reinforcing feedback loop.

The effects of battery capacities and range anxiety have to be further elaborated in future research. ASTRA and TE3 showed different predominant effects in our sensitivity analyses of battery capacity development, which might be not only due to the specific market

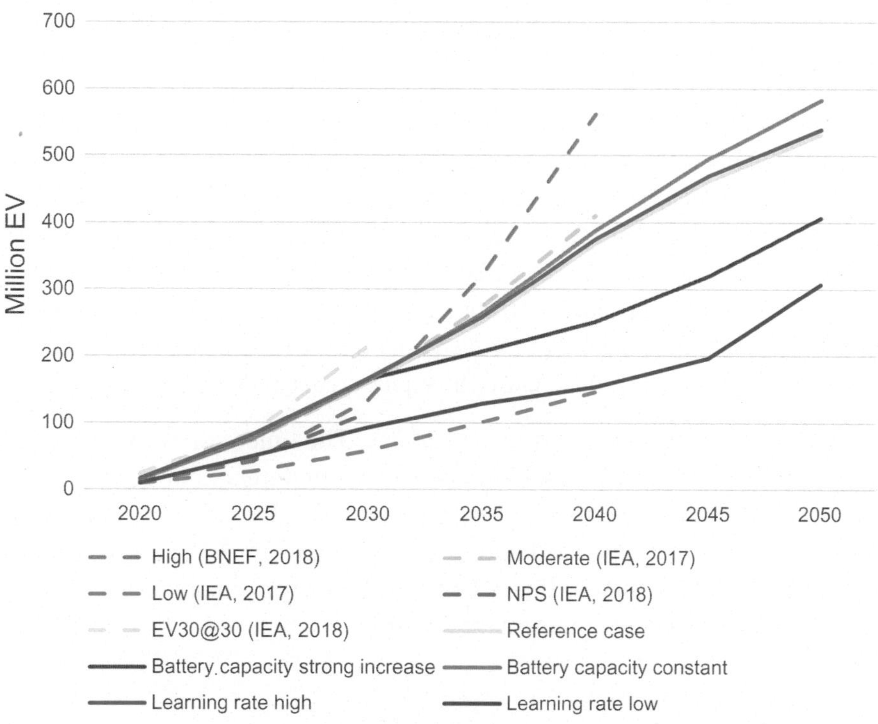

Figure 15.13
EV stock development over time for our scenarios and other studies (BNEF, 2018; International Energy Agency, 2017; International Energy Agency, 2018).

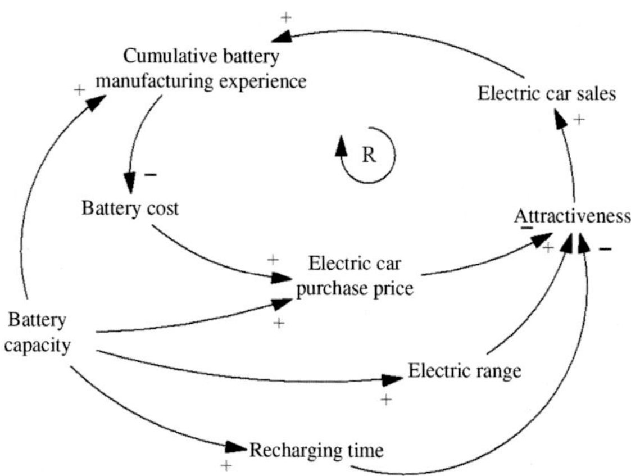

Figure 15.14
Battery capacity and electric car sales: causal loop diagram. Source: *Own work using Vensim.*

characteristics but also due to modeling assumptions and implementation. While ASTRA considers ranges within a fuel procurement cost factor that includes distances to available filling and charging stations and the number of required filling/charging actions, the attractiveness of EVs in the TE3 model depends also on the recharging time that increases with higher battery capacity.

Concerning solutions for the range anxiety issue, Funke et al. (2019) conclude that the deployment of fast-charging infrastructure is preferable to rising battery capacities when comparing the cost efficiency. Hence, the development of battery capacities for EVs and the deployment of fast-charging infrastructure as an alternative to influence the perceived range of EVs are also decisive for future market success. Further differentiation of available battery capacities in the EV product portfolio, and car segments with different battery sizes in the models, respectively, might help to overcome this contradiction of range anxiety versus vehicle price by meeting different user requirements and preferences.

Finally, our modeling work shows further limitations. The time horizon of more than 30 years is very long for such new technologies. Changing preferences of customers and other policy instruments can have a significant impact on market development. Furthermore, influences on decisive prices, such as fuel or other resources for EVs, for instance lithium or cobalt, may influence the market shares considerably. Even supply chain bottlenecks due to a scarcity of required raw materials might constrain EV uptake in the future.

15.6 Conclusion

The market share of EVs is increasing in all simulated scenarios. In 2050, the number of registered EVs in all considered countries amounts to 300–600 million. These numbers are in line with several other studies. The key non-European countries have a higher impact on the development of battery prices through a high contribution to the cumulative battery capacity due to their higher overall car stock compared to the European countries. In the Reference scenario, only 16% of EV car stock is located in Europe, while the main part of the overall EV car stock belongs to the four countries Japan, China, India, and the US. The sensitivity analyses showed clearly that the assumed learning rates have the highest impact on battery prices. Depending on the mechanisms triggered in the models, the effects of global learning on battery prices and EV sales were different. Coupling the models can be considered valuable for the simulation of the battery price as scenario assumptions for countries not covered in a single model can be explicitly defined, and feedback mechanisms between countries can take effect. In order to further contribute to uncertainties in scenarios for EV diffusion, global impacts could be further investigated by simulating strong policies such as sales bans for pure ICEVs. Moreover, by implementing further policies and measures such as stricter CO_2 standards and pricing schemes or more extensive deployment of charging infrastructure for Europe, it could be analyzed how strong the impact of global learning on EV sales would then remain.

References

Abbas, K.A., Bell, M.G.H., 1994. System dynamics applicability to transportation modeling. Transp. Res., A: Policy Pract. 28 (5), 373–390.

Abt, D., 1998. Die Erklärung der Technikgenese des Elektromobils', Frankfurt am Main (Ph.D. thesis). University of Frankfurt (Main), Germany.

BNEF, 2018. Long-Term Electric Vehicle Outlook 2018 (EVO 2018).

Capros, P., de Vita, A., Tasios, N., Siskos, P., Kannavou, M., Petropoulos, A., et al., 2016. EU Reference Scenario 2016—Energy, Transport and GHG Emissions Trends to 2050. [Online]. Available from: <https://ec.europa.eu/energy/sites/ener/files/documents/ref2016_report_final-web.pdf>.

Chatzikomis, C.I., Spentzas, K.N., Mamalis, A.G., 2014. Environmental and economic effects of widespread introduction of electric vehicles in Greece. Eur. Transp. Res. Rev. 6 (4), 365–376.

Creutzig, F., Jochem, P., Edelenbosch, O.Y., Mattauch, L., van Vuuren, D.P., McCollum, D., et al., 2015. Transport: a roadblock to climate change mitigation? Science 350 (6263), 911–912.

Fermi, F., Fiorello, D., Krail, M., Schade, W., 2014. Description of the ASTRA-EC model and of the user interface: deliverable D4.2 of ASSIST (assessing the social and economic impacts of past and future sustainable transport policy in Europe). In: Deliverable D4, vol. 2.

Ford, A., 1995. Simulating the controllability of feebates. Syst. Dyn. Rev. 11 (1), 3–29.

Forrester, J.W., 1995. The beginning of system dynamics. McKinsey Q. 4–17. [Online]. Available from: <http://static.clexchange.org/ftp/documents/system-dynamics/SD1989-07BeginningofSD.pdf>.

Forrester, J.W., 1997. Industrial dynamics. J. Oper. Res. Soc. 48 (10), 1037–1041.

Freunberger, S.A., 2017. True performance metrics in beyond-intercalation batteries. Nat. Energy 2, 17091.

Funke, S.Á., Plötz, P., Wietschel, M., 2019. Invest in fast-charging infrastructure or in longer battery ranges? A cost-efficiency comparison for Germany. Appl. Energy 235, 888–899.

Gnann, T., Plötz, P., Funke, S., Wietschel, M., 2015. What is the market potential of plug-in electric vehicles as commercial passenger cars? A case study from Germany. Transp. Res., D: Transp. Environ. 37, 171–187.

Gómez Vilchez, J.J., 2016. Car technologies and their impact on climate – a system dynamics approach. In: Presented at the 14th World Conference on Transport Research (WCTR).

Gómez Vilchez, J.J., 2019. The Impact of Electric Cars on Oil Demand and Greenhouse Gas Emissions in Key Markets (Ph.D. thesis). KIT Scientific Publishing, Karlsruhe.

Gómez Vilchez, J., Jochem, P., Fichtner, W. (Eds.), 2015. Energy use and emissions impacts from car technologies market scenarios: a multi-country system dynamics model. [Online]. Available from: <https://www.systemdynamics.org/assets/conferences/2015/proceed/papers/P1252.pdf>.

Gómez Vilchez, J., Jochem, P., Fichtner, W. (Eds.), 2016. Car technology market evolution and emissions impacts an example of energy policy scenarios under uncertainty. [Online]. Available from: <https://www.systemdynamics.org/assets/conferences/2016/proceed/papers/P1260.pdf>.

Gómez Vilchez and Jochem, 2019. Simulating Vehicle Fleet Composition: A Review of System Dynamics Models [under review].

Guarnieri, M., 2012. Looking back to electric cars. In: 2012 Third IEEE History of Electro-Technology Conference. 5–7 September 2012, Pavia, Italy. [Place of publication not identified], IEEE, pp. 1–6.

Haasz, T., Gómez Vilchez, J.J., Kunze, R., Deane, P., Fraboulet, D., Fahl, U., et al., 2018. Perspectives on decarbonizing the transport sector in the EU-28. Energy Strategy Rev. 20, 124–132.

Hardman, S., Jenn, A., Tal, G., Axsen, J., Beard, G., Daina, N., et al., 2018. A review of consumer preferences of and interactions with electric vehicle charging infrastructure. Transp. Res., D: Transp. Environ. 62, 508–523.

Harrison, G., Thiel, C., Jones, L., 2016. Powertrain Technology Transition Market Agent Model (PTTMAM): an introduction. In: Publications Office of the European Union, EUR-Scientific and Technical Research Reports.

Hawkins, T.R., Gausen, O.M., Strømman, A.H., 2012. Environmental impacts of hybrid and electric vehicles—a review. Int. J. Life Cycle Assess. 17 (8), 997–1014.

International Energy Agency, 2017. Energy Technology Perspectives 2017 Catalysing Energy Technology Transformations. Organisation for Economic Co-Operation and Development, Paris.

International Energy Agency, 2018. Global EV outlook 2018. [Online]. Available from: <https://webstore.iea.org/global-ev-outlook-2018> (accessed 19.06.19.).

Jochem, P., Babrowski, S., Fichtner, W., 2015. Assessing CO_2 emissions of electric vehicles in Germany in 2030. Transp. Res., A: Policy Pract. 78, 68–83.

Jochem, P., Gómez Vilchez, J.J., Ensslen, A., Schäuble, J., Fichtner, W., 2018. Methods for forecasting the market penetration of electric drivetrains in the passenger car market. Transp. Rev. 38 (3), 322–348.

Keith, D.R., 2012. Essays on the Dynamics of Alternative Fuel Vehicle Adoption: Insights From the Market for Hybrid-Electric Vehicles in the United States. Massachusetts Institute of Technology. [Online]. Available from: <https://dspace.mit.edu/handle/1721.1/79546>.

Kieckhäfer, K., 2013. Marktsimulation zur strategischen Planung von Produktportfolios: Dargestellt am Beispiel innovativer Antriebe in der Automobilindustrie. Springer-Verlag.

Kieckhäfer, K., Volling, T., Spengler, T.S., 2014. A hybrid simulation approach for estimating the market share evolution of electric vehicles. Transp. Sci. 48 (4), 651–670.

Kim, H.C., Wallington, T.J., Arsenault, R., Bae, C., Ahn, S., Lee, J., 2016. Cradle-to-gate emissions from a commercial electric vehicle Li-ion battery: a comparative analysis. Environ. Sci. Technol. 50 (14), 7715–7722.

Kotler, P., Armstrong, G., Wong, V., Saunders, J., Kotler, P., Armstrong, G. (Eds.), 2008. Principles of Marketing, fifth ed. Pearson Education, Inc, Upper Saddle River, NJ.

Lieven, T., Mühlmeier, S., Henkel, S., Waller, J.F., 2011. Who will buy electric cars? An empirical study in Germany. Transp. Res., D: Transp. Environ. 16 (3), 236–243.

Louwen A., Krishnan S., Derks M., Junginger M., 2018. D3.2 Comprehensive Report on Experience Curves. Utrecht, The Netherlands.

Ma, S.-C., Fan, Y., Feng, L., 2017. An evaluation of government incentives for new energy vehicles in China focusing on vehicle purchasing restrictions. Energy Policy 110, 609–618.

Meadows, D.H., Wright, D., 2008. Thinking in Systems: A Primer. Chelsea Green Publishing.

Munoz F., 2019. Global SUV boom continues in 2018 but growth moderates. [Online]. Available from: < https://www.jato.com/global-suv-boom-continues-in-2018-but-growth-moderates > (accessed 28.08.19).

Nelson, P.A., Gallagher, K.G., Bloom, I.D., Dees, D.W., 2012. Modeling the Performance and Cost of Lithium-Ion Batteries for Electric-Drive Vehicles. Argonne National Lab. (ANL), Argonne, IL. [Online]. Available from: <http//www.osti.gov/bridge>.

Rezvani, Z., Jansson, J., Bodin, J., 2015. Advances in consumer electric vehicle adoption research: a review and research agenda. Transp. Res., D: Transp. Environ. 34, 122–136.

Rogers, E.M., 2003. Diffusion of Innovations. Free Press, New York, London, Toronto, Sydney.

Schmidt, O., Hawkes, A., Gambhir, A., Staffell, I., 2017. The future cost of electrical energy storage based on experience rates. Nat. Energy 2 (8), 17110.

Shafiei, E., Stefansson, H., Asgeirsson, E.I., Davidsdottir, B., Raberto, M., 2013. Integrated agent-based and system dynamics modelling for simulation of sustainable mobility. Transp. Rev. 33 (1), 44–70.

Shafiei, E., Leaver, J., Davidsdottir, B., 2017. Cost-effectiveness analysis of inducing green vehicles to achieve deep reductions in greenhouse gas emissions in New Zealand. J. Cleaner Prod. 150, 339–351.

She, Z.-Y., Sun, Q., Ma, J.-J., Xie, B.-C., 2017. What are the barriers to widespread adoption of battery electric vehicles? A survey of public perception in Tianjin, China. Transp. Policy 56, 29–40.

Shen, W., Han, W., Wallington, T.J., 2014. Current and future greenhouse gas emissions associated with electricity generation in China: implications for electric vehicles. Environ. Sci. Technol. 48 (12), 7069–7075.

Shepherd, S.P., 2014. A review of system dynamics models applied in transportation. Transportmetrica B: Transp. Dyn. 2 (2), 83–105.

Shepherd, S., Bonsall, P., Harrison, G., 2012. Factors affecting future demand for electric vehicles: a model based study. Transp. Policy 20, 62–74.

Sterman, J.D., 2000. Business dynamics: systems thinking and modeling for a complex world. McGraw-Hill Education.

Tsiropoulos, I., Tarvydas, D., Lebedeva, N., 2018. Li-Ion Batteries for Mobility and Stationary Storage Applications – Scenarios for Costs and Market Growth. Publications Office of the European Union, Luxembourg.

Wang, N., Pan, H., Zheng, W., 2017a. Assessment of the incentives on electric vehicle promotion in China. Transp. Res., A: Policy Pract. 101, 177–189.

Wang, S., Li, J., Zhao, D., 2017b. The impact of policy measures on consumer intention to adopt electric vehicles: evidence from China. Transp. Res., A: Policy Pract. 105, 14–26.

Wansart, J., 2012. Analyse von Strategien der Automobilindustrie zur Reduktion von CO2-Flottenemissionen und zur Markteinführung alternativer Antriebe: Ein systemdynamischer Ansatz am Beispiel der kalifornischen Gesetzgebung. Springer-Verlag.

de Wolff, P., 1938. The demand for passenger cars in the United States. Econometrica: J. Econometric Soc. 6 (2), 113–129. Available from: < https://www.jstor.org/stable/i332567?refreqid = excelsior%3A5d4993cbed9f7db47980ca0b26bdb593>.

Xinhua, 2017. China to quadruple new energy vehicle production by 2020. [Online]. Available from: <http://english.gov.cn/state_council/ministries/2017/01/16/content_281475543045788.htm> (accessed 28.06.19.).

ZSW, 2019. ZSW: Datenservice, Bestand Elektro-Pkw weltweit, Neuzulassungen Elektro-Pkw weltweit. [Online]. Available from: <https://www.zsw-bw.de/mediathek/datenservice.html> (accessed 20.06.19.).

PART IV

Final words

CHAPTER 16

Synthesis, conclusions, and recommendations

Martin Junginger[1] and Atse Louwen[1,2]

[1]*Copernicus Institute of Sustainable Development, Utrecht University, Utrecht, The Netherlands,* [2]*Institute for Renewable Energy, Eurac Research, Bolzano, Italy*

Abstract

The past 15 chapters provide an overview of the technological development and cost reductions achieved from a number of major energy technologies that are expected to be deployed as part of the ongoing energy transition. At the same time, these chapters highlight how future cost reductions and subsequent deployment of these technologies may shape the future mix of the electricity, heat, and transport sectors. In this final chapter, we discuss both methodological issues that appeared throughout the book and present a synthesis of the outlook of the technologies investigated. We discuss amongst others general lessons and recommendations for policy makers, industry, and academics, focusing on what technologies may require further policy support in the short term to have a major impact later on, which investments will be needed, and what scientific knowledge gaps remain for future research.

Chapter outline

16.1 Introduction

The past 15 chapters provide an overview of the technological development and cost reductions achieved from a number of major energy technologies that are expected to be

Technological Learning in the Transition to a Low-Carbon Energy System.
DOI: https://doi.org/10.1016/B978-0-12-818762-3.00016-9

deployed as part of the ongoing energy transition. At the same time, these chapters highlight how future cost reductions and subsequent deployment of these technologies may shape the future mix of the electricity, heat, and transport sectors.

In this final chapter, we discuss both methodological issues that appeared throughout the book and present a synthesis of the outlook of the technologies investigated. We discuss amongst others general lessons and recommendations for policy makers, industry, and academics, focusing on what technologies may require further policy support in the short term to have a major impact later on, which investments will be needed, and what scientific knowledge gaps remain for future research.

16.2 Methodological considerations

Many papers have been published in the past on the methodological limitations of experience curves. This section does not aim to summarize these findings, but mainly focuses on issues emerging from the current book.

16.2.1 Cost of capacity versus LCOE and other metrics

A common theme emerging from various technology chapters is the need for metrics that focus on the feature of a technology for which optimization is carried out. This is for most energy supply technologies the ability to deliver energy (electricity, heat, or transport fuels) at the lowest possible cost. For demand-side technologies, other factors typically play a role as well (e.g., safety, reliability, and comfort of use, see the example of LED lamps later). Still, for most technologies, the upfront investment cost was used as a proxy to reflect the technological learning and associated cost reductions of the energy delivered. While this yielded acceptable results in the past [especially for situations where the capital expenditures (Capex) remained a substantial part of the levelized cost of electricity (LCOE), such as photovoltaics (PV)] for many other technologies, increasingly the need is emerging to focus more on the assessment of LCOE in order to accurately capture and forecast cost trends. In this book, this need was particularly identified for onshore and offshore wind, where the increasing capacity factor but also lower weighted average cost of capital (WACC) and Opex have contributed to the overall reduction of LCOE, and experience curves solely based on Capex are increasingly less suitable to provide accurate trends.

Also for hydrogen production the levelized cost of hydrogen (LCOH) would be a more appropriate metric than stack costs. For electric cars, studies beyond battery pack costs focusing on total cost of ownership and cost per passenger-kilometer traveled will also help better understand diffusion and adoption of electric vehicles. Similarly, for heat pumps, the increases in coefficient of performance (COP) have led to substantial reductions in the cost of heat delivered compared to the reductions in the investment cost alone. However, as discussed above, data availability limits the possibilities to take these developments adequately into account.

Last but not the least, for LED lamps (as a typical consumer product), a correct assessment depends on many factors: the rapid evolution in LED lighting products also introduced a wide variety of new product features that also affect price, posing a challenge for determining a single, well-defined price at any given point in time, and confounding efforts to measure the underlying learning dynamics for the base technology. For instance, the earliest LED lighting products intended for general illumination had relatively low light output, while over time, products with higher output were introduced in the market at a substantial price premium (that eased with time). It is thus essential for any price-trend analysis to control for lumen output, to account for the varying maturity, and market penetration of bulbs with different output. Additional features that can impact the price of LED lighting products, of which relative market penetration varied significantly during the 2010s, include lifetime, color temperature, color rendering, dimmability, color tunability, remote controllability, and the esthetic appearance of the light bulb itself.

16.2.2 Component-based assessments

As was highlighted in Chapter 5, for photovoltaic systems, experience curves for a system, which is an aggregate of several components, should ideally be based on separate experience curves for each component of this system. We observed that the learning rate for PV modules was substantially higher than that for the so-called balance-of-system components. Future price extrapolations should be made on the basis of using separate experience curves for these components, rather than based on a single experience curve for the whole system. As we highlighted in Chapter 5, using this single system-based experience curve will likely lead to an overestimation of future cost reductions. A similar discussion is also made in Chapter 8, where an overview is given of the components of a battery storage system. From this discussion, it is clear that the potential for cost reduction varies for each component; hence, it is argued that it is feasible to assume the aggregate learning rate for battery systems will decrease over time as the relative cost shares of components with high learning rates decrease more quickly.

16.2.3 Two- and multifactor experience curves

In Chapters 5 and 8, the concept of multifactor experience curves is discussed. Multifactor experience curves attempt to expand the single-factor experience curve, by including additional parameters aside from cumulative production. These parameters commonly include R&D activities, by means of a variety of proxy datasets, and input material prices. We observed in Chapter 5, that the price of silicon is highly correlated with the observed cost developments of crystalline silicon PV modules, while it is difficult to separate the effects of cumulative production and R&D efforts, as the developments over time of these two parameters are highly correlated. For battery storage systems (Chapter 8), material input prices show little correlation with the observed cost reductions, a result of battery designs being highly diversified in terms of material compositions and having design features that are resilient to strong short-term price changes in, for instance, lithium and cobalt.

The use of multifactor experience curves in price extrapolations and energy modeling activities would address several issues that arise from the application of single-factor experience curves, for example, it would be possible to take into account changes in input material prices, the effects of policy measure that enhance R&D efforts, and possibly also take into account knowledge (and production experience) spillover activities from one product to the other. This would especially be true if a combination is made of component-based and multifactor experience curves. Endogenous implementation of multifactor curves in energy modeling is however still a developing field. The successful implementation faces some tough methodological challenges, as the data requirements for multifactor and component-based experience curves are much larger compared to single-factor curves. First, much more detailed data needs to be collected, validated, and verified. Second, the endogenous application of multifactor and component-based experience curves in energy models requires that these models produce a much larger set of input data for the curves, such as raw material prices, R&D activities, and cumulative production for each separate component of a product. Still, if further research were to be successful in addressing these issues, multifactor experience curves have the potential to improve the accuracy of modeling future cost trajectories of technologies.

16.2.4 Environmental experience curves and social learning

As pointed out both in Chapter 4, and in the chapter on electricity storage (Chapter 8), monitoring cost developments may not be the only application of the experience curve. There are clear indications that with decreasing use of materials, next to cost, also the environmental impacts during the production phase of, for example, solar cells or wind turbines, decrease. Likewise, higher efficiencies of demand-side technologies reduce the demand for fuels or electricity, and thus lower again the environmental impacts in the use phase of many technologies. So far, deploying the experience curve concept to both assess the historical environmental impacts and extrapolate such trends for future projections has been very limited, partly also due to data limitations. One promising technology where a historical analysis may be promising is onshore wind (given data availability), but also other technologies could be scrutinized. After all, with massive deployment of these technologies, assessing environmental impacts is equally important as describing the overall cost of deployment.

Another application of the concept of technological learning is discussed in Chapter 4, where social learning mechanisms are investigated in a case study on the market diffusion of electric vehicles. By taking into account a set of social dynamics, a more realistic model can be derived on the uptake of these novel technologies in society, which includes more than only fully rational decisions a consumer would make based on total cost of ownership of a technology. Here, social learning represents the change in risk perception consumers have of novel technologies, for instance, due to first adopters buying into these novel technologies and essentially demonstrating the technologies' value and capability to replace incumbent technologies.

16.2.5 Application in energy and climate models

Implementation of technological learning processes in energy modeling is nowadays common, yet not without its issues and drawbacks. As discussed in Chapter 3, a variety of model characteristics can hamper endogenous implementation. Many energy models are restricted in geographical scope, while technological progress is most often considered a global process. In these cases, it is likely that an (at least partly) exogenous cost trajectory based on experience curves is necessary, but a feedback between development within the model under study and technological learning is in this case only possible to a limited extent.

Other issues are encountered relate to technical or practical considerations in energy modeling, such as the mathematical layout of the model, computation time, or the ability of models to produce the required input parameters for endogenous (multifactor) experience curves. When comparing an endogenous versus exogenous approach of implementing technological learning in different energy system models, it appears that especially top-down models allow easier implementation.

Testing the impact of the uncertainty of learning rates in in three different bottom-up models (see Chapter 14) revealed that the diffusion of different technologies is not impacted equally: heat pump diffusion for residential heating is only moderately affected, as installation rates are also dependent on, for example, technology preferences independent of pure cost parameters and policy preferences (e.g., support of centralized vs decentralized systems). On the other hand, assuming higher learning rates for batteries may significantly determine the diffusion of electric vehicles (see Chapter 15) in the transport sector and may shift new investments from gas turbines to redox-flow batteries (see Chapter 14) in the power sector. Similarly, the assumption whether CCS technologies do or do not learn largely determines investment in CCS plants by 2050. Unfortunately, there is considerable uncertainty for many of the learning rates applied in these models, amongst others, due to the limited availability of reliable learning rates and experience curves for many new energy technologies such as CCS. Thus these model results should also be handled with care, and sensitivity of the models to variation in learning rates should always be tested.

16.2.6 Data availability and future data collection

Based on the nine technology chapters, we conclude that data availability and quality differ strongly between the individual technologies investigated. For some technologies, data availability is excellent, such as onshore wind, offshore wind, solar PV, and batteries. Especially for the electricity supply technologies, data availability is high for the United States and Europe, often due to excellent long-term publicly funded bodies that systematically collect these data.

For other technologies, there is surprisingly little public data available. For example, LEDs have been around for many years, are currently rapidly gaining market share, and are generally considered as *the* lighting technology for the coming decades. Yet, other than the data presented in this chapter, there are surprisingly little time series on LED price developments available—possibly due to the large amount of data to be collected in order to correctly assess and compare LED lamps. Likewise, heat pumps for space heating and cooling have been deployed for decades (and in the form of air conditioning units on massive global scale) and are generally seen as one of the most promising technologies to provide low-temperature heat for residential buildings—yet, systematic collection of data on capital costs and COP is largely missing. Hydrogen production through electrolysis has been carried out on a large scale in the middle of the 20th century, but documentation of the declining cost of hydrogen has been minimal. Given the fact that these technologies are expected to play a major role in coming years, more comprehensive data collections on Capex and other variables (see earlier) is of vital importance to better monitor and asses future cost trends.

Similarly, there is also an actual lack of experience and data to make quality forecasts for electricity storage technologies. The rapid pace of advances on the battery chemistry front introduces new challenges that are novel and cannot be compared with other technologies such as hydropower dams or natural gas combined cycle plants. Technological learning studies should also incorporate alternative indices related to the life cycle of greenhouse gas emissions from storage options, materials availability of emerging battery chemistries, and cost indicators that incorporate multiple services and applications provided by storage. Also here, we call for transparency, and public access of data remains key to validating new learning curve models.

On green hydrogen production through alkaline electrolysis—one of the key technologies to decarbonize the energy system—relatively little public data is available, and often the data is incomplete or unclear (e.g., what parts are included, the size of the system). For future data collection, we recommend collecting data for the separate components of electrolyzers (stack, gas dryer, compressor, etc.) and generating experience curves for each component that makes up the system, similarly as performed for PV (see Chapter 5). This may be particularly useful for proton exchange membrane (PEM) electrolyzers. PEM—due to is flexibility with dynamic operation—might play an important role in hydrogen production in the future.

16.3 Technology outlook till 2030

This section provides an outlook for the various technologies covered in this book until 2030, including likely cost reduction levels.

At the time of writing, three out of four *electricity production technologies* covered in this book (solar PV, onshore wind, and offshore wind) reported that production cost levels could (at least in some instances) outcompete the fossil reference technology:

- Given the already low cost of PV systems, especially for large-scale systems, the LCOE from PV is already competitive with fossil generation in high irradiance locations and has achieved grid parity for private consumers years ago in a much larger geographical region. Chapter 5, shows that there is still substantial room for further system cost reductions, and so it is likely that electricity generation with PV will be cost-competitive in many more locations, even those with relatively low solar irradiance.
- Also, onshore wind is rapidly gaining market share and pushing out incumbent fossil generation in many parts of the world. As shown in Chapter 6, onshore wind has shown cost reductions for more than three decades, but the importance of underlying factors has varied over time. Next to lower upfront Capex, the capacity factor has also increased significantly. While Capex and LCOE have also temporarily increased between 2005 and 2011, the overall learning rate for LCOE for data between 1990 and 2017 is 11.4%. Combining this learning rate with anticipated growth in global onshore wind deployment yields a projected LCOE of about 33€/MWh by 2030, a reduction of approximately 25% from 2018 levels, making it highly competitive with expected prices of new coal and natural gas generation.
- After an increase between 2000 and 2015, the LCOE of offshore wind has declined dramatically from 190€/MWh in 2015 to about 100€/MWh at the end of 2018, with average projections for 2021 reaching as low as 70€/MWh. Especially, the increase in capacity factor has been a major driver in reducing the LCOE. Given the strong fluctuations in the past and many factors influencing the LCOE of offshore wind projects, it was not possible to derive meaningful one-factor experience curves and

learning rates that would allow extrapolation for future cost projections, but similarly to onshore wind and PV, it is expected that in the coming years, offshore wind will increasingly be able to compete with fossil fuels without direct economic support.
- In contrast, while concentrated solar power has also displayed cost decline in the past decades, current LCOE is still at an average of 210€/MWh; it cannot currently compete with PV, wind, and natural gas for electricity production.

Overall, with further opportunities to reduce LCOE, these technologies are set to deliver the large-scale diffusion needed in many energy scenarios (especially those with ambitious climate–change mitigation targets), delivering a surplus of electricity to provide energy for the mobility and heating sector. However, due to the increasingly intermittent availability of electricity in such systems, storage options will likely play a vital role.

Therefore past and future cost reductions of e*lectric mobility and electricity storage options* (EV, batteries, and H_2) have also been assessed in this book.

- For storage technologies, by 2030, stationary systems may cost between 200 and 440€/kWh, with pumped hydro and an electrolysis-fuel cell combination as minimum and maximum value, respectively. When accounting for experience rate uncertainty, the price range expands to 150–520€/kWh (min: utility-scale lithium-ion, max: electrolysis-fuel cell). For battery-based storage technologies, this means typically a cost reduction of more than 50% between 2018 and 2030.
- Similarly, the price of battery packs for transport applications is also expected to decline in a similar fashion from 50 to 190€/kWh in 2030 (40–200€/kWh with uncertainty), partly also depending on the level of diffusion of electric cars, which might reach between 100 and 240 million vehicles on the road by 2030. With the anticipated strong growth in uptake, our chapter suggests that by 2040, the cost could drop an additional 50% from today's level, ultimately reaching 50€/kWh. Such trajectories are feasible based on costs of new lithium-ion cathode chemistries and other battery pack materials. Meeting policy goals such as the EU's Strategic Energy Technology Plan cost target of 75€/kWh is feasible in both high and moderate growth scenarios This result, based on experience rates, indicates that aggressive targets may not be so difficult to meet, which can help as a transportation decarbonization strategy.
- For hydrogen production, a less clear picture emerged. The experience curve for alkaline electrolysis system between 1956 and 2016 shows a learning rate of 16% ± 6% with Capex decreasing from 2100 to $750€_{2017}/kW_{input}$ in 1956 to a range between 900 and $500€_{2017}/kW_{input}$ in 2016 but with a poor R^2 of 0.307, which can be attributed to discrepancies in the Capex composition of the gathered data and the wide spread in capacity (1–100 MW).

With regard to heating and cooling technologies, this book investigated both condensing natural gas boiler (the dominant fossil fuel–based heating technology in many EU countries) and heat pumps. Based on Swiss and Dutch case studies, the learning rate of the main cost components of heat pumps is 12%–22% and the one of the system as a whole about 20%, while their utility, in terms of less noise emission, system integration, and energy efficiency, improved over the past decades. This is equal or higher as compared to the LR of condensing gas boiler (13%). This holds for both countries assessed, although it should be kept in mind that the time series is quite short in the Dutch case. Moreover, a learning rate for the coefficient of performance was found, which is 5% in the case of ground-source heat pumps and 9% for air-to-water heat pump (HP). Thus heat pumps offer a high potential to improve their cost-effectiveness relative to reference systems in terms of cost of delivered heating energy, which depend on both specific investment costs and on the energy efficiency. Given the historical production and sales of heat pumps and condensing boilers, the cost reduction of heat pumps will likely be more dynamic as the cumulative production (or sales) will double faster, particularly in a climate mitigation scenario. Moreover, there is still a considerable technical potential to improve peak performance and energy efficiency (as opposed to gas condensing boilers where the technical potential is basically tapped). Therefore competitiveness particularly will be improved from a life-cycle-cost perspective, and from this perspective, heat pumps are already competitive for different use cases in many countries, also depending on the framework conditions.

Last but not the least, the progress achieved with LED lamps has been scrutinized. The 2010 decade saw a steady and rapid decline in price for LED lighting products, with prices falling several folds from the high initial market-entry prices observed in 2010. For LED A-line lamps sold in the US market, a steady decline of 20%–30% per year was observed through the first half of the decade, in conjunction with fast growth in consumer uptake, resulting in a manyfold increase in cumulative production in the same period. This situation presented an unusual opportunity to observe significant technological learning effects in near real time as they occurred over a period of only a few years. For A-line lamps in the United States, a learning rate of 18% was found using only 2–3 years of data on price and lamp sales. With such a limited time period, this learning rate should be handled with care: manufacturers projected a 40% decline in costs from 2015 to 2020, while the price-based forecasts point to a fourfold price drop over the same period. The discrepancy may partly also be explained by changing margins for manufacturers. On the other hand, learning rates for PV modules (another modular technology) of between 18% and 21% have been observed for a period of over 50 years (see earlier); hence, this learning rate does not seem overly optimistic.

16.4 Final conclusions and recommendations

When reading through the previous section, it becomes clear that technological learning and associated cost reductions for most technologies covered have been impressive over the past two decades and are now important drivers for further large-scale diffusion as part of the transition to a low-carbon energy system. In many cases (onshore and offshore wind energy, PV, and LED lamps), these technologies have reached the stage where they are able to directly compete with some (e.g., coal, nuclear) or virtually all fossil/traditional energy technologies and thus are now entering the phase where society will reap learning benefits, that is, both lower costs (and thus economic benefits), lower greenhouse gas (GHG) emission, and thus lower environmental (and in consequence again economic) impacts. Even in the absence of stringent climate policies, these technologies now have the potential to rapidly displace fossil-based and inefficient technologies. This is mainly due to persistent learning investment by countries such as Denmark, Germany, the United States and Japan, but more recently also China.

In other cases, there is still a long way to go before this stage is reached. Green hydrogen production is seen by many as the ultimate way to decarbonize large parts of our energy system and has shown clear progress over the past decades, yet the production costs are still significantly higher than those of gray or blue hydrogen, and significant investments in both R&D and deployment will be needed to bring down the production costs, much like those of solar PV 50 years ago. The heat pump, as the technology to provide low-temperature heat using electricity, is already cost-competitive in some market segments; but the need for additional building insulation and the sheer size of the building stock to be covered implies that this will still be a process over decades.

Thus we do repeat a key lesson from the LED-lighting chapter: an appreciation for the effects of technological learning is essential for sound decision-making with regard to emerging technologies, both for market actors and for policymakers. Decisions that may seem bold, or even foolhardy, in the context of status quo market conditions may in fact appear wise and beneficial once the full effects of technological learning are considered.

Last but not the least, in this book, a selected number of technologies, which are deemed crucial for the ongoing transition to a low-carbon energy system, were highlighted. However, not all relevant technologies were covered: conversion technologies using fossil fuels and nuclear energy were barely touched upon, even though they will continue to play a major role for decades, and are also still learning. Carbon capture, utilization, and storage technologies of both fossil and biogenic carbon may be crucial in keeping global mean temperature increases below 2°C but were not included either. Advanced biofuels and solar fuels may provide a renewable fuel for aviation and shipping where little alternatives exist on the short term, but again did not feature in this book. This was partly based on time and resource constraints, partly on the fact that the deployment of fossil fuels will hopefully be phased out, but largely also because for many of these technologies, there is barely any public data available and/or little actual progress and deployment has been achieved. For example, both carbon cature, utilisation and storage (CCUS) and advanced biofuels have only seen marginal deployment in the past decade (for differing reasons), making it difficult to deploy experience-curve-based assessments. While many may favor technologies, such as wind, solar, and energy savings through efficiency measures, it remains very likely that we will have to rely on a wide portfolio of technological options to fully transition to a low-carbon energy system. Thus investing in these technologies to "push them down the experience curve" is likely equally important as pursuing deployment of those that have reached commercial maturity.

Index

Note: Page numbers followed by "*f*" and "*t*" refer to figures and tables, respectively.

E

F

G

M

N

O

P